| DATE DUE | | | |
|---|---|---|---|
| APR 1 '88'S | | | |
| JAN 26 91'S | | | |
| FEB 1 5 '93'S | | | |
| MAR 06 1995 S | | | |
| MAY 0 9 | | | |
| | | | |
| | | | |
| | | | |
| | | | |
| | | | |
| | | | |
| | | | |

# An Introduction to

# Chaotic Dynamical Systems

# An Introduction to

# Chaotic Dynamical Systems

ROBERT L. DEVANEY

Department of Mathematics
Boston University

**Addison-Wesley Publishing Company, Inc.**
*The Advanced Book Program*
Redwood City, California • Menlo Park, California • Reading, Massachusetts
New York • Amsterdam • Don Mills, Ontario • Sydney
Bonn • Madrid • Singapore • Tokyo • Bogotá • Santiago
San Juan • Wokingham, United Kingdom

Sponsoring Editor: Richard W. Mixter
Production Editor: Karen Gulliver
Cover Designer: Daved Garza

Cover:

*The cover image shows a computer graphics display
of the Julia set of the complex exponential function
(Chapter 3, Section 8).*

Library of Congress Cataloging-in-Publication Data

Devaney, Robert L., 1948-
   An introduction to chaotic dynamical systems.

   (Global analysis series, pure and applied ; no. 3)
   Includes index.
   1. Differentiable dynamical systems.   2. Chaotic behavior
in systems.   I. Title.   II. Series.
QA614.8.D48  1986   515.3'52   85-15801
ISBN 0-8053-1601-9


   DEFGHIJ-MA-8987

TO KATH AND MEGGIE

# Preface

The last twenty five years have seen an explosion of interest in the study of nonlinear dynamical systems. Scientists in all disciplines have come to realize the power and the beauty of the geometric and qualitative techniques developed during this period. More importantly, they have been able to apply these techniques to a number of important nonlinear problems ranging from physics and chemistry to ecology and economics. The results have been truly exciting: systems which once seemed completely intractable from an analytic point of view can now be understood in a geometric or qualitative sense rather easily. Chaotic and random behavior of solutions of deterministic systems is now understood to be an inherent feature of many nonlinear systems, and the geometric theory developed over the past few decades handles this situation quite nicely.

Modern dynamical systems theory has a relatively short history. It begins with Poincaré (of course), who revolutionized the study of nonlinear differential equations by introducing the qualitative techniques of geometry and topology rather than strict analytic methods to discuss the *global* properties of solutions of these systems. To Poincaré, a global understanding of the gross behavior of *all* solutions of the system was more important than the local behavior of particular, analytically-precise solutions. Poincaré's point of view was enthusiastically adopted and furthered by Birkhoff in the first part of the twentieth century. Birkhoff realized the importance of the study of mappings and emphasized discrete dynamics as a means of understanding the more difficult dynamics arising from differential equations.

The infusion of geometric and topological techniques during this period gradually led mathematicians away from the study of the dynamical systems themselves and to the study of the underlying geometric structures. Manifolds, the natural "state spaces" of dynamical systems, became objects of study in their own right. Fields such as differential topology and algebraic topology were born and eventually flourished. Rapid advances in these fields gave mathematicians new and varied techniques for attacking geometric problems. Meanwhile, the study of the dynamical systems themselves languished in relative disfavor, except in the Soviet Union, where mathematicians such as Liapounov, Pontryagin, Andronov and others, continued to study dynamics from various points of view.

All of this changed around 1960, due mainly to the influence of Moser and Smale in the United States, Peixoto in Brazil and Kolmogorov, Arnol'd and Sinai in the Soviet Union. Differential topological techniques enabled Smale, Peixoto and their followers to understand the chaotic behavior of a large class of dynamical systems known as hyperbolic or Axiom *A* systems. Geometry combined with hard analysis allowed Kolmogorov, Arnol'd and Moser to push through their celebrated KAM theory. Smooth ergodic theory, topological dynamics, Hamiltonian mechanics, and the qualitative theory of ordinary differential equations all developed as disciplines in their own right.

More recently, dynamical systems has benefited from an infusion of interest and techniques from a variety of fields. Physicists such as Feigenbaum have rekindled interest in low dimensional discrete dynamical systems. Breakthroughs in mathematical biology and economics have attracted a diverse group of scientists to the field. The discovery of stably chaotic systems such as the Lorenz system from meteorology have convinced scientists that there are many more stable types of dynamical behavior than just stable equilibrium points and limit cycles. And, by no means least of all, computer graphics has shown that the dynamics of simple systems can be at once beautiful and alluring.

All of these developments have made dynamical systems theory an attractive and important branch of mathematics of interest to scientists in many disciplines. Unfortunately, because of the background of many of the contemporary researchers in such advanced fields as differential topology, algebraic topology, and differential geometry, the available introductions to this subject presuppose a familiarity on the part of the student with several of these fields. It is our feeling that the elements of dynamical systems theory can be introduced without prerequisites such as the theory of differentiable manifolds, advanced analysis, etc. Dynamical systems on simple spaces like the real line or the plane exhibit all of the chaotic and interesting behavior that occur on more general manifolds. Without these unnecessary prerequisites, the basic ideas of the field should be accessible to junior and senior mathematics majors as well as to graduate students and scientists in other disciplines. This is the basic goal of this text.

The field of dynamical systems and especially the study of chaotic systems has been hailed as one of the important breakthroughs in science in this century. While the field is still relatively young, there is no question that the field is becoming more and more important in a variety of scientific disciplines. We hope that this text serves to excite and to lure many others into this dynamic field.

## A NOTE TO THE READER

This is first of all a Mathematics text. Throughout, we emphasize the mathematical aspects of the theory of discrete dynamical systems, not the many and diverse applications of this theory. The text begins at a relatively unsophisticated level and, by the end, has progressed so as to require not much more than the typical mathematics education of an engineer or a physicist. Fully three quarters of the text is accessible to students with only a solid advanced calculus and linear algebra background. Of course, a good dose of mathematical sophistication is useful throughout.

The first chapter, one-dimensional dynamics, is by far the longest. It is the author's belief that virtually all of the important ideas and techniques of nonlinear dynamics can be introduced in the setting of the real line or the circle. This has the obvious advantage of minimizing the topological complications of the system and the algebraic machinery necessary to handle them. In particular, the only real prerequisite for this chapter is a good calculus course. (O.K., we do multiply a $2 \times 2$ matrix once or twice in §1.14 and we use the Implicit Function Theorem in two variables in §1.12, but these are exceptions.) With only these tools, we manage to introduce such important topics as structural stability, topological conjugacy, the shift map, homoclinic points, and bifurcation theory. To emphasize the point that chaotic dynamics occurs in the simplest of systems, we carry out most of our analysis in this section on a basic model, the quadratic mapping given by $F_\mu(x) = \mu x(1 - x)$. This map has the advantage of being perhaps the simplest nonlinear map yet one which illustrates virtually every concept we wish to introduce. A few topological ideas, such as the notion of a dense set or a Cantor set, are introduced in detail when needed.

The second chapter is devoted to higher dimensional dynamical systems. With many of the prerequisites already introduced in the first chapter, the discussion of such higher dimensional maps as Smale's horseshoe, the hyperbolic toral automorphisms, and the solenoid become especially simple. This chapter assumes that the student is familiar with some multi-dimensional calculus as well as linear algebra, including the notion of eigenvalues and eigenvectors for $3 \times 3$ matrices. One of the major differences between one dimensional and higher dimensional dynamics, the possibility of both contraction and expansion at the same time, is treated at length in a section devoted to the proof of the Stable Manifold Theorem. We end the chapter with a lengthy set of exercises all centered on the important Hénon map of the plane. This section serves as a summary of many of the previous topics in the section as well as a good "final" project for the reader.

The last chapter should be regarded as a "special topics" chapter in that we presuppose a working knowledge of complex analysis. In this chapter we describe some of the fascinating and beautiful recent work on the dynamics of complex analytic maps and, in particular, the structure of the Julia set of polynomials. This gives a complementary view of the dynamics of maps such as the quadratic map, which receives so much attention in chapter one.

Each of the chapters is self-contained, assuming familiarity with the basic concepts of dynamics as outlined in the first chapter. Accordingly, we have numbered the Theorems, Figures, etc. consecutively within each subsection, without reference to the chapter number. As there is very little cross-referencing between chapters, this should cause no confusion.

There are many themes developed in this book. We have tried to present several different dynamical concepts in their most elementary formulation in chapter one and to return to these subjects for further refinement at later stages in the book. One such topic is bifurcation theory. We introduce the most elementary bifurcations, the saddle-node and the period- doubling bifurcations, early in chapter one. Later in the same chapter we treat the accumulation points of such bifurcations which occur when a homoclinic point develops. In chapter two, we return to bifurcation theory to discuss the Hopf bifurcation as well as the elliptic bifurcations which occur in area-preserving systems. Finally, in the last chapter, we explore several types of bifurcations that occur in analytic dynamics, including a discussion of the global aspects of the saddle-node bifurcation.

Another recurrent theme is symbolic dynamics. We think of symbolic dynamics as a tool whereby complicated dynamical systems are reduced to seemingly quite different systems which enjoy the advantage that they can be analyzed quite easily. Symbolic dynamics appears quite early in chapter one when we first discuss the quadratic map. It is clear that the most elementary setting for the phenomena associated with the Smale horseshoe mapping occurs in one dimension and we fully exploit this idea. Later, symbolic dynamics is extended to the case of subshifts of finite type via another quadratic example. And finally the related concepts of Markov partitions and inverse limits are introduced in the second chapter.

Examples abound in the text. We often motivate new concepts by working through them in the setting of a specific dynamical system. In fact, we have often sacrificed generality in order to concentrate on a specific system or class of systems. Many of the results throughout the text are stated in a form that is nowhere near full generality. We feel that the general theory is best left to more advanced texts which presuppose more advanced Mathematics.

Much of what many researchers consider dynamical systems has been deliberately left out of this text. For example, we do not treat continuous systems or differential equations at all. There are several reasons for this. First, as is well known, computations with specific nonlinear ordinary differential equations are next to impossible. Secondly, the study of differential equations necessitates a much higher level of sophistication on the part of the student, certainly more than that necessary for chapter one of this text. We adopt instead the attitude that any dynamical phenomena that occurs in a continuous system also occurs in a discrete system, and so we might as well make life easy and study maps first. There are many texts currently available that treat continuous systems almost exclusively. We hope that this book presents an solid introduction to the topics treated in these more advanced texts.

Another topic that has been excluded is ergodic theory. It is our feeling that measure theory would take us too far afield in an elementary text. Of course, it can be argued that measure theory is no more advanced than the complex analysis necessary for chapter three. However, we feel that the topological approach adopted throughout this text is inherently easier to understand, at least for an undergraduate in Mathematics. There is no question, however, that ergodic theory would provide an ideal sequel to the material presented here, as would a course in nonlinear differential equations.

This text has benefited from the suggestions and comments of many people. I would like to thank Clark Robinson, Guido Sandri, Harvey Keynes, Phil Boyland, Paul Blanchard, Dick Hall, and Elwood Devaney for helpful comments on portions of the manuscript. Richard Millman, Chris Golé, and Steve Batterson read the entire text and made many useful suggestions regarding content and organization (and incorrect proofs.) Finally, this text never would have been completed without the constant advice and encouragement of Phil Holmes. The book owes much to his experience and expertise.

The book was produced using TₑX at Boston University by Tom Orowan. Tom's near-perfect typing and formatting of the text made the production of the book effortless and fun. Thanks are also due Chris Mayberry for his help designing the figures. And finally, it is a pleasure to thank Rick Mixter and his staff at Benjamin-Cummings for their enthusiastic support for the duration of this project.

Robert L. Devaney
*Boston, Mass.*
*April, 1985*


This printing has benefited immensely from the suggestions of many mathematicians, including Susan Dabros, Jenny Harrison, Roger Kraft, Tyre Newton, John Milnor, Connie Overzet, Charles Pugh, Joe Silverman, and Mary Lou Zeeman. Elwood Devaney again digested the entire manuscript and returned many stylistic suggestions. And thanks are especially due to Gary Meisters and his class at the University of Nebraska for their enthusiastic response to this book, but I'm afraid that I'm not the coach they think I am!

Robert L. Devaney
*Boston, Mass.*
*September, 1986*

# Table of Contents

# An Introduction to

# Chaotic Dynamical Systems

# Chapter One

# One-Dimensional Dynamics

The goal of this first chapter is to introduce many of the basic techniques from the theory of dynamical systems in a setting that is as simple as possible. Accordingly, all of the dynamical systems that we will encounter take place in one dimension, either on the real line or on the unit circle in the plane. For that reason. much of this chapter can be read with only a solid background in calculus

We regard the first twelve sections of this chapter as central to the theory of dynamical systems. Here we introduce such topics as hyperbolicity, symbolic dynamics, topological conjugacy, structural stability, and chaos. These form the essential background for all that follows. Indeed, the last two chapters of this text may be regarded as extensions and refinements of the material presented in these introductory sections.

Our main thrust in this chapter is to understand what it means for a dynamical system to be chaotic. We feel that this is best understood in light of examples. Hence most of our initial effort revolves around a single family of examples, the family of quadratic maps $F_\mu(x) = \mu x(1 - x)$. Later, using such tools as Sarkovskii's Theorem and the Schwarzian derivative, we will show that the seemingly specialized results for the quadratic map actually hold for a large variety of dynamical systems.

The next four sections in this chapter present material which is somewhat more technical than the preceding material. The concepts introduced – subshifts of finite type, Morse-Smale maps, the rotation number, and homoclinic bifurcation – are important in the sequel, however. The final three sections on the kneading theory and the period-doubling route to chaos should be regarded as special topics. They will not be used in what follows. However, for the reader interested in recent work on the transition to chaotic dynamics, these sections should provide an introduction to many of the topics in the current literature.

## §1.1 EXAMPLES OF DYNAMICAL SYSTEMS

This brief section is intended merely as motivation for the succeeding sections. Our aim is to give a couple of simple examples of dynamical systems. These examples show how dynamical systems occur in the "real world" and how some very simple phenomena from nature yield rather complicated dynamical systems.

First, what is a dynamical system? The answer is quite simple: take a scientific calculator and input any number whatsoever. Then start striking one of the function keys over and over again. This iterative procedure is an example of a *discrete dynamical system*. For example, if we repeatedly strike the "exp" key, given an initial input $x$, we are computing the sequence of numbers

$$x, \ e^x \ e^{e^x}, \ e^{e^{e^x}} \ \ldots$$

That is, we are iterating the exponential function. If this experiment is performed over and over again, it becomes apparent that any choice of initial $x$ leads rather quickly to an "overflow" message from the calculator: that is, successive iterations of $\exp(x)$ tend to $\infty$. This is, in fact, the main question we will ask in the sequel: given a function $f$ and an initial value $x_0$, what ultimately happens to the sequence of iterates

$$x_0, \ f(x_0), \ f(f(x_0)), \ f(f(f(x_0))), \ldots$$

As another example, consider $\sin x$. A few keystrokes on the calculator will be enough to convince the reader that any initial $x_0$ leads to a sequence of iterates which tends to 0. Similarly, for cos $x$, any $x_0$ yields a sequence which converges fairly rapidly to .73908 ... (in radians, or to .99984 ... in degrees). The reader may begin to suspect that iteration of a given function on a given initial value always yields a sequence that converges to a fixed limit (maybe $x_0$, maybe 0, in any case a unique limit). Actually, nothing could be further from the truth. Very simple functions, even the simplest quadratic functions on the real line, lead to bizarre and unpredictable results when iterated. For example, program a computer or calculator to iterate the simple function $f(x) = 4x(1 - x)$. Input a random number between 0 and 1 and watch the results of the iteration. One gets dramatically different behaviors depending upon which initial $x$ is input. Sometimes the values repeat; other times they do not. Most often they wander aimlessly about the unit interval with no discernible pattern. Now change the parameter from 4 to 3.839, i.e. iterate the function $f(x) = 3.839x(1 - x)$. For a random entry between 0 and 1, one observes that the iterates of this point eventually settle down to a repeating cycle of three numbers, .149888 ..., .489172 ..., and .959299 ..., repeated over and over again in succession. Two comments are in order. The first example illustrates the phenomenon of chaos or unpredictability that forms one of the major themes of this book. Despite its complexity, we will see how to analyze this unpredictability completely. Second, chaos occurs in many, many dynamical systems. The second example above, which seems comparatively rather tame, also admits a set of initial $x$ values which behave just as unpredictably as in the first example. However, due to roundoff or "experimental" error, we do not see this randomness at first glance. Nevertheless, as we shall see, it lurks in the background and has an increasingly important effect on the system as the accuracy of the computations is increased.

At this juncture, we should note that there are many other types of dynamical systems besides iterated functions. For example, differential equations are examples of *continuous*, as opposed to discrete, dynamical systems. In this book, we will not deal with these types of systems at all. These types of systems are much easier to understand once the basic behavior of discrete systems has been mastered.

Let us now consider several "applied" examples. Dynamical systems occur in all branches of science, from the differential equations of classical mechanics in physics to the difference equations of mathematical economics and biology. We will first describe a simple model from population biology which will serve as motivation for all of the succeeding chapter.

Population biologists are interested in the long-term behavior of the population of a certain species or collection of species. Given certain observed or experimentally determined parameters (number of predators, severity of climate, availability of food, etc.), the biologist sets up a mathematical model to describe the fluctuations in the population. This may take the form of a differential equation or a difference equation, depending upon whether the population is assumed to change continuously or discretely, such as when the population is measured once a year or once a generation. In either case, the population biologist is interested in what happens to an initial population of $P_0$ members. Does the population tend to zero as time goes on, leading to extinction of the species? Does the population become arbitrarily large, indicating eventual overcrowding? Or does the population fluctuate periodically or even randomly? Thus the problem facing the population biologist is a typical dynamical systems question: given $P_0$, can one predict the long-term behavior of the population?

Several simple biological models are encountered in elementary calculus courses. For example, the differential equation of exponential growth or decay is often the first differential equation a student is exposed to. In this model, we assume that the population of a single species changes at a rate that is directly proportional to the population present at the given time. This is, of course, an extremely naive model, which does not take into account obvious factors such as overcrowding, the death rate, etc. However, this model does produce an especially simple differential equation which is readily solved. If $P(t)$ denotes the population at time $t$, the assumptions above may be translated into

$$\frac{dP}{dt} = kP.$$

The solution to this equation is $P(t) = P_0 e^{kt}$ where $P_0 = P(0)$ is the initial population of the species. Hence, if the constant of proportionality is positive, $P(t) \rightarrow \infty$ as $t \rightarrow \infty$ leading to population explosion. If $k < 0$, then $P(t) \rightarrow 0$ as $t \rightarrow \infty$, leading to extinction.

This procedure illustrates (in an exceedingly simple situation) the typical application of dynamical systems in science. A population biologist sets up a mathematical model for which the mathematician is asked to provide some idea about the long-term behavior of the solutions.

This simple model can also be studied as a difference equation. Let us write $P_n =$ population after $n$ generations, where $n$ is a natural number. The simplest growth law one can imagine is that the population in the next generation is directly proportional to that in the present generation. That is

$$P_{n+1} = k P_n$$

where again $k$ is a constant.

We have

$$P_1 = kP_0$$
$$P_2 = kP_1 = k^2 P_0$$
$$P_3 = kP_2 = k^3 P_0$$

$$\vdots$$

$$P_n = kP_{n-1} = k^n P_0$$

so that the ultimate fate of the population is again easy to decide. If $k >$ 1, $P_n \to \infty$, whereas if $0 < k < 1$, then $P_n \to 0$.

For later use, let us recast this difference equation as a function. Let $x = P_0$ and set $f(x) = kx$. Note that, in the above terms, $f(x) = P_1$, $f(f(x)) = k^2 x = P_2$, $f(f(f(x))) = P_3$, etc. Hence the ultimate behavior of the population is intimately related to the asymptotic behavior of the iteration of the function $f$.

In either of the above models, one has a rather idealized situation. There are essentially only two possibilities: unchecked growth or extinction. Experience tells the population biologist that more complicated patterns arise in nature. So the biologist tries to incorporate additional constraints or parameters in the model, hoping for a better reflection of reality. One such approach again often encountered in calculus is to assume that there is some limiting value $L$ for the population. If $P(t)$ exceeds $L$, the population should tend to decrease (there is overcrowding, not enough food, etc.). On the other hand, if $P(t) < L$, there is room for more of the species so $P(t)$ should increase. The simplest biological model leading to this behavior is

$$\frac{dP}{dt} = kP(L - P)$$

Note that we have simply tacked on the factor $L - P$ to the previous model.

Let us assume that $k > 0$, the case that previously led to unlimited growth. Here we note that

1. if $P = L$, $\dfrac{dP}{dt} = 0$

2. if $P > L$, $\dfrac{dP}{dt} < 0$

3. if $P < L$, $\dfrac{dP}{dt} > 0$

Thus elementary calculus shows that this model behaves according to our expectations. The population remains constant, decreases, or increases depending upon whether $P = L, P > L$, or $P < L$. In fact, one can explicitly solve the above differential equation via separation of variables and integration by partial fractions. One finds that

$$P(t) = \frac{LP_0 e^{Lkt}}{L - P_0 + P_0 e^{Lkt}}.$$

Using this formula, one may easily sketch the solutions of this system.

While this model conforms more to reality than the exponential growth model, nevertheless we see no cyclic behavior or other fluctuations in the population. One might naively expect that the corresponding difference equation behaves similarly. However, we are in for a great surprise: the analogous difference equation leads to one of the most complicated dynamical systems imaginable. To this day, the dynamics of this system are not completely understood. Moreover, this system exhibits many of the pathologies of higher dimensional systems and for this reason may be considered as one of the most basic nonlinear dynamical systems. We will return to it throughout the chapter as it provides a rich source of illustrative examples.

Let us make a simplification in our model. Let us assume that $L = 1$ is the limiting value. Obviously, we are not now talking about populations but rather percentage of population. $P_n$ represents the percentage of the limiting population present in generation $n$. The population is then assumed to satisfy the following difference equation.

$$P_{n+1} = kP_n(1 - P_n)$$

where again $k$ is a positive constant. As before, we may write $x = P_0$ and $f(x) = kx(1-x)$. This, of course, is the quadratic function mentioned above. We have

$$P_1 = f(x)$$
$$P_2 = f(f(x))$$
$$P_3 = f(f(f(x)))$$

and so on. Thus to determine the fate of a population for a given constant $k$, we must determine the asymptotic behavior of the function $kx(1 - x)$. This function, known as the logistic function, and its dynamics have been the subject of much contemporary mathematical research. In the following chapters, we will only begin to describe the complications and pathologies that arise in this simple system.

Another example of a dynamical system which arises in practical applications is Newton's method for finding the roots of a polynomial. Let

$$Q(x) = a_n x^n + a_{n-1} x^{n-1} + \ldots + a_0$$

be a polynomial. In general, it is impossible to factor $Q$ if the degree of $Q$ is high. Nevertheless, it is often important in applications to find a root of $Q$. One such procedure for doing this is the classical recursion scheme of Newton. Let $x_0$ be a real number. Consider the recursion

$$x_1 = x_0 - \frac{Q(x_0)}{Q'(x_0)}$$

$$x_2 = x_1 - \frac{Q(x_1)}{Q'(x_1)}$$

$$\vdots$$

$$x_n = x_{n-1} - \frac{Q(x_{n-1})}{Q'(x_{n-1})}.$$

For most choices of the initial value $x_0$, it is well known from calculus that the sequence of values $x_0, x_1, x_2, \ldots$ converges to one of the roots of $Q$.

Given the polynomial $Q$, we thus see that Newton's method determines a dynamical system. Let

$$N(x) = x - \frac{Q(x)}{Q'(x)}.$$

As long as $Q'(x) \neq 0$, this function is well defined. As in our population model, Newton's method reduces to the iteration of $N$. Again we ask the same question: given $x$, what happens as we compute successively higher iterates of $N$ at $x$ ?

We remark that Newton's method does not always converge. For certain initial values $x_0$, the iterative scheme does not yield convergence to a root of $Q$. The structure of the set where $N$ fails to converge is extremely interesting (especially in the complex plane) and leads to unpredictable behavior similar to the logistic function. We will take up this topic in chapter three.

## §1.2 PRELIMINARIES FROM CALCULUS

In this section, we recall some elementary (and not-so-elementary) notions from single variable and multivariable calculus. In the sequel, we will also need a few notions from point-set topology, so we include them here as well. First, we fix some notation. $\mathbf{R}$ denotes the real numbers. $I$ or $J$ will always denote closed intervals in $\mathbf{R}$, i.e., all points $x$ satisfying $a \leq x \leq b$ for some $a$ and $b$. $\mathbf{R}^2$ denotes the Cartesian plane.

Let $f: \mathbf{R} \to \mathbf{R}$ be a function. We denote the derivative of $f$ at $x$ by $f'(x)$, the second derivative by $f''(x)$, and higher derivatives by $f^{(r)}(x)$. We say that $f$ is of class $C^r$ on $I$ if $f^{(r)}(x)$ exists and is continuous at all $x \in I$. A function is said to be *smooth* if it is of class $C^1$. The function $f(x)$ is $C^\infty$ if *all* derivatives exist and are continuous. Throughout this book, function means $C^\infty$ function; occasionally we will use functions which are continuous but non-differentiable as examples, but in general, when we say function, we mean $C^\infty$ function.

There are other classes of functions which are commonly studied in calculus. For example, analytic functions (i.e., those with convergent power series representations) are often encountered. For our purposes in this chapter, these types of functions are too rigid in the following sense. We wish to allow small changes in or perturbations of the functions which will change the function in a certain interval but not everywhere. This is accomplished by the use of *bump functions* which we will introduce in the Exercises. These small changes are impossible if we are restricted to analytic functions, for a small change in any of the coefficients of the power series affects the behavior of the function *everywhere*. Later, in chapter three, when we discuss complex analytic dynamical systems, we will restrict our attention solely to these types of functions.

There are some special classes of functions that often arise. The function $f(x)$ is linear if $f(x) = ax$ for some constant $a$; $f(x)$ is affine if $f(x) = ax + b$; $f(x)$ is piecewise linear if $f(x)$ is affine on a collection of intervals. For example, $f(x) = |x|$ is piecewise linear, the "pieces" being the positive and negative reals on each of which $f(x)$ is linear.

**Definition 2.1.** $f(x)$ is one-to-one if $f(x) \neq f(y)$ whenever $x \neq y$.

Clearly, increasing or decreasing functions are the only types of continuous one-to-one functions of a real variable. If $f: I \to J$ is one-to-one, then we may define the inverse of $f$, written $f^{-1}(x)$, by the rule $f^{-1}(x) = y$ if and only if $f(y) = x$. For example, if $f(x) = x^3$, then $f^{-1}(x) = \sqrt[3]{x}$ and if $g(x) = \tan x$, then $g^{-1}(x) = \arctan x$. Here $g: (-\pi/2, \pi/2) \to \mathbf{R}$ so $g^{-1}: \mathbf{R} \to (-\pi/2, \pi/2)$.

**Definition 2.2.** Let $I$ and $J$ be intervals and $f: I \to J$. The function $f$ is *onto* if for any $y$ in $J$ there is an $x \in I$ such that $f(x) = y$. See Fig. 2.1.

**Definition 2.3.** Let $f: I \to J$. The function $f(x)$ is a *homeomorphism* if $f(x)$ is one-to-one, onto, and continuous, and $f^{-1}(x)$ is also continuous.

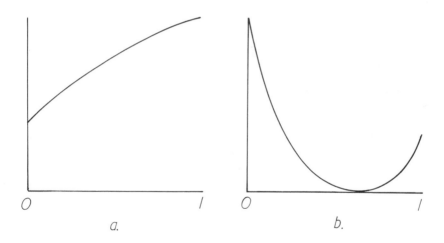

**Fig. 2.1.** In a. $f(x)$ is one-to-one on the interval $[0, 1]$; in b. $f(x)$ is onto the interval $[0, 1]$.

For example, $\tan x$ is a homeomorphism between $(-\pi/2, \pi/2)$ and $\mathbf{R}$. Thus we say the open interval $(-\pi/2, \pi/2)$ is *homeomorphic* to $\mathbf{R}$. Functions which are one-to-one are also said to be *injective*, while functions which are onto are also called *surjective*.

**Definition 2.4.** Let $f: I \to J$. The function $f(x)$ is a $C^r$-*diffeomorphism* if $f(x)$ is a $C^r$-homeomorphism such that $f^{-1}(x)$ is also $C^r$.

For example, it is easy to see that $\tan x$ is a $C^\infty$ diffeomorphism from $(-\pi/2, \pi/2)$ to $\mathbf{R}$, whereas $f(x) = x^3$ is a homeomorphism which is *not* a diffeomorphism since $f^{-1}(x) = x^{1/3}$ and $(f^{-1})'(0)$ does not exist.

We will see in subsequent chapters that diffeomorphisms on the real line are extremely simple, dynamically speaking. Therefore, in this chapter, we will primarily consider non-invertible functions. In higher dimensions, diffeomorphisms become much more interesting and therefore become the focal point for dynamical systems theory.

We denote the *composition* of two functions by $f \circ g(x) = f(g(x))$. The $n$-fold composition of $f$ with itself recurs over and over again in the sequel. We denote this function by $f^n(x) = \underbrace{f \circ \ldots \circ f(x)}_{n\ times}$. Note that $f^n$ does not mean $f(x)$ raised to the $n^{th}$ power, a function which we will *never* use, nor does it mean the $n^{th}$ derivative of $f(x)$, which we denote by $f^{(n)}(x)$. If $f^{-1}(x)$ exists, we write $f^{-n}(x) = f^{-1} \circ \ldots \circ f^{-1}(x)$.

Perhaps the most important feature from elementary calculus that we will use is the Chain Rule:

**Proposition 2.5.** *If $f$ and $g$ are functions, then*

$$(f \circ g)'(x) = f'(g(x))g'(x)$$

*In particular, if $h(x) = f^n(x)$, then*

$$h'(x) = f'(f^{n-1}(x)) \cdot f'(f^{n-2}(x)) \cdot \ldots \cdot f'(x).$$

Another important notion from elementary calculus is the Mean Value Theorem:

**Theorem 2.6.** *Suppose $f: [a, b] \to \mathbf{R}$ is $C^1$. Then there exists $c \in [a, b]$ such that*

$$f(b) - f(a) = f'(c)(b - a).$$

Fig. 2.2 illustrates the content of the Mean Value Theorem. The third important result from calculus is the Intermediate Value Theorem:

**Theorem 2.7.** *Suppose $f: [a, b] \to \mathbf{R}$ is continuous. Suppose that $f(a) = u$ and $f(b) = v$. Then for any $z$ between $u$ and $v$, there exists $c$, $a \leq c \leq b$, such that $f(c) = z$.*

One of the most abstract and seemingly useless theorems from multi-variable calculus is the Implicit Function Theorem. Most beginning students have no appreciation of the power of this Theorem when they encounter it in

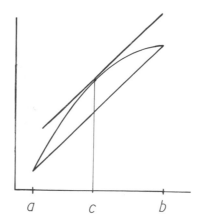

**Fig. 2.2.** The Mean Value Theorem.

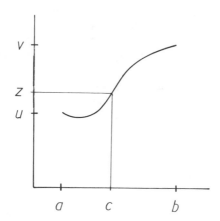

**Fig. 2.3.** The Intermediate Value Theorem.

their first analysis course. We hope that the geometric results in bifurcation theory that we will encounter later will help dispel any misconceptions about the usefulness of this theorem.

**Theorem 2.8.** *Suppose $G: \mathbf{R}^2 \rightarrow \mathbf{R}^1$ is a $C^1$-function (i.e., both partial derivatives of $G$ exist and are continuous). Suppose further that*

    *1. $G(x_0, y_0) = 0$*

    *2. $\dfrac{\partial G}{\partial y}(x_0, y_0) \neq 0$*

*Then there exists open intervals $I$ about $x_0$ and $J$ about $y_0$ and a $C^1$-function*

$p: I \rightarrow J$ *satisfying*

    1. $p(x_0) = y_0$
    2. $G(x, p(x)) = 0$ *for all* $x \in I$.

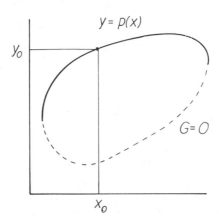

**Fig. 2.4.** The Implicit Function Theorem.

    Rather than prove the Implicit Function Theorem, we give several examples of how to apply it. While these examples are obviously concocted, they nevertheless are typical, as we shall see later.

**Example 2.9.** Let $G(x, y) = x^2 + y^2 - 1$. The level sets of $G$ are clearly circles, and $G = 0$ defines the unit circle in the plane.

    Suppose $G(x_0, y_0) = 0$ and $y_0 > 0$, i.e., $(x_0, y_0)$ is a point on the upper or lower semicircle. Clearly,

$$\frac{\partial G}{\partial y}(x_0, y_0) = 2y_0 \neq 0$$

so the Implicit Function Theorem applies. The result is a function $p(x)$ which satisfies $G(x, p(x)) = 0$ for all $x$ sufficiently close to $x_0$. What is $p(x)$? In this case, we can construct $p(x)$ explicitly. Clearly, $p(x) = \sqrt{1 - x^2}$, which is $C^{\infty}$ as long as $x \neq \pm 1$ (when $y = 0$ ). We have $G(x, \sqrt{1 - x^2}) = 0$ for $|x| < 1$, as the Implicit Function Theorem guarantees. If $y_0 < 0$, then we must choose $p(x) = -\sqrt{1 - x^2}$.

    It is important to realize that, in practice, one cannot very often solve for the function $p(x)$ as we did here. Nevertheless, the Implicit Function

Theorem guarantees its existence (whether or not we can explicitly write it down), and that is often exactly what we need.

**Example 2.10.** $G(x,y) = x^5 y^4 - xy^5 - yx^2 + 1$ satisfies $G(1,1) = 0$ and

$$\frac{\partial G}{\partial y}(1,1) = -2.$$

Hence there is a function $p(x)$ defined in some interval about $x = 1$ and which satisfies $G(x, p(x)) = 0$. Solving $G(x,y) = 0$ for $y = p(x)$ is impossible, however.

Fixed points for functions are points $x$ which satisfy $f(x) = x$. These points will play a dominant role in the theory of dynamical systems. The following easy application of the Intermediate Value Theorem gives an important criterion for the existence of a fixed point. See Fig. 2.5.

**Proposition 2.11.** *Let* $I = [a,b]$ *be an interval and let* $f: I \to I$ *be continuous. Then* $f$ *has at least one fixed point in* $I$.

*Proof.* Let $g(x) = f(x) - x$. Clearly, $g(x)$ is continuous on $I$. Suppose $f(a) > a$ and $f(b) < b$ (otherwise, one of $a$ or $b$ is fixed). We thus have $g(a) > 0$ and $g(b) < 0$, so the Intermediate Value Theorem gives the existence of $c$ between $a$ and $b$ for which $g(c) = 0$. Therefore, $f(c) = c$ and we are done.

q.e.d.

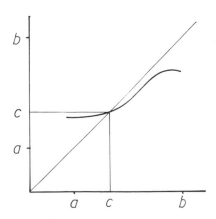

**Fig. 2.5.** $f: I \to I$ has at least one fixed point.

This theorem is a special case of a much more general theorem called the Brouwer Fixed Point Theorem, which gives a similar sufficient condition for the existence of fixed points in higher dimensions. One can actually do better with a little differentiability. The following result is a special case of the Contraction Mapping Theorem.

**Proposition 2.12.** *Let $f: I \to I$ and assume that $|f'(x)| < 1$ for all $x$ in $I$. Then there exists a unique fixed point for $f$ in $I$. Moreover*

$$|f(x) - f(y)| < |x - y|$$

*for all $x, y \in I, x \neq y$.*

*Proof.* Proposition 2.11 guarantees at least one fixed point for $f$, so we suppose that both $x$ and $y$ are fixed points, $x \neq y$. By the Mean Value Theorem, there is a $c$ between $x$ and $y$ such that

$$f'(c) = \frac{f(y) - f(x)}{y - x} = 1$$

But this contradicts our assumption that $|f'(c)| < 1$ for all $c$ in $I$. Hence $x = y$.

To establish the second assertion of the Proposition, we again use the Mean Value Theorem to assert that for any $x, y \in I, x \neq y$

$$|f(y) - f(x)| = |f'(c)||y - x| < |y - x|$$

as required.

q.e.d.

We close this section with a few notions from general topology. In general, these notions are beyond the scope of elementary calculus courses. However, many of them occur in the simplest possible setting on the real line, and this is precisely the setting in which we will work.

**Definition 2.13.** Let $S \subset \mathbf{R}$. A point $x \in \mathbf{R}$ is a limit point of $S$ if there is a sequence of points $x_n \in S$ converging to $x$. $S$ is a closed set if it contains all of its limit points.

Clearly, closed intervals of the form $a \leq x \leq b$ are closed sets. Any finite union of closed sets is also closed. Infinite unions of closed sets, however, need not be closed, as the following example shows.

**Example 2.14.** Let $I_n = [\frac{1}{n}, 1]$. Then

$$\bigcup_{n=1}^{\infty} I_n = (0, 1]$$

which is not closed, since 0 is a limit point of $S$ which is not in $S$.

Intersections of closed sets yield closed sets, however (the empty set is, by definition, a closed set). Moreover, if $I_n$ is a closed and bounded interval for each $n$ and $I_{n+1} \subset I_n$, then $\cap_{n=1}^{\infty} I_n$ is a closed, *non-empty* set. The crucial word here is, of course, non-empty.

**Definition 2.15.** Let $S \subset \mathbf{R}$. $S$ is an open set if, for any $x \in S$, there is an $\epsilon > 0$ such that all points $t$ in the open interval $x - \epsilon < t < x + \epsilon$ are contained in $S$.

It is clear that the complement of a closed set is open and vice versa. Unlike closed sets, infinite unions of open intervals are open sets in $\mathbf{R}$. However, infinite intersections of open intervals are not open sets. For example, if $J_n = (-\frac{1}{n}, \frac{1}{n})$, then $\cap_{n=1}^{\infty} J_n = \{0\}$ which is closed.

For any set $S$, we denote the closure of $S$ by $\overline{S}$. $\overline{S}$ consists of all points in $S$ together with all limit points of $S$. For example, if $S$ is the open interval $(0, 1)$, then $\overline{S}$ is the closed interval $[0, 1]$. Clearly, if $S$ is closed, then $\overline{S} = S$.

**Definition 2.16.** A subset $U$ of $S$ is *dense* in $S$ if $\overline{U} = S$.

For example, any open set $S$ is dense in its closure $\overline{S}$. A more interesting example is the set of rational numbers $Q$, which is dense in $\mathbf{R}$. Similarly, the irrationals are dense in $\mathbf{R}$. We caution the reader against thinking that dense subsets are necessarily large. Even open and dense sets may be quite small in the sense of total length. Here is an example in the unit interval $I$ given by $0 \leq x \leq 1$. Since the rationals form a countable set in $I$, we may list them in some order. One such ordering is

$$0, 1, \frac{1}{2}, \frac{1}{3}, \frac{2}{3}, \frac{1}{4}, \frac{3}{4}, \frac{1}{5}, \frac{2}{5}, \frac{3}{5}, \frac{4}{5}, \frac{1}{6}, \cdots$$

Now let $\epsilon > 0$ be small. Consider the open interval of length $\epsilon^n$ about the $n^{th}$ rational in this list. The union of all of these intervals is an open set in $I$ which is clearly dense since it contains all of the rationals in $I$. However, the total length of this set is quite small. Indeed, the length is given by

$$\sum_{n=1}^{\infty} \epsilon^n = \frac{\epsilon}{1 - \epsilon}.$$

This example shows clearly the difference between the topological approach to dynamics that we will adopt in the sequel and the measure theoretic approach. In a topological sense, an open, dense subset is considered "large". These sets may or may not be large in a measure theoretic sense, i.e., in the sense of total length.

**Exercises**

**1.**   Decide whether each of the following functions are one-to-one, onto, homeomorphisms, or diffeomorphisms on their domains of definition.

   a. $f(x) = x^{5/3}$

   b. $f(x) = x^{4/3}$

   c. $f(x) = 3x + 5$

   d. $f(x) = e^x$

   e. $f(x) = 1/x$

   f. $f(x) = 1/x^2$

**2.**   Identify which of the following subsets of $\mathbf{R}$ are closed, open, or neither.

   a. $\{x \mid x \text{ is an integer }\}$

   b. $\{x \mid x \text{ is a rational number }\}$

   c. $\{x \mid x = \frac{1}{n} \text{ for some natural number } n\}$

   d. $\{x \mid \sin(\frac{1}{x}) = 0\}$

   e. $\{x \mid x \sin(\frac{1}{x}) = 0\}$

   f. $\{x \mid \sin(\frac{1}{x}) > 0\}$

**3.**   Prove that the set of rational numbers of the form $p/2^n$ for $p, n \in \mathbf{Z}$ is dense in $\mathbf{R}$.

The goal of the next few exercises is to construct special functions which will be useful later when we perturb or change slightly a given function. These functions are called "bump functions." Define

$$B(x) = \begin{cases} \exp(-1/x^2) & \text{if } x > 0 \\ 0 & \text{if } x \leq 0 \end{cases}$$

**4.**   Sketch the graph of $B(x)$.

**5.**   Prove that $B'(0) = 0$.

**6.**   Inductively prove that $B^{(n)}(0) = 0$ for all $n$. Conclude that $B(x)$ is a $C^\infty$ function.

**7.**  Modify $B(x)$ to construct a $C^\infty$ function $C(x)$ which satisfies

    a.  $C(x) = 0$ if $x \leq 0$.

    b.  $C(x) = 1$ if $x \geq 1$.

    c.  $C'(x) > 0$ if $0 < x < 1$.

**8.**  Modify $C(x)$ to construct a $C^\infty$ *bump function* $D(x)$ on the interval $[a, b]$, i.e., $D(x)$ satisfies

    a.  $D(x) = 1$ for $a \leq x \leq b$.

    b.  $D(x) = 0$ for $x < \alpha$ and $x > \beta$ where $\alpha < a$ and $\beta > b$.

    c.  $D'(x) \neq 0$ on the intervals $(\alpha, a)$ and $(b, \beta)$.

**9.**  Use a bump function to construct a diffeomorphism $f: [a, b] \rightarrow [c, d]$ which satisfies $f'(a) = f'(b) = 1$ and $f(a) = c$, $f(b) = d$.

## §1.3 ELEMENTARY DEFINITIONS

The basic goal of the theory of dynamical systems is to understand the eventual or asymptotic behavior of an iterative process. If this process is a differential equation whose independent variable is time, then the theory attempts to predict the ultimate behavior of solutions of the equation in either the distant future $(t \rightarrow \infty)$ or the distant past $(t \rightarrow -\infty)$. If the process is a discrete process such as the iteration of a function, then the theory hopes to understand the eventual behavior of the points $x, f(x), f^2(x), \ldots, f^n(x)$ as $n$ becomes large. That is, dynamical systems asks the somewhat non-mathematical sounding question: where do points go and what do they do when they get there? In this chapter, we will attempt to answer this question at least partially for one of the simplest classes of dynamical systems, functions of a single real variable. Functions which determine dynamical systems are also called *mappings*, or *maps*, for short. This terminology connotes the geometric process of taking one point to another. As much of the sequel will in fact be geometric, we will use all of these terms synonymously.

**Definition 3.1.**  The forward orbit of $x$ is the set of points $x, f(x), f^2(x), \ldots$ and is denoted by $O^+(x)$. If $f$ is a homeomorphism, we may define the full orbit of $x$, $O(x)$, as the set of points $f^n(x)$ for $n \in \mathbf{Z}$, and the backward orbit of $x$, $O^-(x)$, as the set of points $x, f^{-1}(x), f^{-2}(x), \ldots$.

Thus our basic goal is to understand all orbits of a map. Orbits and forward orbits of points can be quite complicated sets, even for very simple nonlinear mappings. However, there are some orbits which are especially simple and which will play a central role in the study of the entire system.

**Definition 3.2.** The point $x$ is a fixed point for $f$ if $f(x) = x$. The point $x$ is a periodic point of period $n$ if $f^n(x) = x$. The least positive $n$ for which $f^n(x) = x$ is called the prime period of $x$. We denote the set of periodic points of (not necessarily prime) period $n$ by $\text{Per}_n(f)$, and the set of fixed points by $\text{Fix}(f)$. The set of all iterates of a periodic point form a periodic orbit.

Maps may have many fixed points. For example, the identity map $id(x) = x$ fixes all points in $\mathbf{R}$, whereas the map $f(x) = -x$ fixes the origin, while all other points have period 2. These, however, are atypical dynamical systems; maps with intervals of fixed or periodic points are rare in a sense which will be made precise later. Most of the dynamical systems we will encounter will have isolated periodic points.

**Example 3.3.** The map $f(x) = x^3$ has $0, 1$, and $-1$ as fixed points and no other periodic points. The map $P(x) = x^2 - 1$ has fixed points at $(1 \pm \sqrt{5})/2$, while the points $0$ and $-1$ lie on a periodic orbit of period 2.

**Example 3.4.** Let $S^1$ denote the unit circle in the plane. We denote a point in $S^1$ by its angle $\theta$ measured in radians in the standard manner. Hence a point is determined by any angle of the form $\theta + 2k\pi$ for an integer $k$. Now let $f(\theta) = 2\theta$. (Note that $f(\theta + 2\pi) = f(\theta)$ on the circle so this map is well defined). Now $f^n(\theta) = 2^n\theta$, so that $\theta$ is periodic of period $n$ if and only if $2^n\theta = \theta + 2k\pi$ for some integer $k$, i.e., if and only if $\theta = 2k\pi/(2^n - 1)$ where $0 \le k \le 2^n$ is an integer. Hence the periodic points of period $n$ for $f$ are the $(2^n - 1)^{th}$ roots of unity. It follows that the set of periodic points are dense in $S^1$. See Exercise 10.

**Definition 3.5.** A point $x$ is eventually periodic of period $n$ if there exists $m > 0$ such that $f^{n+i}(x) = f^i(x)$ for all $i \ge m$. That is, $f^i(x)$ is periodic for $i \ge m$.

**Example 3.6.** Let $f(x) = x^2$. Then $f(1) = 1$ is fixed, while $f(-1) = 1$ is eventually fixed.

**Example 3.7.** Let $f(\theta) = 2\theta$ on the circle. Note that $f(0) = 0$ is fixed. If $\theta = 2k\pi/2^n$ then $f^n(\theta) = 2k\pi$ so that $\theta$ is eventually fixed. It follows that eventually fixed points are also dense in $S^1$. See Exercise 11.

We remark that eventually periodic points cannot occur if the map is a homeomorphism.

**Definition 3.8.** Let $p$ be periodic of period $n$. A point $x$ is forward asymptotic to $p$ if $\lim_{i \to \infty} f^{in}(x) = p$. The stable set of $p$, denoted by $W^s(p)$, consists of all points forward asymptotic to $p$.

If $p$ is non-periodic, we may still define forward asymptotic points by requiring $|f^i(x) - f^i(p)| \to 0$ as $i \to \infty$. Also, if $f$ is invertible, we may consider *backward asymptotic* points by letting $i \to -\infty$ in the above definition. The set of points backwards asymptotic to $p$ is called the *unstable set* of $p$ and is denoted by $W^u(p)$.

**Example 3.9.** Let $f(x) = x^3$. Then $W^s(0)$ is the open interval $-1 < x < 1$. $W^u(1)$ is the positive real axis, whereas $W^u(-1)$ is the negative real axis.

**Definition 3.10.** A point $x$ is a critical point of $f$ if $f'(x) = 0$. The critical point is non-degenerate if $f''(x) \neq 0$. The critical point is degenerate if $f''(x) = 0$.

For example $f(x) = x^2$ has a non-degenerate critical point at 0, but $f(x) = x^n$ for $n > 2$ has a degenerate critical point at 0. Note that degenerate critical points may be maxima, minima, or saddle points (as in the case of $f(x) = x^3$ ). But non-degenerate critical points must be either maxima or minima. Critical points cannot occur for diffeomorphisms, but their existence for non-invertible maps is one reason why these kinds of maps are more complicated.

The goal of dynamical systems is to understand the nature of all orbits, and to identify the set of orbits which are periodic, eventually periodic, asymptotic, etc. Generally, this is an impossible task. For example, if $f(x)$ is a quadratic polynomial, then finding explicitly the periodic points of period $n$ necessitates solving the equation $f^n(x) = x$, which is a polynomial equation of degree $2^n$. A computer does not help matters much, for numerical computations of periodic points are often misleading. Round-off errors tend to accumulate and make many periodic points invisible to the computer. Therefore we are left with only qualitative or geometric techniques to understand the dynamics of a given system. This means that we should look for a geometric picture of the behavior of all orbits of a system. This geometric picture is provided by the phase portrait which we now discuss.

The graph of a function on the reals provides information about its first iterate, but gives very little information about subsequent iterates. To understand higher iterates, we could attempt to sketch each of their graphs, but this is a cumbersome procedure. There is a much more efficient, geometric method for describing the orbits of a dynamical system, the *phase portrait*. This is a picture, on the real line itself, as opposed to the plane, of all orbits of a system. For example, to indicate that all non-zero orbits of $f(x) = -x$ have period 2, we could sketch the phase portrait as in Fig. 3.1.a. This figure also depicts the phase portraits of some other simple maps.

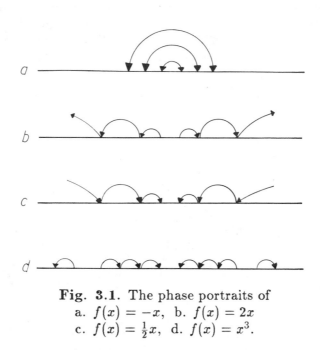

**Fig. 3.1.** The phase portraits of
a. $f(x) = -x$,  b. $f(x) = 2x$
c. $f(x) = \frac{1}{2}x$,  d. $f(x) = x^3$.

The graph of $f(x)$ does of course contain information about the first iteration of $f$. We may use it to gain insight into higher iterations and hence the phase portrait via the following procedure which we call *graphical analysis*. Identify the diagonal $\Delta = \{(x,x)|x \in \mathbf{R}\}$ with $\mathbf{R}$ in the obvious way. A vertical line from $(p,p)$ to the graph of $f$ meets the graph at $(p, f(p))$. Then a horizontal line from $(p, f(p))$ to $\Delta$ meets the diagonal at $(f(p), f(p))$. Hence a vertical line to the graph followed by a horizontal line back to $\Delta$ yields the image of the point $p$ under $f$ on the diagonal. We may thus visualize the phase portrait of a map as taking place on the diagonal rather than on the $x$-axis. Then an orbit is given by repeatedly drawing line segments vertically from $\Delta$ to the graph and then horizontally from the graph to $\Delta$. Fig. 3.2 illustrates this procedure for $f(x) = x^3$ and $g(x) = 2x - x^2$.

Diffeomorphisms of the circle form an interesting class of maps which are

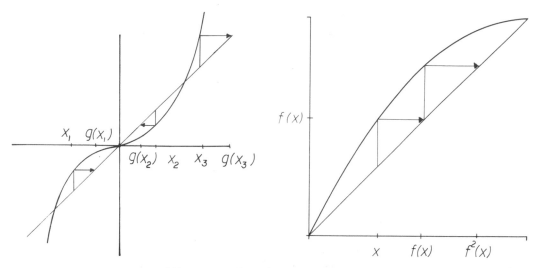

**Fig. 3.2.** Graphical analyses of
a. $g(x) = x^3$ and b. $f(x) = 2x - x^2$.

somewhat different from maps of **R**. The following example is typical.

**Example 3.11.** Let $f(\theta) = \theta + \epsilon \sin(2\theta)$ for $0 < \epsilon < 1/2$. Note that $f$ has fixed points at $0, \pi/2, \pi$, and $3\pi/2$. We compute $f'(0) = f'(\pi) = 1 + 2\epsilon > 1$ whereas $f'(\pi/2) = f'(3\pi/2) = 1 - 2\epsilon < 1$. Hence $0$ and $\pi$ are repelling fixed points and $\pi/2$ and $3\pi/2$ are attracting. More generally, $f(\theta) = \theta + \epsilon \sin(N\theta)$ has $N$ attracting and $N$ repelling fixed points arranged alternately around the circle as long as $0 < \epsilon < 1/N$.

The phase portraits of these maps may be sketched as in Fig. 3.3. Another important class of circle maps are the translation maps.

**Example 3.12.** Translations of the circle. Let $\lambda \in \mathbf{R}$ and $T_\lambda(\theta) = \theta + 2\pi\lambda$. The maps $T_\lambda$ behave quite differently depending upon the rationality or irrationality of $\lambda$. If $\lambda = p/q$, where $p$ and $q$ are integers, then $T_\lambda^q(\theta) = \theta + 2\pi p = \theta$ so that all points are fixed by $T_\lambda^q$. When $\lambda$ is irrational, the situation is quite different. The following result is known as Jacobi's Theorem.

**Theorem 3.13.** *Each orbit $T_\lambda$ is dense in $S^1$ if $\lambda$ is irrational.*

*Proof.* Let $\theta \in S^1$. The points on the orbit of $\theta$ are distinct for if $T_\lambda^n(\theta) = T_\lambda^m(\theta)$ we would have $(n - m)\lambda \in \mathbf{Z}$, so that $n = m$. Any infinite set of

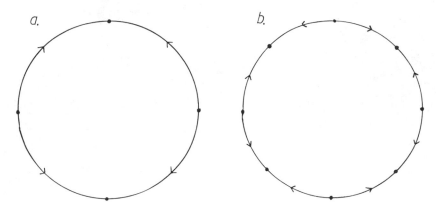

**Fig. 3.3.** The phase portraits of
a. $f(\theta) = \theta + \epsilon \sin(2\theta)$ and
b. $f(\theta) = \theta + \epsilon \sin(4\theta)$.

points on the circle must have a limit point. Thus, given any $\epsilon > 0$, there must be integers $n$ and $m$ for which $|T_\lambda^n(\theta) - T_\lambda^m(\theta)| < \epsilon$. Let $k = n - m$. Then $|T_\lambda^k(\theta) - \theta| < \epsilon$.

Now $T_\lambda$ preserves lengths in $S^1$. Consequently, $T_\lambda^k$ maps the arc connecting $\theta$ to $T_\lambda^k(\theta)$ to the arc connecting $T_\lambda^k(\theta)$ and $T_\lambda^{2k}(\theta)$ which has length less than $\epsilon$. In particular it follows that the points $\theta, T_\lambda^p(\theta), T_\lambda^{2p}(\theta), \ldots$ partition $S^1$ into arcs of length less than $\epsilon$. Since $\epsilon$ was arbitrary, this completes the proof.

q.e.d.

## Exercises

**1.** Use a calculator to iterate each of the following functions (using an arbitrary initial value).

    a. $C(x) = \cos(x)$
    b. $S(x) = \sin(x)$
    c. $E(x) = e^x$
    d. $F(x) = \frac{1}{e}e^x$
    e. $A(x) = \arctan(x)$

Explain the results.

**2.** Using the graph of the function, identify the fixed points for each of the maps in the previous Exercise.

**3.**   List all periodic points for each of the following maps. Then use the graph of $f(x)$ to sketch the phase portrait of $f(x)$ on the indicated interval.

   a. $f(x) = -\frac{1}{2}x$    $-\infty < x < \infty$

   b. $f(x) = -3x$    $-\infty < x < \infty$

   c. $f(x) = x - x^2$    $0 \le x \le 1$

   d. $f(x) = \frac{\pi}{2}\sin x$    $0 \le x \le \pi$

   e. $f(x) = -x^3$    $-\infty < x < \infty$

   f. $f(x) = \frac{1}{2}(x^3 + x)$    $-1 \le x \le 1$

**4.**   Identify the stable sets of each of the fixed points for the maps in the previous Exercise.

**5.**   For each of the following functions, list all critical points and decide whether each is degenerate or non-degenerate.

   a. $f(x) = x^3 - x$

   b. $S(x) = \sin(x)$

   c. $f(x) = x^4 - 2x^2$

   d. $g(x) = x^3 + x^4$

**6.**   Describe the phase portrait of the map of the circle given by

$$f(\theta) = \theta + \frac{\pi}{n} + \epsilon \sin(n\theta)$$

for $0 < \epsilon < 1/n$.

**7.**   Prove that a homeomorphism of $\mathbf{R}$ can have no periodic points with prime period greater than 2. Give an example of a homeomorphism that has a periodic point of period 2.

**8.**   Prove that a homeomorphism cannot have eventually periodic points.

**9.**   Let $S: S^1 \to S^1$ be given by $S(\theta) = \theta + \omega + \epsilon \sin(\theta)$ where $\omega$ and $\epsilon$ are constants. Prove that $S$ is a homeomorphism of the circle if $|\epsilon| < 1$.

**10.**   Let $f(\theta) = 2\theta$ be the map of $S^1$ discussed in Example 3.4. Prove that periodic points of $f$ are dense in $S^1$.

**11.**   Prove that eventually fixed points for the map in Exercise 10 are also dense in $S^1$.

## §1.4 HYPERBOLICITY

Simple maps like $id(x) = x$ and $f(x) = -x$ are, unfortunately, atypical among dynamical systems. There are many reasons why this is so, but perhaps the most unusual feature of these maps is the fact that all points are periodic under iteration of these maps. Most maps do not have this type of behavior. Periodic points tend to be more spread out on the line. In this section we will introduce one of the main themes of this book, hyperbolicity. Maps with hyperbolic periodic points are the ones that occur typically in many dynamical systems and, moreover, they provide the simplest types of periodic behavior to analyze.

**Definition 4.1.** Let $p$ be a periodic point of prime period $n$. The point $p$ is hyperbolic if $|(f^n)'(p)| \neq 1$. The number $(f^n)'(p)$ is called the multiplier of the periodic point.

**Example 4.2.** Consider the diffeomorphism $f(x) = \frac{1}{2}(x^3 + x)$. There are 3 fixed points: $x = 0, 1$, and $-1$. Note that $f'(0) = 1/2$ and $f'(\pm 1) = 2$. Hence each fixed point is hyperbolic. The graph and phase portrait of $f(x)$ are depicted in Fig. 4.1.

**Example 4.3.** Let $f(x) = -\frac{1}{2}(x^3 + x)$. 0 is a hyperbolic fixed point, with $f'(0) = -\frac{1}{2}$. The points $\pm 1$ now lie on a periodic orbit of period 2. We compute $(f^2)'(\pm 1) = f'(1) \cdot f'(-1) = 4$ by the chain rule. Hence this periodic point is hyperbolic, and the phase portrait is depicted in Fig. 4.2. Note that points in the interval $(-1, 1)$ spiral toward 0 and away from $\pm 1$.

We observe that, in the above two examples, we have $|f'(0)| < 1$ and that points close to 0 are forward asymptotic to 0. This situation occurs often:

**Proposition 4.4.** *Let $p$ be a hyperbolic fixed point with $|f'(p)| < 1$. Then there is an open interval $U$ about $p$ such that if $x \in U$, then*

$$\lim_{n \to \infty} f^n(x) = p$$

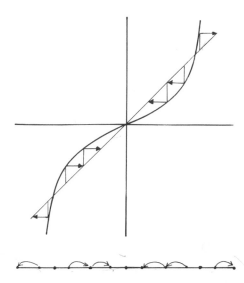

**Fig. 4.1.** The graph and phase portraits of
$$f(x) = \tfrac{1}{2}(x^3 + x).$$

*Proof.* Since $f$ is $C^1$, there is $\epsilon > 0$ such that $|f'(x)| < A < 1$ for $x \in [p - \epsilon, p + \epsilon]$. By the Mean Value Theorem

$$|f(x) - p| = |f(x) - f(p)| \le A|x - p| < |x - p| \le \epsilon.$$

Hence $f(x)$ is contained in $[p - \epsilon, p + \epsilon]$ and, in fact, is closer to $p$ than $x$ is. Via the same argument

$$|f^n(x) - p| \le A^n|x - p|$$

so that $f^n(x) \to p$ as $n \to \infty$.

q.e.d.

**Remarks.**

**1.**  It follows that the interval $[p - \epsilon, p + \epsilon]$ is contained in the stable set associated to $p$, $W^s(p)$.

**2.**  A similar result is true for hyperbolic periodic points of period $n$. In this case, we get an open interval $U$ about $p$ which is mapped inside itself by $f^n$. Of course, the assumption in this case is that $|(f^n)'(p)| < 1$.

**Definition 4.5.**  Let $p$ be a hyperbolic periodic point of period $n$ with $|(f^n)'(p)| < 1$. The point $p$ is called an *attracting periodic point* (an *attractor*) or a *sink*.

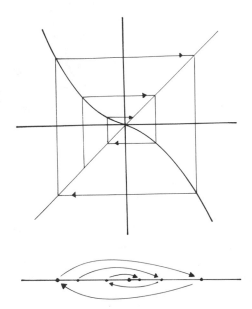

**Fig. 4.2.** The graph and phase portraits of
$$f(x) = -\tfrac{1}{2}(x^3 + x).$$

Attracting periodic points of period $n$ therefore have neighborhoods which are mapped inside themselves by $f^n$. Such a neighborhood is called the *local stable set* and is denoted by $W^s_{loc}$. We may actually distinguish three different types of attracting fixed points, namely those where $f'(p) =$ . $0$, $0 < f'(p) < 1$, and $-1 < f'(p) < 0$. The behavior near these types of fixed points is illustrated in Fig. 4.3.

The behavior of a map near periodic points where the derivative is larger than one in absolute value is quite different from that of sinks.

**Proposition 4.6.** *Let $p$ be a hyperbolic fixed point with $|f'(p)| > 1$. Then there is an open interval $U$ of $p$ such that, if $x \in U$, $x \neq p$, then there exists $k > 0$ such that $f^k(x) \notin U$.*

The proof is similar to the proof of the preceding proposition and is therefore left as an exercise. Graphically, the result is quite clear; see Fig. 4.4.

**Definition 4.7.** A fixed point $p$ with $|f'(p)| > 1$ is called a *repelling fixed point* (a *repellor*) or *source*. The neighborhood described in the Proposition

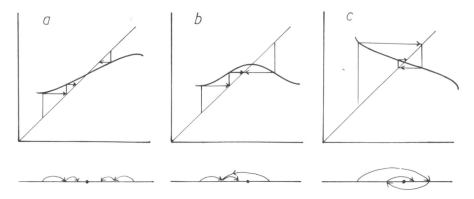

**Fig. 4.3.** The phase portraits near an attracting fixed point $p$. in case a. $0 < f'(p) < 1$, b. $f'(p) = 0$, c. $-1 < f'(p) < 0$.

is called the local unstable set and denoted $W^u_{loc}$.

We remark that periodic points of period $n$ exhibit similar behavior when $|(f^n)'(p)| > 1$.

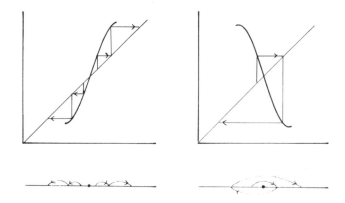

**Fig. 4.4.** The phase portraits near a repelling fixed point.

Hyperbolic periodic points therefore have local behavior which is governed by the derivative at the periodic point. This is not true when the point is indifferent or non-hyperbolic, as the following example shows.

**Example 4.8.** Each of the maps in Fig. 4.5 satisfy $f(0) = 0$ and $f'(0) = 1$, but each have vastly different phase portraits near 0. In a., the map $f(x) = x + x^3$ has a *weakly* repelling fixed point at 0. In b., the map $f(x) = x - x^3$

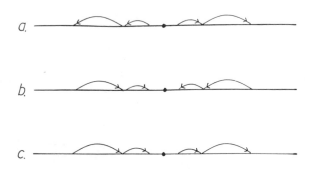

**Fig. 4.5.** The phase portraits of a. $f(x) = x + x^3$,
b. $f(x) = x - x^3$, c. $f(x) = x + x^2$.

has a *weakly* attracting fixed point at 0. In c., the map $f(x) = x + x^2$ is weakly repelling from the right but weakly attracting from the left.

Most maps have only hyperbolic periodic points, as we shall see later. However, non-hyperbolic periodic points often occur in families of maps. When this happens, the periodic point structure often undergoes a *bifurcation*. We will deal with bifurcation theory more extensively later, but for now we give several examples.

**Example 4.9.** Consider the family of quadratic functions $Q_c(x) = x^2 + c$, where $c$ is a parameter. The graphs of $Q_c$ assume three different positions relative to the diagonal depending upon whether $c > 1/4$, $c = 1/4$, or $c < 1/4$. See Fig. 4.6. Note that $Q_c$ has no fixed points for $c > 1/4$. When $c = 1/4$, $Q_c$ has a unique non-hyperbolic fixed point at $x = 1/2$. And when $c < 1/4$, $Q_c$ has a pair of fixed points, one attracting and one repelling. Thus the phase portrait of $Q_c$ changes as $c$ decreases through $1/4$. This change is an example of a bifurcation.

**Example 4.10.** Let $F_\mu(x) = \mu x(1 - x)$ with $\mu > 1$. $F_\mu$ has two fixed points: one at 0 and the other at $p_\mu = (\mu - 1)/\mu$. Note that $F_\mu'(0) = \mu$ and $F_\mu'(p_\mu) = 2 - \mu$. Hence 0 is a repelling fixed point for $\mu > 1$ and $p_\mu$ is attracting for $1 < \mu < 3$. When $\mu = 3$, $F_\mu'(p_\mu) = -1$. We sketch the graphs of $F_\mu^2$ for $\mu$ near 3. See Fig. 4.7. Note that 2 new fixed points for $F_\mu^2$ appear as $\mu$ increases through 3. These are new periodic points of period 2. Another bifurcation has occurred: this time we have a change in $\mathrm{Per}_2(F_\mu)$.

This quadratic family actually exhibits many of the phenomena that are crucial in the general theory. The next section is devoted entirely to this function.

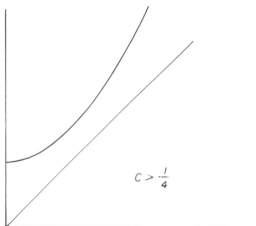
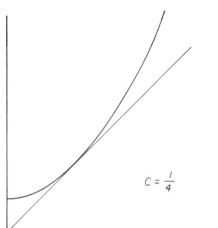

$$C > \frac{1}{4}$$

$$C = \frac{1}{4}$$

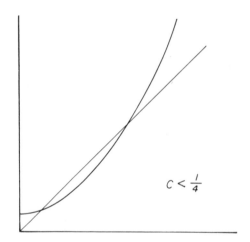

$$C < \frac{1}{4}$$

**Fig. 4.6.** The graphs of $Q_c(x) = x^2 + c$ for $c > 1/4$
$c = 1/4$ and $c < 1/4$.

## Exercises

**1.** Find all periodic points for each of the following maps and classify them as attracting, repelling, or neither. Sketch the phase portraits.

    a. $f(x) = x - x^2$

    b. $f(x) = 2(x - x^2)$

    c. $f(x) = x^3 - \frac{1}{9}x$

    d. $f(x) = x^3 - x$

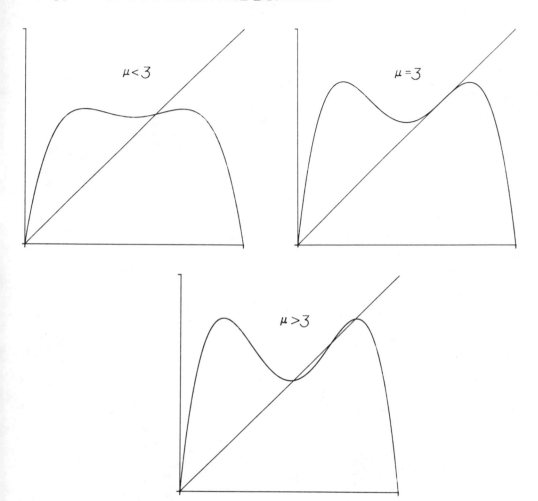

**Fig. 4.7.** The graphs of $F_\mu^2(x)$ where
$F_\mu(x) = \mu x(1-x)$ for
$\mu < 3, \mu = 3$, and $\mu > 3$.

e.  $S(x) = \frac{1}{2}\sin(x)$

f.  $S(x) = \sin(x)$

g.  $E(x) = e^{x-1}$

h.  $E(x) = e^x$

i.  $A(x) = \arctan x$

j.  $A(x) = \frac{\pi}{4}\arctan x$

 k. $A(x) = -\frac{\pi}{4} \arctan x$

**2.** Discuss the bifurcations which occur in the following families of maps for the indicated parameter value

 a. $S_\lambda(x) = \lambda \sin x, \quad \lambda = 1$

 b. $E_\lambda(x) = \lambda e^x, \quad \lambda = 1/e$

 c. $E_\lambda(x) = \lambda e^x, \quad \lambda = -e$

 d. $Q_c(x) = x^2 + c, \quad c = -3/4$

 e. $F_\mu(x) = \mu x(1 - x), \quad \mu = 1$

 f. $A_\lambda(x) = \lambda \arctan x, \quad \lambda = 1$

 g. $A_\lambda(x) = \lambda \arctan x, \quad \lambda = -1$

**3.** Suppose $f$ is a diffeomorphism. Prove that all hyperbolic periodic points are isolated.

**4.** Show via an example that hyperbolic periodic points need not be isolated.

**5.** Find an example of a $C^1$ diffeomorphism with a non-hyperbolic fixed point which is an accumulation point of other hyperbolic fixed points.

**6.** Discuss the dynamics of the family $f_\alpha(x) = x^3 - \alpha x$ for $-\infty < \alpha \le 1$. Find all parameter values where bifurcations occur. Describe how the phase portrait of $f_\alpha$ changes at these points.

**7.** Consider the linear maps $f_k(x) = kx$. Show that there are four open sets of parameters for which the phase portraits of $f_k$ are similar. The exceptional cases are $k = 0, \pm 1$.

## §1.5 AN EXAMPLE: THE QUADRATIC FAMILY

In this section, we will continue the discussion of the quadratic family $F_\mu(x) = \mu x(1 - x)$. Actually, we will return to this example repeatedly throughout the remainder of this chapter, since it illustrates many of the most important phenomena that occur in dynamical systems.

**Proposition 5.1.**

 1. $F_\mu(0) = F_\mu(1) = 0$ and $F_\mu(p_\mu) = p_\mu$, where $p_\mu = \dfrac{\mu - 1}{\mu}$.

2. $0 < p_\mu < 1$ *if* $\mu > 1$

The proof of this Proposition is straightforward. From now on we will concentrate on the case $\mu > 1$. The following Proposition shows that most points behave rather tamely under iteration of $F_\mu$: all points which do not lie in the interval $[0, 1]$ tend to $-\infty$.

**Proposition 5.2.** *Suppose $\mu > 1$. If $x < 0$, then $F_\mu^n(x) \to -\infty$ as $n \to \infty$. Similarly, if $x > 1$, then $F_\mu^n(x) \to -\infty$ as $n \to \infty$.*

*Proof.* If $x < 0$, then $\mu x(1-x) < x$ so $F_\mu(x) < x$. Hence $F_\mu^n(x)$ is a decreasing sequence of points. This sequence cannot converge to $p$, for then we would have $F_\mu^{n+1}(x) \to F_\mu(p) < p$, whereas $F_\mu^n(x) \to p$. Hence $F_\mu^n(p) \to -\infty$ as required. If $x > 1$, then $F_\mu(x) < 0$ so $F_\mu^n(x) \to -\infty$ as well.

q.e.d.

Graphical analysis yields the above results easily, as shown in Fig. 5.1. As a consequence of this Proposition, all of the interesting dynamics of the quadratic family occur in the unit interval $I = \{x \mid 0 \leq x \leq 1\}$. For low values of $\mu$, the dynamics of $F_\mu$ are not too complicated.

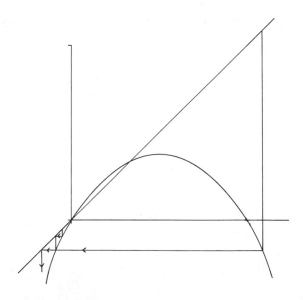

**Fig. 5.1.** Graphical analysis of $F_\mu(x) = \mu x(1 - x)$
when $\mu > 1$.

**Proposition 5.3.** *Let* $1 < \mu < 3$.

1. $F_\mu$ *has an attracting fixed point at* $p_\mu = (\mu - 1)/\mu$ *and a repelling fixed point at* $0$.

2. *If* $0 < x < 1$, *then*

$$\lim_{n \to \infty} F_\mu^n(x) = p_\mu$$

*Proof.* Part 1 was proved in Example 4.10 at the end of the last section. For part 2, we first deal with the case $1 < \mu < 2$. Suppose $x$ lies in the interval $(0, 1/2]$. Then graphical analysis immediately shows that

$$|F_\mu(x) - p_\mu| < |x - p_\mu|.$$

if $x \neq p_\mu$. See Fig. 5.2. Consequently, $F_\mu^n(x) \to p_\mu$ as $n \to \infty$. If, on the other hand, $x$ lies in the interval $(1/2, 1)$, then $F_\mu(x)$ lies in $(0, 1/2)$, so that the previous argument implies

$$F_\mu^n(x) = F_\mu^{n-1}(F_\mu(x)) \to p_\mu$$

as $n \to \infty$.

The case when $2 < \mu < 3$ is more difficult. Graphical analysis shows what is different in this case. See Fig. 5.2. Note that $1/2 < p_\mu < 1$. Let $\hat{p}_\mu$ denote the unique point in the interval $(0, 1/2)$ that is mapped onto $p_\mu$ by $F_\mu$. Then the reader may easily check that $F_\mu^2$ maps the interval $[\hat{p}_\mu, p_\mu]$ inside $[1/2, p_\mu]$. It follows that $F_\mu^n(x) \to p_\mu$ as $n \to \infty$ for all $x \in [\hat{p}_\mu, p_\mu]$. Now suppose $x < \hat{p}_\mu$. Again graphical analysis shows that there exists $k > 0$ such that $F_\mu^k(x) \in [\hat{p}_\mu, p_\mu]$. Thus $F_\mu^{k+n}(x) \to p_\mu$ as $n \to \infty$ in this case as well. Finally, as before, $F_\mu$ maps the interval $(p_\mu, 1)$ onto $(0, p_\mu)$, so the result follows here as well. Since $(0, 1) = (0, \hat{p}_\mu) \cup [\hat{p}_\mu, p_\mu] \cup (p_\mu, 1)$, we are finished. We leave the intermediate case $\mu = 2$ to the reader. See Exercise 1.

q.e.d.

Hence for $1 < \mu < 3$, $F_\mu$ has only two fixed points and all other points in $I$ are asymptotic to $p_\mu$. Thus the dynamics of $F_\mu$ are completely understood for $\mu$ in this range. The phase portraits of $F_\mu$ are depicted in Fig. 5.3.

As we showed in Example 4.10 in the previous section, as $\mu$ passes through 3, the dynamics of $F_\mu$ become slightly more complicated: a new periodic point of period 2 is born. This is the beginning of a long story: as $\mu$ continues to increase the dynamics of $F_\mu$ become increasingly more complicated until the phase portrait of $F_\mu$ is dramatically different from the above picture. This is a scenario that we will investigate in much more detail later.

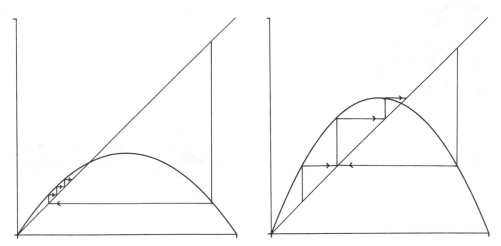

**Fig. 5.2.** Graphical analysis of $F_\mu(x) = \mu x(1-x)$
when a. $1 < \mu < 2$, and b. $2 < \mu < 3$.

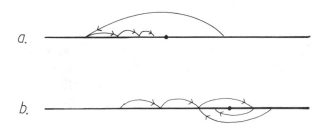

**Fig. 5.3.** The phase portraits for $F_\mu(x) = \mu x(1-x)$
when a. $1 < \mu < 2$, and b. $2 < \mu < 3$.

We now turn to the case when $\mu > 4$. For the remainder of this section, we will drop the subscript $\mu$ and write $F$ instead of $F_\mu$. As above, all of the interesting dynamics of $F$ occur in the unit interval $I$. Note that, since $\mu > 4$, the maximum value $\mu/4$ of $F$ is larger than one. Hence certain points leave $I$ after one iteration of $F$. Denote the set of such points by $A_0$. Clearly, $A_0$ is an open interval centered at $\frac{1}{2}$ and has the property that, if $x \in A_0$, then $F(x) > 1$, so $F^2(x) < 0$ and $F^n(x) \to -\infty$. $A_0$ is the set of points which immediately escape from $I$. All other points in $I$ remain in $I$ after one iteration of $F$.

Let $A_1 = \{x \in I \mid F(x) \in A_0\}$. If $x \in A_1$, then $F^2(x) > 1$, $F^3(x) < 0$, and so, as before, $F^n(x) \to -\infty$. Inductively, let $A_n = \{x \in A_{n-1} \mid F^n(x) \in A_0\}$.

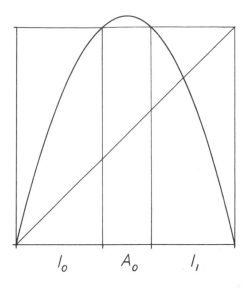

$$I_0 \qquad A_0 \qquad I_1$$

**Fig. 5.4.**

That is, $A_n = \{x \in I | F^i(x) \in I$ for $i \le n$ but $F^{n+1}(x) \notin I\}$, so that $A_n$ consists of all points which escape from $I$ at the $n+1^{st}$ iteration. As above, if $x$ lies in $A_n$, it follows that the orbit of $x$ tends eventually to $-\infty$. Since we therefore know the ultimate fate of any point which lies in the $A_n$, it therefore remains only to analyze the behavior of those points which never escape from $I$, i.e., the set of points which lie in

$$I - \left( \bigcup_{n=0}^{\infty} A_n \right).$$

Let us denote this set by $\Lambda$. Our first question is: what precisely is this set of points? To understand $\Lambda$, we describe more carefully its recursive construction.

Since $A_0$ is an open interval centered at $1/2$, $I - A_0$ consists of two closed intervals, $I_0$ on the left and $I_1$ on the right. See Fig. 5.4.

Note that $F$ maps both $I_0$ and $I_1$ monotonically onto $I$; $F$ is increasing on $I_0$ and decreasing on $I_1$. Since $F(I_0) = F(I_1) = I$, there are a pair of open intervals, one in $I_0$ and one in $I_1$, which are mapped into $A_0$ by $F$. Therefore this pair of intervals is precisely the set $A_1$.

Now consider $I - (A_0 \cup A_1)$. This set consists of 4 closed intervals and $F$ maps each of them monotonically onto either $I_0$ or $I_1$. Consequently $F^2$ maps each of them onto $I$. We therefore see that each of the four intervals in

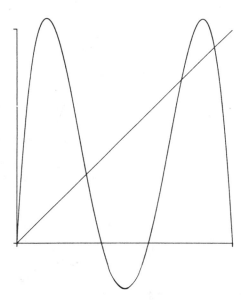

**Fig. 5.5.** The graph of $F^2$.

$I - (A_0 \cup A_1)$ contains an open subinterval which is mapped by $F^2$ onto $A_0$. Therefore, points in these intervals escape from $I$ upon the third iteration of $F$. This is the set we called $A_2$. For later use, we observe that $F^2$ is alternately increasing and decreasing on these four intervals. It follows that the graph of $F^2$ must therefore have two humps as shown in Fig. 5.5.

Continuing in this manner we note two facts. First, $A_n$ consists of $2^n$ disjoint open intervals. Hence $I - (A_0 \cup \ldots \cup A_n)$ consists of $2^{n+1}$ closed intervals since

$$1 + 2 + 2^2 + \ldots + 2^n = 2^{n+1} - 1.$$

Secondly, $F^{n+1}$ maps each of these closed intervals monotonically onto $I$. In fact, the graph of $F^{n+1}$ is alternately increasing and decreasing on these intervals. Thus the graph of $F^{n+1}$ has exactly $2^n$ humps on $I$, and it follows that the graph of $F^n$ crosses the line $y = x$ at least $2^n$ times. This implies that $F^n$ has at least $2^n$ fixed points or, equivalently, $\text{Per}_n(F)$ consists of $2^n$ points in $I$. Clearly, the structure of $\Lambda$ is much more complicated when $\mu > 4$ than the earlier case $\mu < 3$.

The construction of $\Lambda$ is reminiscent of the construction of the Cantor Middle Thirds set: $\Lambda$ is obtained by successively removing open intervals from the "middles" of a set of closed intervals.

**Definition 5.4.** A set $\Lambda$ is a Cantor set if it is a closed, totally disconnected, and perfect subset of $I$. A set is totally disconnected if it contains no intervals; a set is perfect if every point in it is an accumulation point or limit point of other points in the set.

**Example 5.5.** *The Cantor Middle-Thirds Set.* This is the classical example of a Cantor set. Start with $I$ but remove the open "middle third," i.e., the interval $(\frac{1}{3}, \frac{2}{3})$. Next, remove from what remains the two middle thirds again, i.e., the pair of intervals $(\frac{1}{9}, \frac{2}{9})$ and $(\frac{7}{9}, \frac{8}{9})$. Continue removing middle thirds in this fashion; note that $2^n$ open intervals are removed at the $n^{th}$ stage of this process. Thus, this procedure is entirely analogous to our construction above. Exercise 7 shows that the Cantor Middle-Thirds set is indeed a Cantor set as defined in 5.4.

**Remark.** The Cantor Middle-Thirds set is an example of a *fractal*. Intuitively, a fractal is a set which is self-similar under magnification. In the Cantor Middle-Thirds set, suppose we look only at those points which lie in the left-hand interval $[0, \frac{1}{3}]$. Under a microscope which magnifies this interval by a factor of three, the "piece" of the Cantor set in $[0, \frac{1}{3}]$ looks exactly like the original set. More precisely, the linear map $L(x) = 3x$ maps the portion of the Cantor set in $[0, \frac{1}{3}]$ homeomorphically onto the entire set. See Exercise 10. This process does not stop at the first level: one may magnify any piece of the Cantor set at the $n^{th}$ stage of the construction by a factor of $3^n$ and obtain the original set. See Exercise 11.

To guarantee that our set $\Lambda$ is a Cantor set, we need an additional hypothesis on $\mu$. Let us assume that $\mu$ is large enough so that $|F'(x)| > 1$ for all $x \in I_0 \cup I_1$. The reader may check that $\mu > 2 + \sqrt{5}$ suffices. Hence, for these values of $\mu$, there exists $\lambda > 1$ such that $|F'(x)| > \lambda$ for all $x \in \Lambda$. By the chain rule, it follows that $|(F^n)'(x)| > \lambda^n$ as well. We claim that $\Lambda$ contains no intervals. Indeed, if this were so, we could choose $x, y \in \Lambda$, $x \neq y$, with the closed interval $[x, y] \subset \Lambda$. But then, $|(F^n)'(\alpha)| > \lambda^n$ for all $\alpha \in [x, y]$. Choose $n$ so that $\lambda^n |y - x| > 1$. By the Mean Value Theorem, it then follows that $|F^n(y) - F^n(x)| \geq \lambda^n |y - x| > 1$, which implies that at least one of $F^n(y)$ or $F^n(x)$ lies outside of $I$. This is a contradiction, and so $\Lambda$ is totally disconnected.

Since $\Lambda$ is a nested intersection of closed intervals, $\Lambda$ is closed. We now prove that $\Lambda$ is perfect. First note that any endpoint of an $A_k$ is in $\Lambda$: indeed, such points are eventually mapped to the fixed point at 0, and so they stay in $I$ under iteration. Now if $p \in \Lambda$ were isolated, every nearby point must leave $I$ under iteration of $F$. Such points must belong to some $A_k$. Either there is a sequence of endpoints of the $A_k$ converging to $p$, or else all points in a deleted neighborhood of $p$ are mapped out of $I$ by some power of $F$. In the former case, we are done as the endpoints of the $A_k$ map to 0 and hence are in $\Lambda$. In the latter, we may assume that $F^n$ maps $p$ to 0 and all other points in a neighborhood of $p$ into the negative real axis. But then $F^n$ has a maximum at $p$ so that $(F^n)'(p) = 0$. By the chain rule, we must have $F'(F^i(p)) = 0$ for some $i < n$. Hence $F^i(p) = 1/2$. But then $F^{i+1}(p) \notin I$ and so $F^n(p) \to -\infty$, contradicting the fact that $F^n(p) = 0$.

Hence we have proved

**Theorem 5.6.** *If $\mu > 2 + \sqrt{5}$, then $\Lambda$ is a Cantor set.*

**Remark.** The theorem is true for $\mu > 4$, but the proof is more delicate.

We have now succeeded in understanding the gross behavior of orbits of $F_\mu$ when $\mu > 4$. Either a point tends to $-\infty$ under iteration of $F_\mu$, or else its entire orbit lies in $\Lambda$. Hence we understand the orbit of a point under $F_\mu$ perfectly well as long as the point does not lie in $\Lambda$. In the next section, we will complete the analysis of the dynamics of $F_\mu$ by analyzing the dynamics of $F_\mu$ on $\Lambda$.

When $\mu > 2 + \sqrt{5}$, we have shown that $|F_\mu'(x)| > 1$ on $I_0 \cup I_1$. This implies that $|F_\mu'(x)| > 1$ on $\Lambda$. This is a condition similar to the hyperbolicity condition of §3, except that we require $|F_\mu'(x)| \neq 1$ on a whole set, not just at a periodic point. This motivates the definition of a hyperbolic set:

**Definition 5.9.** A set $\Gamma \subset \mathbf{R}$ is a repelling (resp. attracting) hyperbolic set for $f$ if $\Gamma$ is closed, bounded and invariant under $f$ and there exists an $N > 0$ such that $|(f^n)'(x)| > 1$ (resp. $< 1$) for all $n \geq N$.

The Cantor set $\Lambda$ for the quadratic map when $\mu > 2 + \sqrt{5}$ is of course a repelling hyperbolic set with $N = 1$.

**Exercises**

**1.** Prove that $F_2(x) = 2x(1 - x)$ satisfies: if $0 < x < 1$, then $F_2^n(x) \to 1/2$ as $n \to \infty$.

**2.** Sketch the graph of $F_4^n(x)$ on the unit interval, where $F_4(x) = 4x(1-x)$. Conclude that $F_4$ has at least $2^n$ periodic points of period $n$.

**3.** Sketch the graph of the tent map

$$T_2(x) = \begin{cases} 2x & 0 \leq x \leq 1/2 \\ 2 - 2x & \frac{1}{2} \leq x \leq 1 \end{cases}$$

on the unit interval. Use the graph of $T^n$ to conclude that $T$ has exactly $2^n$ periodic points of period $n$.

4.  Prove that the set of all periodic points of $T(x)$ are dense in $[0, 1]$.

5.  Sketch the graph of the baker map

$$B(x) = \begin{cases} 2x & 0 \le x \le 1/2 \\ 2x - 1 & 1/2 < x \le 1 \end{cases}.$$

How many periodic points of period $n$ does $B$ have?

6.  The following exercises deal with the family of functions $F(x) = x^3 - \lambda x$ for $\lambda > 0$.

    a.  Find all periodic points and classify them when $0 < \lambda < 1$.

    b.  Prove that, if $|x|$ is sufficiently large, then $|f^n(x)| \to \infty$.

    c.  Prove that if $\lambda$ is sufficiently large, then the set of points which do not tend to infinity is a Cantor set.

7.  Prove that the Cantor Middle-Thirds set described in Example 5.5 is closed, nonempty, perfect, and totally disconnected.

8.  Show that, at the $n^{th}$ stage of the construction of the Cantor Middle-Thirds set, the sum of the lengths of the remaining intervals is

$$1 - \frac{1}{3}\left(\sum_{i=0}^{n-1} \left(\frac{2}{3}\right)^i\right).$$

Conclude that the sum of the lengths of these intervals tends to 0 as $n \to \infty$.

9.  Construct a Middle-Fifths Cantor set in which the middle fifth of each remaining subinterval of the unit interval is removed. What can be said about the sum of the lengths of the remaining intervals in this case?

10.  Let $\Gamma$ be the Cantor Middle-Thirds set. Prove that the linear map $L(x) = 3x$ maps $\Gamma \cap [0, \frac{1}{3}]$ homeomorphically onto $\Gamma$.

11.  Generalize Exercise 10 to show that the portion of $\Gamma$ contained in an interval remaining at the $n^{th}$ stage of the construction of $\Gamma$ is homeomorphic to $\Gamma$.

## §1.6 SYMBOLIC DYNAMICS

    Our goal in this section is to give a model for the rich dynamical structure of the quadratic map on the Cantor set $\Lambda$ discussed in the previous section.

To do this we will set up a model mapping which is completely equivalent to $F$. At first, this model may seem artificial and unintuitive. But, as we go along, it will become clear that such symbolic models describe the dynamics of $F$ completely and also in the simplest possible way.

We need a "space" on which our model map will act. The points in this space will be infinite sequences of 0's and 1's. We don't worry about convergence of these sequences; rather, the difficult notion here is to imagine such an infinite sequence as representing a single "point" in space.

**Definition 6.1.** $\Sigma_2 = \{ s = (s_0 s_1 s_2 \ldots) | s_j = 0 \text{ or } 1 \}$.

$\Sigma_2$ is called the *sequence space* on the two symbols 0 and 1. More generally, we can consider the space $\Sigma_n$ consisting of infinite sequences of integers between 0 and $n - 1$. Elements of $\Sigma_2$ are infinite strings of integers, like $(000\ldots)$ or $(0101\ldots)$. We may make $\Sigma_2$ into a metric space as follows. For two sequences $s = (s_0 s_1 s_2 \ldots)$ and $t = (t_0 t_1 t_2 \ldots)$, define the distance between them by

$$d[s, t] = \sum_{i=0}^{\infty} \frac{|s_i - t_i|}{2^i}$$

Since $|s_i - t_i|$ is either 0 or 1, this infinite series is dominated by the geometric series

$$\sum_{i=0}^{\infty} \frac{1}{2^i} = 2$$

and therefore it converges.

For example, if $s = (000\ldots)$ and $t = (111\ldots)$, then $d[s, t] = 2$. If $r = (1010\ldots)$, then

$$d[s, r] = \sum_{i=0}^{\infty} \frac{1}{2^{2i}} = \frac{1}{1 - \frac{1}{4}} = \frac{4}{3}.$$

**Proposition 6.2.** *d is a metric on $\Sigma_2$.*

*Proof.* Clearly, $d[s, t] \geq 0$ for any $s, t \in \Sigma_2$, and $d[s, t] = 0$ iff $s_i = t_i$ for all $i$. Since $|s_i - t_i| = |t_i - s_i|$, it follows that $d[s, t] = d[t, s]$. Finally, if $r, s$, and $t \in \Sigma_2$, then $|r_i - s_i| + |s_i - t_i| \geq |r_i - t_i|$ from which we deduce that $d[r, s] + d[s, t] \geq d[r, t]$.

q.e.d.

The metric $d$ allows us to decide which subsets of $\Sigma_2$ are open and which are closed, as well as which sequences are close to each other.

**Proposition 6.3.** *Let* $s, t \in \Sigma_2$ *and suppose* $s_i = t_i$ *for* $i = 0, 1, \ldots, n$. *Then* $d[s, t] \leq 1/2^n$. *Conversely, if* $d[s, t] < 1/2^n$, *then* $s_i = t_i$ *for* $i \leq n$.

*Proof.* If $s_t = t_i$ for $i \leq n$, then

$$d[s, t] = \sum_{i=0}^{n} \frac{|s_i - s_i|}{2^i} + \sum_{i=n+1}^{\infty} \frac{|s_i - t_i|}{2^i}$$

$$\leq \sum_{i=n+1}^{\infty} \frac{1}{2^i} = \frac{1}{2^n}$$

On the other hand, if $s_j \neq t_j$ for some $j \leq n$, then we must have

$$d[s, t] \geq \frac{1}{2^j} \geq \frac{1}{2^n}$$

consequently, if $d[s, t] < 1/2^n$, then $s_i = t_i$ for $i \leq n$.

q.e.d.

The importance of this result is that we can decide quickly whether or not two sequences are close to each other. Intuitively, this result says that two sequences in $\Sigma_2$ are close provided their first few entries agree. We now define the most important ingredient in symbolic dynamics, the shift map on $\Sigma_2$.

**Definition 6.4.** The shift map $\sigma: \Sigma_2 \to \Sigma_2$ is given by $\sigma(s_0 s_1 s_2 \ldots) = (s_1 s_2 s_3 \ldots)$.

The shift map simply "forgets" the first entry in a sequence, and shifts all other entries one place to the left. Clearly, $\sigma$ is a two-to-one map of $\Sigma_2$, as $s_0$ may be either 0 or 1. Moreover, in the metric defined above, $\sigma$ is a continuous map.

**Proposition 6.5.** $\sigma: \Sigma_2 \to \Sigma_2$ *is continuous.*

*Proof.* Let $\epsilon > 0$ and $s = s_0 s_1 s_2 \ldots$. Pick $n$ such that $1/2^n < \epsilon$. Let $\delta = 1/2^{n+1}$. If $t = t_0 t_1 t_2 \ldots$ satisfies $d[s, t] < \delta$, then by Proposition 6.3 we have $s_i = t_i$ for $i \leq n + 1$. Hence the $i^{th}$ entries of $\sigma(s)$ and $\sigma(t)$ agree for $i \leq n$. Therefore $d[\sigma(s), \sigma(t)] \leq 1/2^n < \epsilon$.

q.e.d.

In the next section, we will show that the shift map is an exact model for the quadratic map $F_\mu$ when $\mu > 4$. Here we will simply show that the dynamics of $\sigma$ can be understood completely. For example, periodic

points correspond exactly to repeating sequences, i.e. sequences of the form
$s = (s_0 \ldots s_{n-1}, s_0 \ldots s_{n-1}, s_0 \ldots s_{n-1} \ldots)$. Hence there are $2^n$ periodic points
of period $n$ for $\sigma$, each generated by one of the $2^n$ finite sequence of 0's and
1's of length $n$.

Eventually periodic points are equally abundant and easy to recognize.
For example, any sequence of the form $(s_0 \ldots s_n 1111 \ldots)$ is eventually fixed,
while any eventually repeating sequence is eventually periodic for $\sigma$.

Another interesting fact about $\sigma$ is that periodic points form a dense
subset of $\Sigma_2$. Recall that a subset is dense in $\Sigma_2$ provided its closure is the
entire space $\Sigma_2$. To prove that $\mathrm{Per}(\sigma)$ is dense, we must produce a sequence of
periodic points $\tau_n$ which converge to an arbitrary point $s = (s_0 s_1 s_2 \ldots)$ in $\Sigma_2$.
We define the sequence $\tau_n = (s_0 \ldots s_n, s_0 \ldots s_n, \ldots)$, i.e. $\tau_n$ is the repeating
sequence whose entries agree with s up to the $n^{th}$ entry. By Proposition 6.3,
$d[\tau_n, s] \le 1/2^n$, so that we have $\tau_n \to s$.

Of course, not all points in $\Sigma_2$ are periodic or eventually periodic. Any
non-repeating sequence can never be periodic. In fact, the non-periodic
sequences greatly outnumber the periodic sequences in $\Sigma_2$. Moreover, there
are non-periodic orbits in $\Sigma_2$ which wind densely about $\Sigma_2$, i.e. the closure
of the orbit is $\Sigma_2$ itself. Another way to say this is there are points in $\Sigma_2$
whose orbit comes arbitrarily close to any given sequence in $\Sigma_2$. To see this,
consider

$$s^* = \underbrace{(0\ 1)}_{1block} | \underbrace{00\ 01\ 10\ 11}_{2blocks} | \underbrace{000\ 001 \cdots}_{3blocks} | \underbrace{\cdots}_{4blocks} .$$

$s^*$ is constructed by successively listing all blocks of 0's and 1's of length $n$,
then length $n+1$, etc. Clearly, some iterate of $\sigma$ applied to $s^*$ yields a sequence which agrees with any given sequence in an arbitrarily large number
of places. Mappings which have dense orbits are called *topologically transitive*.

Let us list these properties of $\sigma$:

**Proposition 6.6.**
  *1. Card* $\mathrm{Per}_n(\sigma) = 2^n$.
  *2. Per $(\sigma)$ is dense in $\Sigma_2$.*
  *3. There exists a dense orbit for $\sigma$ in $\Sigma_2$.*

In the next section, we will show that the shift map on $\Sigma_2$ is in fact the
"same" map as $f$ on $\Lambda$.

Symbolic dynamics is one of the main themes of this book. It will appear in various guises throughout, including later in this chapter when we

introduce subshifts of finite type and also the kneading theory to describe
the dynamics of $F_\mu$ when $\mu < 4$.


**Exercises**

1.   Let
$$s = (001\,001\,001\ldots)$$
$$t = (01\,01\,01\ldots)$$
$$r = (10\,10\,10\ldots).$$

Compute:
   a. $d[s,t]$
   b. $d[t,r]$
   c. $d[s,r]$.

2.   Identify all sequences in $\Sigma_2$ which are periodic points of period 3 for $\sigma$.
Which sequences lie on the same orbit under $\sigma$?

3.   Rework Exercise 2 for periods four and five.

4.   Let $\Sigma'$ consist of all sequences in $\Sigma_2$ satisfying: if $s_j = 0$ then $s_{j+1} = 1$.
In other words, $\Sigma'$ consists of only those sequences in $\Sigma_2$ which never have
two consecutive zeros.
   a. Show that $\sigma$ preserves $\Sigma'$ and that $\Sigma'$ is a closed subset of $\Sigma$.
   b. Show that periodic points of $\sigma$ are dense in $\Sigma'$.
   c. Show that there is a dense orbit in $\Sigma'$.
   d. How many fixed points are there for $\sigma, \sigma^2, \sigma^3$ in $\Sigma'$ ?
   e. Find a recursive formula for the number of fixed points of $\sigma^n$ in terms
      of the number of fixed points of $\sigma^{n-1}$ and $\sigma^{n\,2}$.

5.   Let $\Sigma_N$ consist of all sequences of natural numbers $1, 2, \ldots, N$. There is
a natural shift on $\Sigma_N$.
   a. How many periodic points does $\sigma$ have in $\Sigma_N$ ?
   b. Show that $\sigma$ has a dense orbit in $\Sigma_N$.

6.   Let $s \in \Sigma_2$. Define the stable set of s, $W^s(s)$, to be the set of sequences
t such that $d[\sigma^i(s)\,\sigma^i(t)] \to 0$ as $i \to \infty$. Identify all of the sequences in
$W^s(s)$.

## §1.7 TOPOLOGICAL CONJUGACY

The goal of this section is to relate the shift map discussed in the previous section to the quadratic map $F_\mu(x) = \mu x(1 - x)$ when $\mu$ is sufficiently large. Recall that all points in $\mathbf{R}$ tend to $-\infty$ under iteration of $F_\mu$ with the exception of those points in the Cantor set $\Lambda$. In order to complete the description of the dynamics of $F_\mu$, we must then understand the restriction of $F_\mu$ to $\Lambda$.

Recall first that $\Lambda \subset I_0 \cup I_1$. If $x \in \Lambda$, then the entire orbit of $x$ lies in $\Lambda$ and hence in one of these two intervals. We can thus get a rough idea of the behavior of the orbit by noting in which of these intervals the various iterates of $x$ fall. Accordingly, we make the following definition.

**Definition 7.1.** The *itinerary* of $x$ is a sequence $S(x) = s_0 s_1 s_2 \ldots$ where $s_j = 0$ if $F_\mu^j(x) \in I_0$, $s_j = 1$ if $F_\mu^j(x) \in I_1$.

Thus the itinerary of $x$ is an infinite sequence of 0's and 1's. That is, $S(x)$ is a point in the sequence space $\Sigma_2$. We think of $S$ as a map from $\Lambda$ to $\Sigma_2$. This map has several interesting properties.

**Theorem 7.2.** *If $\mu > 2 + \sqrt{5}$, then $S$ is a homeomorphism.*

*Proof.* We first show that $S$ is one-to-one. Let $x, y \in \Lambda$ and suppose $S(x) = S(y)$. Then, for each $n$, $F_\mu^n(x)$ and $F_\mu^n(y)$ lie on the same side of $1/2$. This implies that $F_\mu$ is monotonic on the interval between $F_\mu^n(x)$ and $F_\mu^n(y)$. Consequently, all points in this interval remain in $I_0 \cup I_1$. This contradicts the fact that $\Lambda$ is totally disconnected.

To see that $S$ is onto, we first introduce the following notation. Let $J \subset I$ be a closed interval. Let

$$F_\mu^{-n}(J) = \{x \in I | F_\mu^n(x) \in J\}$$

In particular, $F_\mu^{-1}(J)$ denotes the preimage of $J$. Observe that, if $J \subset I$ is a closed interval, then $F_\mu^{-1}(J)$ consists of two subintervals, one in $I_0$ and one in $I_1$. See Fig. 7.1.

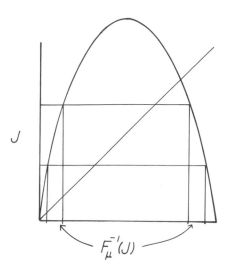

$J$

$F_\mu^{-1}(J)$

**Fig. 7.1.** The preimage of a closed interval $J$ is a
pair of closed intervals, one in $I_0$ and one in $I_1$.

Now let $\mathbf{S} = s_0 s_1 s_2 \ldots$ . We must produce $x \in \Lambda$ with $S(x) = \mathrm{s}$. To that
end we define

$$I_{s_0 s_1 \ldots s_n} = \{x \in I | x \in I_{s_0}, F_\mu(x) \in I_{s_1}, \ldots, F_\mu^n(x) \in I_{s_n}\}$$
$$= I_{s_0} \cap F_\mu^{-1}(I_{s_1}) \cap \ldots \cap F_\mu^{-n}(I_{s_n})$$

We claim that the $I_{s_0 \ldots s_n}$ form a nested sequence of nonempty closed intervals
as $n \to \infty$. Note that

$$I_{s_0 s_1 \ldots s_n} = I_{s_0} \cap F_\mu^{-1}(I_{s_1 \ldots s_n})$$

By induction, we may assume that $I_{s_1 \ldots s_n}$ is a nonempty subinterval, so that,
by the observation above, $F_\mu^{-1}(I_{s_1 \ldots s_n})$ consists of two closed intervals, one
in $I_0$ and one in $I_1$. Hence $I_{s_0} \cap F_\mu^{-1}(I_{s_1 \ldots s_n})$ is a single closed interval.

These intervals are nested because

$$I_{s_0 \ldots s_n} = I_{s_0 \ldots s_{n-1}} \cap F_\mu^{-n}(I_{s_n}) \subset I_{s_0 \ldots s_{n-1}} .$$

Therefore we conclude that

$$\bigcap_{n \geq 0} I_{s_0 s_1 \ldots s_n}$$

is nonempty. Note that if $x \in \cap_{n \geq 0} I_{s_0 s_1 \ldots s_n}$, then $x \in I_{s_0}$, $F_\mu(x) \in I_{s_1}$, etc.
Hence $S(x) = (s_0 s_1 \ldots)$. This proves that $S$ is onto.

Note that $\cap_{n \geq 0} I_{s_0 s_1 \ldots s_n}$ consists of a unique point. This follows immediately from the fact that $S$ is one-to-one. In particular, we have that diam $I_{s_0 s_1 \ldots s_n} \to 0$ as $n \to \infty$.

Finally, to prove continuity of $S$, we choose $x \in \Lambda$ and suppose that $S(x) = s_0 s_1 s_2 \ldots$ . Let $\epsilon > 0$. Pick $n$ so that $1/2^n < \epsilon$. Consider the closed subintervals $I_{t_0 t_1 \ldots t_n}$ defined above for all possible combinations $t_0 t_1 \ldots t_n$. These subintervals are all disjoint, and $\Lambda$ is contained in their union. There are $2^{n+1}$ such subintervals, and $I_{s_0 s_1 \ldots s_n}$ is one of them. Hence we may choose $\delta$ such that $|x - y| < \delta$ and $y \in \Lambda$ implies that $y \in I_{s_0 s_1 \ldots s_n}$. Therefore, $S(y)$ agrees with $S(x)$ in the first $n + 1$ terms. Hence, by Proposition 6.3,

$$d[S(x), S(y)] < \frac{1}{2^n} < \epsilon.$$

This proves the continuity of $S$. It is easy to check that $S^{-1}$ is also continuous. Thus, $S$ is a homeomorphism.

q.e.d.

This theorem shows that, as sets, $\Lambda$ and $\Sigma_2$ are the same. More importantly, the coding $S$ also gives an equivalence between the dynamics of $F_\mu$ on $\Lambda$ and $\sigma$ on $\Sigma_2$. This is the content of the following theorem.

**Theorem 7.3.** $S \circ F_\mu = \sigma \circ S.$

*Proof.* A point $x$ in $\Lambda$ may be defined uniquely by the nested sequence of intervals

$$\bigcap_{n \geq 0} I_{s_0 s_1 \ldots s_n \ldots}$$

determined by the itinerary $S(x)$. Now

$$I_{s_0 \ldots s_n} = I_{s_0} \cap F_\mu^{-1}(I_{s_1}) \cap \ldots \cap F_\mu^{-n}(I_{s_n})$$

so that $F_\mu(I_{s_0 \ldots s_n})$ may be written

$$I_{s_1} \cap F_\mu^{-1}(I_{s_2}) \cap \ldots F_\mu^{-n+1}(I_{s_n}) = I_{s_1 \ldots s_n} \, ,$$

since $F_\mu(I_{s_0}) = I$. Hence

$$SF_\mu(x) = SF_\mu \left( \cap_{n=0}^{\infty} I_{s_0 s_1 \ldots s_n} \right)$$
$$= S \left( \cap_{n=1}^{\infty} I_{s_1 \ldots s_n} \right)$$
$$= s_1 s_2 \ldots = \sigma S(x).$$

q.e.d.

**Definition 7.4.** Let $f: A \to A$ and $g: B \to B$ be two maps. $f$ and $g$ are said to be topologically conjugate if there exists a homeomorphism $h: A \to B$ such that, $h \circ f = g \circ h$. The homeomorphism $h$ is called a topological conjugacy.

Mappings which are topologically conjugate are completely equivalent in terms of their dynamics. For example, if $f$ is topologically conjugate to $g$ via $h$, and $p$ is a fixed point for $f$, then $h(p)$ is fixed for $g$. Indeed, $h(p) = hf(p) = gh(p)$. Similarly, $h$ gives a one-to-one correspondence between $\mathrm{Per}_n(f)$ and $\mathrm{Per}_n(g)$. One may also check that eventually periodic and asymptotic orbits for $f$ go over via $h$ to similar orbits for $g$, and that $f$ is topologically transitive if and only if $g$ is. In particular, since $F_\mu$ on $\Lambda$ is topologically conjugate to the shift, we have now proved that the quadratic map enjoys the striking properties we uncovered so easily for $\sigma$ in the last section. These may be summarized as follows.

**Theorem 7.5.** *Let* $F_\mu(x) = \mu x(1-x)$ *with* $\mu > 2 + \sqrt{5}$. *Then*
  1. *Card* $\mathrm{Per}_n(F_\mu) = 2^n$.
  2. $F_\mu$ *has a dense orbit in* $\Lambda$.

This Theorem shows the power of symbolic dynamics and topological conjugacy. Actually computing the $2^n$ periodic points of period $n$ for $F_\mu$ is a hopeless task. But topological conjugacy guarantees that these orbits are there, and, moreover, symbolic dynamics gives a rough measure of the complexity of the orbits in $\Lambda$. Thus these two notions provide justification for our statement that the shift map is an accurate model for the quadratic map.

**Exercises**

**1.**  Let $Q_c(x) = x^2 + c$. Prove that if $c < 1/4$, there is a unique $\mu > 1$ such that $Q_c$ is topologically conjugate to $F_\mu(x) = \mu x(1-x)$ via a map of the form $h(x) = \alpha x + \beta$.

**2.**  A point $p$ is a *non-wandering* point for $f$, if, for any open interval $J$ containing $p$, there exists $x \in J$ and $n > 0$ such that $f^n(x) \in J$. Note that we do not require that $p$ itself return to $J$. Let $\Omega(f)$ denote the set of non-wandering points for $f$.
  a. Prove that $\Omega(f)$ is a closed set.
  b. If $F_\mu$ is the quadratic map with $\mu > 2 + \sqrt{5}$, show that $\Omega(F_\mu) = \Lambda$.
  c. Identify $\Omega(F_\mu)$ for each $\mu$ satisfying $0 < \mu \leq 3$.

**3.**   A point $p$ is *recurrent* for $f$ if. for any open interval $J$ about $p$, there exists $n > 0$ such that $f^n(p) \in J$. Clearly. all periodic points are recurrent.

   a. Give an example of a non-periodic recurrent point for $F_\mu$ when $\mu > 2 + \sqrt{5}$.

   b. Give an example of a non-wandering point for $F_\mu$ which is not recurrent.

**4.**   *Order of the periodic points.* Let $\Gamma_n$ denote the set of repeating sequences of period $n$ in $\Sigma_2$. Identify such a sequence with a finite string $s_1, \ldots, s_n$ of 0's and 1's in the natural way. Under the topological conjugacy, each element of $\Gamma_n$ corresponds to a unique point in $I$ for a given value of $\mu > 2 + \sqrt{5}$.

   a. Prove that the order of these points in $I$ is independent of $\mu > 2 + \sqrt{5}$. Let $N(s_1, \ldots, s_n)$ denote the integer between 0 and $2^n - 1$ corresponding to this order, numbering from left to right, i.e. $N(0, \ldots, 0) = 0$. Let $B(s_1, \ldots, s_n)$ denote $N$ in binary form. That is, $B(s_1, \ldots, s_n) = (a_1, \ldots, a_n)$ where $a_j = 0$ or 1 and

$$N(s_1, \ldots, s_n) = a_1 \cdot 2^{n-1} + a_2 \cdot 2^{n-2} + \ldots + a_n \cdot 2^0.$$

   b. Use induction to prove that $B$ is given by the following formula:

$$a_j = \sum_{i=1}^{j} s_j \bmod 2.$$

   For example, the fixed point $1, 1, 1 \in \Gamma_3$ occupies position 5 on the real line since

$$a_1 = s_1 = 1$$
$$a_2 = s_1 + s_2 = 0 \bmod 2$$
$$a_3 = s_1 + s_2 + s_3 = 1 \bmod 2.$$

   c. List all points in $\Gamma_n$ for $n = 2, 3, 4$ according to this ordering.

   d. Describe an algorithm for ordering the points in $\Gamma_n$ knowing the ordering of $\Gamma_{n-1}$.

## §1.8 CHAOS

   The quadratic map exhibits in stunning fashion a phenomenon which is only partially understood: the chaotic behavior of orbits of a dynamical system. There are many possible definitions of chaos, ranging from measure theoretic notions of randomness in ergodic theory to the topological approach we will adopt here.

**Definition 8.1.** $f: J \rightarrow J$ is said to be topologically transitive if for any pair of open sets $U, V \subset J$ there exists $k > 0$ such that $f^k(U) \cap V \neq \emptyset$.

Intuitively, a topologically transitive map has points which eventually move under iteration from one arbitrarily small neighborhood to any other. Consequently, the dynamical system cannot be decomposed into two disjoint open sets which are invariant under the map. Note that if a map possesses a dense orbit, then it is clearly topologically transitive. The converse is also true (for compact subsets of $\mathbf{R}$ or $S^1$ ), but we will not prove it here since the proof depends on the Baire Category Theorem.

**Definition 8.2.** $f: J \rightarrow J$ has sensitive dependence on initial conditions if there exists $\delta > 0$ such that, for any $x \in J$ and any neighborhood $N$ of $x$, there exists $y \in N$ and $n \geq 0$ such that $|f^n(x) - f^n(y)| > \delta$.

Intuitively, a map possesses sensitive dependence on initial conditions if there exist points arbitrarily close to $x$ which eventually separate from $x$ by at least $\delta$ under iteration of $f$. We emphasize that not *all* points near $x$ need eventually separate from $x$ under iteration, but there must be at least one such point in every neighborhood of $x$. If a map possesses sensitive dependence on initial conditions, then for all practical purposes, the dynamics of the map defy numerical computation. Small errors in computation which are introduced by round-off may become magnified upon iteration. The results of numerical computation of an orbit. no matter how accurate, may bear no resemblance whatsoever with the real orbit.

**Example 8.3.** The quadratic map $\mu x(1 - x)$ with $\mu > 2 + \sqrt{5}$ possesses sensitive dependence on initial conditions on $\Lambda$. To see this, choose $\delta$ less than the diameter of $A_0$, where $A_0$ is the gap between $I_0$ and $I_1$. Let $x, y \in \Lambda$. If $x \neq y$, then $S(x) \neq S(y)$, so the itineraries of $x$ and $y$ must differ in at least one spot, say the $n^{th}$. But this means that $F_\mu^n(x)$ and $F_\mu^n(y)$ lie on opposite sides of $A_0$, so that

$$|F_\mu^n(x) - F_\mu^n(y)| > \delta.$$

**Example 8.4.** An irrational rotation of the circle is topologically transitive but not sensitive to initial conditions, since all points remain the same distance apart after iteration.

We turn now to one of the main themes of this book, the notion of a chaotic dynamical system. There are many possible definitions of chaos in a dynamical system, some stronger and some weaker than ours. We choose this particular definition because it applies to a large number of important examples and because, in many cases, it is easy to verify.

**Definition 8.5.** Let $V$ be a set. $f: V \to V$ is said to be chaotic on $V$ if
1. $f$ has sensitive dependence on initial conditions.
2. $f$ is topologically transitive.
3. periodic points are dense in $V$.

To summarize, a chaotic map possesses three ingredients: unpredictability, indecomposability, and an element of regularity. A chaotic system is unpredictable because of the sensitive dependence on initial conditions. It cannot be broken down or decomposed into two subsystems (two invariant open subsets) which do not interact under $f$ because of topological transitivity. And, in the midst of this random behavior, we nevertheless have an element of regularity, namely the periodic points which are dense.

**Example 8.6.** $f: S^1 \to S^1$ given by $f(\theta) = 2\theta$ is chaotic. As we have seen, the angular distance between two points is doubled upon iteration of $f$. Hence $f$ is sensitive to initial conditions. Topological transitivity also follows from this observation since any small arc in $S^1$ is eventually expanded by some $f^k$ to cover all of $S^1$ and, in particular, any other arc in $S^1$. The density of periodic points was established in §1.3. We remark that this map possesses a strong form of sensitive dependence called expansiveness.

**Definition 8.7.** $f: J \to J$ is expansive if there exists $\nu > 0$ such that, for any $x, y \in J$, there exists $n$ such that $|f^n x - f^n y| > \nu$.

Expansiveness differs from sensitive dependence in that *all* nearby points eventually separate by at least $\mu$.

**Example 8.8.** The quadratic maps $F_\mu(x) = \mu x(1 - x)$ are chaotic on $\Lambda$ when $\mu > 2 + \sqrt{5}$.

This example differs markedly from the previous example in that the chaos is confined to a small subset of $I$, namely the Cantor set $\Lambda$. A much larger chaotic region for a quadratic map is given by the following example.

**Example 8.9.** $F_4(x) = 4x(1 - x)$ is chaotic on the interval $I = [0, 1]$.

*Proof.* Let $g(\theta) = 2\theta$ be the map on $S^1$ discussed in Example 8.6. Define $h_1 : S^1 \to [-1,1]$ by $h_1(\theta) = \cos\theta$. That is, $h_1$ is just projection from $S^1$ to the $x$-axis. Let $q(x) = 2x^2 - 1$. Then we have

$$
\begin{aligned}
h_1 \circ g(\theta) &= \cos(2\theta) \\
&= 2\cos^2\theta - 1 \\
&= q \circ h_1(\theta)
\end{aligned}
$$

so that $h_1$ conjugates $g$ with $q$. Now $q$ is also topologically conjugate to $F_4$. Indeed, if $h_2(t) = \frac{1}{2}(1-t)$, then we have $F_4 \circ h_2 = h_2 \circ q$. Hence we have the following diagram

$$
\begin{array}{ccc}
S^1 & \xrightarrow{\;g\;} & S^1 \\
{\scriptstyle h_1}\downarrow & & \downarrow{\scriptstyle h_1} \\
[-1,1] & \xrightarrow{\;q\;} & [-1,1] \\
{\scriptstyle h_2}\downarrow & & \downarrow{\scriptstyle h_2} \\
[0,1] & \xrightarrow{\;F_4\;} & [0,1].
\end{array}
$$

It follows immediately that $F_4$ is topologically transitive, for if $U$ and $V$ are two open intervals in $I$, we may choose open arcs $\hat{U}$ and $\hat{V}$ in $S^1$ which project onto $U$ and $V$ under $h_2 \circ h_1$. Since there exists $k$ such that $g^k(\hat{U}) \cap \hat{V} \neq \emptyset$, we therefore have $F_4^k(U) \cap V \neq \emptyset$.

To prove sensitive dependence, we note that any neighborhood $U$ of $x \in I$ "lifts" to $\hat{U}$ in $S^1$. There exists $n$ such that $g^n(\hat{U})$ covers $S^1$, so $F_4^n(U)$ covers $I$ as well. Hence there are points in $U$ which move at least $\delta = 1/2$ away from $x$  Finally, density of periodic points for $g$ implies that there is a $g$-periodic point in $\hat{U}$. The projection of this point in $U$ is clearly $F_4$-periodic.

The technique introduced in this Example can also be used to produce other examples of maps which are chaotic on an interval. The so-called Tchebycheff polynomials are important classical examples which feature this type of behavior. See Exercises 1-3.

We remark that the map $h_1$ above is not a homeomorphism since it is a two-to-one at most points. Thus we have *not* shown that $g(\theta) = 2\theta$ on $S^1$ and $F_4$ are topologically conjugate. Rather, we say that these two maps are *semi-conjugate*.

## Exercises

1.   Use the method of Example 8.9 to prove that $F(x) = 4x^3 - 3x$ is chaotic on the interval $[-1,1]$. (Hint: consider $g(\theta) = 3\theta$ on $S^1$.)

**2.** Prove that $F(x) = 8x^4 - 8x^2 + 1$ is chaotic on $[-1, 1]$.

**3.** The polynomial which is given as in Exercises 1 and 2 by projection of $g(\theta) = \cos n\theta$ onto the interval $[-1, 1]$ is called the $n^{th}$ Tchebycheff polynomial, when properly normalized. Show that these polynomials satisfy the differential equation

$$(1 - x^2)y'' - xy' + \alpha^2 y = 0$$

where $\alpha$ is a positive integer.

**4.** Prove that $T(x) = \tan x$ is chaotic on the entire real line, despite the fact that there are a dense set of points at which an iterate of $T$ fails to be defined.

**5.** Prove that the baker map

$$B(x) = \begin{cases} 2x & 0 \le x \le \frac{1}{2} \\ 2x - 1 & \frac{1}{2} \le x \le 1 \end{cases}$$

is chaotic on $[0, 1]$.

The following exercises apply to the tent map

$$T_2(x) = \begin{cases} 2x & 0 \le x \le 1/2 \\ 2(1 - x) & 1/2 \le x \le 1 \end{cases}.$$

Note that the maximum of $T_2$ is 1 and occurs at $x = \frac{1}{2}$. To describe the dynamics of $T_2$ via symbolic dynamics, we thus need to modify $\Sigma_2$ somewhat since there is an ambiguity in the sequence associated to any rational number of the form $p/2^k$ where $p$ is an integer. For example, $1/2$ may be described by either $(11000...)$ or $(01000...)$. To remedy this, we identify any two sequences of the form $(s_0 \ldots s_{k-1}*1000...)$, where $* = 0$ or 1. For example, the sequences $(1101000...)$ and $(1111000...)$ are to be thought of as representing the same point. Let $\Sigma_2'$ denote $\Sigma_2$ with these identifications.

**6.** Prove that $S: I \to \Sigma_2'$ is one-to-one, where $S(x)$ is defined as in §1.8.

**7.** Prove that $\sigma \circ S = S \circ T_2$.

**8.** Prove that $T_2$ has exactly $2^n$ periodic points of period $n$.

**9.** Prove that $T_2$ is chaotic on $I$.

**10.** Prove that $T_2$ is topologically conjugate to the quadratic map $F_4(x) = 4x(1 - x)$.

**11.**    Construct a piecewise linear map on $[0, 1]$ which is topologically conjugate to $F(x) = 4x^3 - 3x$ on $[-1, 1]$.

## §1.9 STRUCTURAL STABILITY

A very important notion in the study of dynamical systems is the stability or persistence of the system under small changes or perturbations. This is the concept of structural stability which we introduce in this section. Briefly, a map $f$ is structurally stable if every "nearby" map is topologically conjugate to $f$ and so has essentially the same dynamics. Clearly, we need to be precise about what nearby means, but the basic idea is simple. If, no matter how we perturb $f$ or change $f$ slightly, we get an equivalent dynamical system, then the dynamical structure of $f$ is stable. Here, equivalent means topologically conjugate. If $f$ and $g$ are topologically conjugate, we will write $f \sim g$.

The notion of structural stability is extremely important in applications. Suppose our dynamical system is the solution of a differential equation or otherwise comes from a real world physical system. Ordinarily, the system itself will be only a model of real world phenomena: certain assumptions will have been made, and certain approximations and experimental errors will be present. Hence the dynamical system itself, albeit a completely accurate solution of the physical model, will nevertheless be only an approximation to reality since the model itself suffers this flaw. Now, if the dynamical system in question is not structurally stable, then the small errors and approximations made in the model have a chance of dramatically changing the structure of the real solution to the system. That is, our "solution" could be radically wrong or unstable. If, on the other hand, the dynamical system in question is structurally stable, then the small errors introduced by approximations and experimental errors may not matter at all: the solution to the model system may be equivalent or topologically conjugate to the actual solution.

This does not mean that the only interesting physical systems are the structurally stable ones. Indeed, most dynamical systems that arise in classical mechanics are not structurally stable. There are also simple examples of systems such as the Lorenz system from meteorology that are "far" from being structurally stable. These systems cannot even be approximated in a sense to be made precise below by stable systems. Nevertheless, the concept of structural stability is an important one in applications of the theory of dynamical systems.

To begin the discussion of structural stability, we need to make precise the notion of "nearness" of two functions.

**Definition 9.1.** Let $f$ and $g$ be two maps. The $C^0$-distance between $f$ and $g$, written $d_0(f,g)$, is given by

$$d_0(f,g) = \sup_{x \in \mathbf{R}} |f(x) - g(x)|.$$

The $C^r$-distance $d_r(f,g)$ is given by

$$d_r(f,g) = \sup_{x \in \mathbf{R}} (|f(x) - g(x)|, \ |f'(x) - g'(x)|, \dots, |f^{(r)}(x) - g^{(r)}(x)|).$$

Intuitively, two maps are $C^r$-close provided they as well as their first $r$ derivatives differ by only a small amount. We may also consider the $C^r$-distance between two maps on an interval $J \subset \mathbf{R}$ by suitably restricting $x$ and $y$. We caution the reader that $d_r$ does not give a useful metric on the set of all functions. Indeed, since the real line is unbounded, two maps can easily be infinitely far apart. Moreover, even if we assume this difficulty away, the resulting topology on the set of functions is nasty. Hence we will use the $C^r$-distance only as a measure of the proximity of two functions and not as a global metric on all maps.

**Example 9.2.** $f(x) = 2x$ and $g(x) = (2+\epsilon)x$ have $C^0$-distance infinity. But $f(x) = 2x$ and $g(x) = 2x + \epsilon$ are $C^r$-$\epsilon$ apart for all $r$. Let $J = [0,10]$. Then $f(x) = 2x$ and $g(x) = (2 + \epsilon)x$ are $C^0$-$10\epsilon$ apart (and, in fact $C^r$-$10\epsilon$ apart) on the interval $J$.

We will be primarily concerned with functions that are $C^1$-close or, at most, $C^2$-close. Fig. 9.1 illustrates the difference graphically between $C^0$-close, $C^1$-close, and $C^2$-close.

We now define $C^r$-structural stability.

**Definition 9.3.** Let $f: J \to J$. $f$ is said to be $C^r$-structurally stable on $J$, if there exists $\epsilon > 0$ such that whenever $d_r(f,g) < \epsilon$ for $g: J \to J$, it follows that $f$ is topologically conjugate to $g$.

**Example 9.4.** Let $L(x) = \frac{1}{2}x$. Then $L$ is $C^1$-structurally stable on $\mathbf{R}$. To see this, we must exhibit an $\epsilon > 0$ such that, if $d_1(L,g) < \epsilon$, then $L$ and $g$ are topologically conjugate. We claim that any $\epsilon < 1/2$ works. For if $d_1(L,g) < \epsilon$, then we must have $0 < g'(x) < 1$ for all $x \in \mathbf{R}$. In particular, $g(x)$ is everywhere increasing. Note also that $g(x)$ has a unique attracting

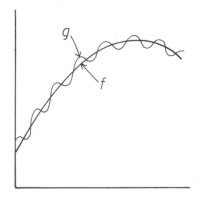 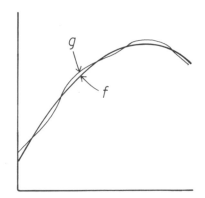

**Fig. 9.1.** In Fig. a, $f$ and $g$ are $C^0$-close
but not $C^1$-close. In Fig. b, $f$ and $g$ are
$C^1$-close but not $C^2$-close.

fixed point $p$ in $\mathbf{R}$ and that all points in $\mathbf{R}$ tend to $p$ under iteration. That $g$ has a unique fixed point follows from the Mean Value Theorem: between any two fixed points of $g$ must lie a point with derivative $= 1$, which cannot happen. Alternatively, since $|g'(x)| < 1$, $g$ is a global contraction.

This shows that $L$ and $g$ have the same dynamics, the basic idea behind structural stability. To be strictly precise, however, we must exhibit a topological conjugacy between $L$ and $g$. To do this, we introduce the notion of a *fundamental domain*. This is best done by example. Consider the pair of intervals $5 < |x| \le 10$. Note that the $L$-orbit of any point in $\mathbf{R}$ (with the exception of 0) enters this set exactly once. For $g$, we may find a similar fundamental domain: indeed, it is easy to check that the intervals $g(10) < x \le 10$ and $-10 \le x < g(-10)$ have the same property (Exercise 1).

We now construct a conjugacy $h$ such that $h \circ L = g \circ h$. First define $h: [5, 10] \to [g(10), 10]$ and $h: [-10, -5] \to [-10, g(-10)]$ to be linear, i.e. with a straight line graph. We require that $h$ be increasing so that $h(\pm 10) = \pm 10$. (We remark that any other increasing homeomorphism works just as well.) We complete the definition of $h$ as follows. Let $x \ne 0$. There is an $n \in \mathbf{Z}$ such that $L^n(x)$ belongs to the fundamental domain for $L$. Hence $h \circ L^n(x)$ is well-defined. We then set $h(x) = g^{-n} \circ h \circ L^n(x)$. Note that $h(x)$ is also well-defined, since $g$ is a homeomorphism and so $g^{-n}$ makes sense. Clearly, we have $g^n \circ h(x) = h \circ L^n(x)$. Moreover, if we apply the same construction to $L(x)$, we find that $g \circ h(x) = h \circ L(x)$, as required. Finally, define $h(0) = $ fixed point of $g$. It is easy to check that $h$ as defined is a homeomorphism.

Intuitively, a fundamental domain is visited exactly once by each orbit, except, of course, the fixed point. Hence we may define a conjugacy in virtually any way we please on the fundamental domain, and then extend in the only way possible by iterating the map. The only question is then whether or not we can extend the conjugacy to points whose orbits never enter the fundamental domain.

We now return to the quadratic map $F_\mu(x) = \mu x(1-x)$. As we saw in §1.5, all points tend to $-\infty$ for this map with the exception of those in a set $\Lambda$ on which $F_\mu$ is topologically conjugate to the shift. We claim that, if $\mu$ is large enough, then $F_\mu$ is $C^2$-structurally stable. This may be proved by another fundamental domain argument.

This is more complicated, but let us sketch the details. We first assume that $\mu > 2 + \sqrt{5}$, so that $|F_\mu'(x)| > 1$ on $I_0 \cup I_1$. We will produce an $\epsilon > 0$ such that if $g$ is $C^2$-$\epsilon$ close to $F_\mu$, then $g$ has the same dynamics as $F_\mu$. Let us first choose $\epsilon_1$ small enough so that if $g$ is $C^2$- $\epsilon_1$ close to $F_\mu$, then $g'' < 0$, i.e., so that the graph of $g$ is concave down. This is clearly possible since $F_\mu'' \equiv -2\mu$. Next choose $\epsilon_2 < \epsilon_1$ small enough so that if $g$ is $C^1$-$\epsilon_2$ close to $F_\mu$, then $g$ has two fixed points, $\alpha$ and $\beta$, which satisfy

1. $\alpha < \beta$.
2. $g'(\alpha) > 1$.
3. $g'(\beta) < -1$.

The fact that $\epsilon_2$ may be chosen so that $g$ has at most two fixed points follows from the concavity of the graph of $g$. The fact that $g$ has at least two fixed points can be guaranteed by making $g$ $C^0$-close to $F_\mu$. Finally, the conditions on $g'$ at the fixed points are controlled by the $C^1$-distance of $g$ from $F_\mu$.

Note that $g$ has a unique critical point $c$ and that there exist points $\alpha'$, $\beta'$ with $g(\alpha') = \alpha$, $g(\beta') = \beta$. The points $\alpha$ and $\alpha'$ play the same role as 0 and 1 do for $F_\mu$.

We may finally choose $\epsilon < \epsilon_2$ such that, if $g$ is $C^1$-$\epsilon$ close to $F_\mu$, then $g^{-1}(\alpha')$ consists of a pair of points, $a_0$ and $a_1$, and moreover, if $x \in [\alpha, a_0] \cup [a_1, \alpha']$, then $|g'(x)| > 1$. Thus, if $g$ is $C^2$-$\epsilon$ close (and therefore $C^0$- and $C^1$-$\epsilon$ close) to $F_\mu$, then the graph of $g$ has all of the qualitative properties of the graph of $F_\mu$ on the interval $[\alpha, \alpha']$. See Fig. 9.2.

More importantly, $F_\mu$ and $g$ have the same dynamics. It follows immediately that if $x < \alpha$, then $g^n(x) \to -\infty$. Similarly, if $x > \alpha'$ or if $x \in (a_0, a_1)$, then $g^n(x) \to -\infty$ as well. A similar inductive procedure on the inverse images of $(a_0, a_1)$ as in §1.5 shows that all points except those in a Cantor set $\Lambda_g$ tend to $-\infty$ eventually under iteration of $g$. On $\Lambda_g$, $g$ is again topologically conjugate to the shift automorphism via arguments as in §1.5. We leave the details to the reader.

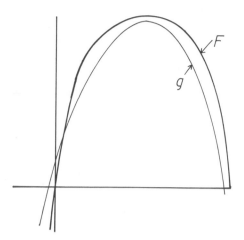

**Fig. 9.2.** The graphs of $F_\mu$ and $g$ are $C^2$-$\epsilon$ close

To prove that $F_\mu \sim g$, we must construct fundamental domains for both $F_\mu$ and $g$ in order to define the conjugacy. This can be accomplished as follows. First choose $x_0 < \min(g^2(c), F_\mu^2(c))$. The intervals $(F_\mu(x_0), x_0)$ and $(g(x_0), x_0)$ are easily seen to be fundamental domains for $F_\mu$ on $\mathbf{R}^-$ and $g$ on $(-\infty, \alpha)$. The conjugacy may then be defined arbitrarily on $(F_\mu(x_0), x_0)$ and extended by $h \circ F_\mu = g \circ h$ to all of $\mathbf{R}^-$.

We then extend $h$ to the interval $(1, \infty)$ and finally to each $A_n$ in the natural way. We remark that care must be taken on $A_0$ since $F_\mu$ is two-to-one on this interval. Once $h$ is defined on all of $\mathbf{R} - \Lambda$, we extend to $\Lambda$ in the only way possible to make $h$ a homeomorphism. We leave the tedious details to the reader.

Alternatively, one can use the fact that both $g$ on $\Lambda_g$ and $F_\mu$ on $\Lambda$ are topologically conjugate to the shift, hence to each other on these sets. The conjugacy may be extended off these sets by the above fundamental domain argument. In either event we have proved

**Theorem 9.5.** *The quadratic map* $F_\mu(x) = \mu x(1 - x)$ *is* $C^2$ *structurally stable if* $\mu > 2 + \sqrt{5}$.

Perhaps more important than the question of when a given map is structurally stable is the converse question: when is it *not* structurally stable? One of the major ways a map can fail to be structurally stable occurs when there is a lack of hyperbolicity.

**Example 9.6.** Let $F_0(x) = x - x^2$. Note that $F_0(0) = 0$ and $F_0'(0) = 1$, so 0

is a non-hyperbolic fixed point. Consider $F_\epsilon(x) = x - x^2 + \epsilon$. Clearly, $F_\epsilon(x)$ is $C^r$-$\epsilon$ close to $F_0$. When $\epsilon > 0$, $F_\epsilon$ is easily seen to have two fixed points, but when $\epsilon < 0$, $F_\epsilon$ has none. Consequently, the $F_\epsilon$ do not have the same dynamics as $F_0$ and therefore $F_0$ is not structurally stable.

**Example 9.7.** Let $T_\lambda(x) = x^3 - \lambda x$. For $-1 < \lambda \le 1$, $T_\lambda$ has three fixed points: at 0 and at $\pm\sqrt{\lambda + 1}$. All points between $\pm\sqrt{\lambda + 1}$ tend to the attracting fixed point at 0. When $\lambda > 1$, this is no longer true. There exists $x$ in the interval $[-\sqrt{\lambda + 1}, \sqrt{\lambda + 1}]$ such that $T_\lambda(x) = -x$, i.e., the graph of $T_\lambda$ crosses the line $y = -x$. Since $T_\lambda(-x) = -T_\lambda(x)$, we also have $T_\lambda(-x) = x$, so that $x$ is a periodic point of period 2. Hence the dynamics of $T_\lambda$ is different for $\lambda \le 1$ and $\lambda > 1$, so that $T_1$ is not structurally stable. We remark that $T_{-1}$ is also not structurally stable. See Exercise 2. Note that $T_1'(0) = -1$, so that the fixed point is again non-hyperbolic when structural stability fails to hold.

A hyperbolic fixed point for $f$ is $C^1$ structurally stable locally. By this we mean there is a neighborhood of the fixed point and an $\epsilon > 0$ such that, if a map $g$ is $C^1$-$\epsilon$ close to $f$ on this neighborhood, then $f$ is topologically conjugate to $g$ on this neighborhood. This fact is established in a series of exercises below. Along the way, we establish the one-dimensional version of Hartman's Theorem:

**Theorem 9.8.** *Let $p$ be a hyperbolic fixed point for $f$ and suppose $f'(p) = \lambda$ with $|\lambda| \ne 0, 1$. Then there are neighborhoods $U$ of $p$ and $V$ of $0 \in \mathbf{R}$ and a homeomorphism $h: U \to \mathbf{R}$ which conjugates $f$ on $U$ to the linear map $L(x) = \lambda x$ on $V$.*

Thus a map near a hyperbolic fixed point is always locally topologically conjugate to its derivative. This allows us to explain why we only require that the conjugacy map in the definition of topological conjugacy be a homeomorphism, not a diffeomorphism. Suppose $f(p) = p$ and $f'(p) = \lambda$. Let $h$ be a diffeomorphism. Then $g = h \circ f \circ h^{-1}$ has a fixed point at $h(p)$, but we have

$$g'(h(p)) = h'(f(p)) \cdot f'(p) \cdot (h^{-1})'(h(p))$$

$$= h'(p) \cdot \lambda \cdot \frac{1}{h'(p)}$$

$$= \lambda.$$

Thus, the multiplier $\lambda$ at the fixed point is preserved by differentiable conjugacies. As we have seen, maps may behave dynamically the same despite having different multipliers at the fixed points. Thus the weakened notion of topological conjugacy is more appropriate for our purposes.

**Exercises**

**1.** Suppose $g(x)$ is as in Example 9.4. Prove that the intervals $g(10) < x \leq 10$ and $-10 \leq x < g(-10)$ form a fundamental domain for $g$.

**2.** Let $T_{-1}(x) = x^3 + x$. Prove that $T_{-1}$ is not structurally stable.

**3.** Let $T_\lambda(x) = x^3 - \lambda(x)$. Prove that $T_\lambda$ is structurally stable if $-1 < \lambda < 0$.

**4.** Prove that $T_{\lambda_0}$ is topologically conjugate to $T_{\lambda_1}$ if $-1 < \lambda_0, \lambda_1 < 0$.

**5.** Prove that $F_4(x) = 4x(1 - x)$ is not structually stable.

**6.** Prove that $S(x) = \sin(x)$ is not structurally stable.

**7.** Prove that, if $f \sim g$ via $h$ and $f$ has a local maximum at $x_0$, then $g$ has either a local maximum or minimum at $h(x_0)$.

**8.** Give an example to show that we may have $f \sim g$ via $h$ and $x_0$ a local maximum for $f$ and $h(x_0)$ a local minimum for $g$.

**9.** Let $S_\lambda(x) = \lambda \sin(x)$. If $0 < \lambda_1 < \lambda_2 < 1$, prove that $S_{\lambda_1} \sim S_{\lambda_2}$.

**10.** Show, however, that neither $S_{\lambda_1}$ nor $S_{\lambda_2}$ is structurally stable.

**11.** We may define a notion of linear structural stability for linear maps by replacing the notion of topological conjugacy by that of linear conjugacy. Two linear maps $T_1, T_2 : \mathbf{R} \to \mathbf{R}$ are linearly conjugate if there is a linear map $L$ such that $T_1 \circ L = L \circ T_2$. $T_1(x) = ax$ is linearly stable if there is a neighborhood $N$ about $a$ such that if $b \in N$, then $T_2(x) = bx$ is linearly conjugate to $T_1$. Find all linearly stable maps and identify all elements of a given conjugacy class.

**12.** (Hartman's Theorem) Let $p$ be a hyperbolic fixed point for $f$ with $f'(p) = \lambda$ and $\lambda \neq 0$. Prove that $f$ is locally topologically conjugate to its derivative map $x \to \lambda x$ as described in Theorem 9.8.

**13.** Combine Exercises 11 and 12 to prove that any small perturbation of a map near a hyperbolic fixed point is locally topologically conjugate to $f$.

**14.** Let $f : [0, 1] \to [0, 1]$ be a diffeomorphism. Prove that, if $f'(x) > 0$, then $f$ has only fixed points and no periodic points. Prove that, if $f'(x) < 0$, then $f$ has a unique fixed point and all other periodic points have period two.

**15.** A diffeomorphism $f : [0, 1] \to [0, 1]$ is called *Morse-Smale* if $f$ has only hyperbolic periodic points. (Note that, since $f$ is onto, the endpoints of $[0, 1]$ are necessarily periodic.) Prove that a Morse-Smale diffeomorphism has only finitely many periodic points.

**16.**    Prove that a Morse-Smale diffeomorphism of $[0, 1]$ is structurally stable.

**17.**    Prove that the map $f(x) = x^3 + \frac{3}{4}x$ is a Morse-Smale diffeomorphism on the interval $[-\frac{1}{2}, \frac{1}{2}]$.

## §1.10 SARKOVSKII'S THEOREM

In this section, we will prove a remarkable theorem due to Sarkovskii. The theorem only holds for maps of the real line, but nevertheless is amazing for its lack of hypotheses ($f$ is only assumed continuous) and strong conclusion. We caution the reader that, as this is our first major theorem, the material in this section is a little "heavier" than in previous sections. As a warmup, and also as a means of highlighting the importance of period three points, we will prove a special case.

**Theorem 10.1.**    *Let* $f : \mathbf{R} \to \mathbf{R}$ *be continuous. Suppose* $f$ *has a periodic point of period three. Then* $f$ *has periodic points of all other periods.*

*Proof.* The proof will depend on two elementary observations. First, if $I$ and $J$ are closed intervals with $I \subset J$ and $f(I) \supset J$, then $f$ has a fixed point in $I$. This is, of course, a simple consequence of the Intermediate Value Theorem. See Fig. 10.1.

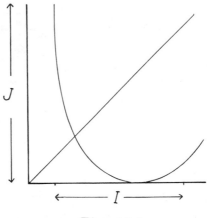

Fig. 10.1

The second observation is the following: suppose $A_0. A_1, \ldots, A_n$ are closed intervals and $f(A_i) \supset A_{i+1}$ for $i = 0, \ldots, n-1$. Then there exists at least one

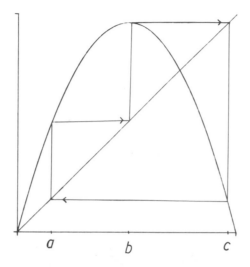

**Fig. 10.2.** The map $F_{3.839}(x) = 3.839x(1-x)$.

subinterval $J_0$ of $A_0$ which is mapped onto $A_1$. There is a similar subinterval in $A_1$ which is mapped onto $A_2$, and thus there is a subinterval $J_1 \subset J_0$ having the property that $f(J_1) \subset A_1$ and $f^2(J_1) = A_2$. Continuing in this fashion, we find a nested sequence of intervals which map into the various $A_i$ in order. Thus there exists a point $x \in A_0$ such that $f^i(x) \in A_i$ for each $i$. We say that $f(A_i)$ *covers* $A_{i+1}$. See Exercise 1.

To prove the Theorem, let $a, b, c \in \mathbf{R}$ and suppose $f(a) = b$, $f(b) = c$, and $f(c) = a$. We assume that $a < b < c$. The only other possibility, $f(a) = c$, is handled similarly. This situation arises in the quadratic map $F_\mu$ for sufficiently large $\mu$, and even for some $\mu < 4$. In fact, we will exploit this fact later when we discuss the case $\mu = 3.839$ in detail in §1.13. See Fig. 10.2

Let $I_0 = [a, b]$ and $I_1 = [b, c]$ and note that our assumptions imply $f(I_0) \supset I_1$ and $F(I_1) \supset I_0 \cup I_1$. The graph of $f$ shows that there must be a fixed point for $f$ between $b$ and $c$. Similarly, $f^2$ must have fixed points between $a$ and $b$, and it is easy to see that at least one of these points must have period two. So we let $n \geq 2$; our goal then is to produce a periodic point of prime period $n > 3$.

Inductively, we define a nested sequence of intervals $A_0, A_1, \ldots, A_{n-2} \subset I_1$ as follows. Set $A_0 = I_1$. Since $f(I_1) \supset I_1$, there is a subinterval $A_1 \subset A_0$ such that $f(A_1) = A_0 = I_1$. Then there is a subinterval $A_2 \subset A_1$ such that $f(A_2) = A_1$, so that $f^2(A_2) = A_0 = I_1$. Continuing, we find a subinterval $A_{n-2} \subset A_{n-3}$ such that $f(A_{n-2}) = A_{n-3}$. According to our second observation above, if $x \in A_{n-2}$, then $f(x), f^2(x), \ldots, f^{n-1}(x) \subset A_0$ and, indeed, $f^{n-2}(A_{n-2}) = A_0 = I_1$.

Now since $f(I_1) \supset I_0$, there exists a subinterval $A_{n-1} \subset A_{n-2}$ such that $f^{n-1}(A_{n-1}) = I_0$. Finally, since $f(I_0) \supset I_1$ we have, $f^n(A_{n-1}) \supset I_1$ so that $f^n(A_{n-1})$ covers $A_{n-1}$. It follows from our first observations that $f^n$ has a fixed point $p$ in $A_{n-1}$.

We claim that $p$ actually has prime period $n$. Indeed, the first $n - 2$ iterations of $p$ lie in $I_1$, the $(n-1)^{st}$ lies in $I_0$, and the $n^{th}$ is $p$ again. If $f^{n-1}(p)$ lies in the interior of $I_0$ then it follows easily that $p$ has prime period $n$. If $f^{n-1}(p)$ happens to lie on the boundary, then $n = 2$ or $3$, and again we are done.

q.e.d.

This theorem is just the beginning of the story. Sarkovskii's Theorem gives a complete accounting of which periods imply which other periods for continuous maps of $\mathbf{R}$. Consider the following ordering of the natural numbers:

$$3 \triangleright 5 \triangleright 7 \triangleright \cdots \triangleright 2 \cdot 3 \triangleright 2 \cdot 5 \triangleright \cdots \triangleright 2^2 \cdot 3 \triangleright 2^2 \cdot 5 \triangleright \cdots$$

$$\triangleright 2^3 \cdot 3 \triangleright 2^3 \cdot 5 \triangleright \cdots \cdots \triangleright 2^3 \triangleright 2^2 \triangleright 2 \triangleright 1.$$

That is, first list all odd numbers except one, followed by 2 times the odds, $2^2$ times the odds, $2^3$ times the odds, etc. This exhausts all the natural numbers with the exception of the powers of two which we list last, in decreasing order. This is the Sarkovskii ordering of the natural numbers. Sarkovskii's Theorem is:

**Theorem 10.2.** *Suppose $f : \mathbf{R} \to \mathbf{R}$ is continuous. Suppose $f$ has a periodic point of prime period $k$. If $k \triangleright \ell$ in the above ordering, then $f$ also has a periodic point of period $\ell$.*

Before proving this Theorem, we note several consequences.

**Remarks.**

**1.** If $f$ has a periodic point whose period is not a power of two, then $f$ necessarily has infinitely many period points. Conversely, if $f$ has only finitely many periodic points, then they all necessarily have periods which are powers of two. This fact will reappear when we discuss the period-doubling route to chaos in a later section.

**2.** Period 3 is the greatest period in the Sarkovskii ordering and therefore implies the existence of all other periods, as we saw above.

**3.**   The converse of Sarkovskii's Theorem is also true! There are maps which have periodic points of period $p$ and no "higher" period points according to the Sarkovskii ordering. We give several examples of this at the end of this section.

We will give an elementary proof of Sarkovskii's Theorem due to Block, Guckenheimer, Misiurewicz and Young. The proof rests mainly on the two observations which we used above. For two closed intervals, $I_1$ and $I_2$, we will introduce the notation $I_1 \rightarrow I_2$ if $f(I_1)$ covers $I_2$. If we find a sequence of intervals $I_1 \rightarrow I_2 \rightarrow \ldots \rightarrow I_n \rightarrow I_1$, then our previous observations show that there is a fixed point of $f^n$ in $I_1$.

We first assume that $f$ has a periodic point $x$ of period $n$ with $n$ odd and $n > 1$. Suppose that $f$ has no periodic points of odd period less than $n$. Let $x_1, \ldots, x_n$ be the points on the orbit of $x$, enumerated from left to right. Note that $f$ permutes the $x_i$. Clearly, $f(x_n) < x_n$. Let us choose the largest $i$ for which $f(x_i) > x_i$. Let $I_1$ be the interval $[x_i, x_{i+1}]$. Since $f(x_{i+1}) < x_{i+1}$, it follows that $f(x_{i+1}) \leq x_i$ and so we have that $f(I_1) \supset I_1$. Therefore, $I_1 \rightarrow I_1$.

Since $x$ does not have period 2, it cannot be that $f(x_{i+1}) = x_i$ and $f(x_i) = x_{i+1}$ so that $f(I_1)$ contains at least one other interval of the form $[x_j, x_{j+1}]$ which we call $I_2$. Hence $I_1 \rightarrow I_2$. Continuing, we choose $I_3, \ldots, I_k$ such that $f(I_j) \supset I_{j+1}$. Since $n$ is odd, there are more $x_i$'s on one side of $I_1$ than on the other, so that some $x_i$'s must change sides under the action of $f$ and some must not. Consequently, $f(I_k) \supset I_1$ for some $k$. We thus have $I_1 \rightarrow I_2 \rightarrow \ldots \rightarrow I_k \rightarrow I_1$. Let us choose the smallest $k$ for which this happens, i.e. $I_1 \rightarrow I_2 \rightarrow \ldots \rightarrow I_k \rightarrow I_1$ is the shortest path from $I_1$ to $I_1$ except, of course, $I_1 \rightarrow I_1$. We therefore find a diagram as in Fig. 10.3.

Now, if $k < n - 1$, then one of the loops $I_1 \rightarrow I_2 \rightarrow \ldots \rightarrow I_k \rightarrow I_1$ or $I_1 \rightarrow \ldots \rightarrow I_k \rightarrow I_1 \rightarrow I_1$ gives a fixed point of $f^m$ with $m$ odd and $m < k$. This point must have prime period $< k$ since $I_1 \cap I_2$ consists of only one point, and that point has period $> m$. Therefore $k = n - 1$.

Since $k$ is the smallest integer that works, we cannot have $I_\ell \rightarrow I_j$ for any $j > \ell + 1$. It follows that the orbit of $x$ must be ordered in $\mathbf{R}$ in one of two possible ways, as depicted in Fig. 10.4.

It follows that we can extend the diagram depicted in Fig. 10.3 to that shown in Fig. 10.5. Sarkovskii's Theorem for the special case of $n$ odd is now immediate. Periods larger than $n$ are given by cycles of the form $I_1 \rightarrow \ldots \rightarrow I_{n-1} \rightarrow I_1 \rightarrow \ldots I_1$. The smaller even periods are given by cycles of the form

$$I_{n-1} \rightarrow I_{n-2} \rightarrow I_{n-1},$$

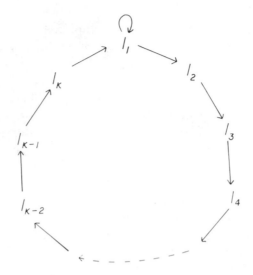

**Fig. 10.3**

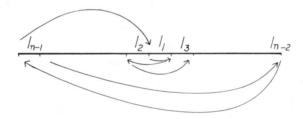

**Fig. 10.4** One possible ordering of the $I_j$.
The other is the mirror image.

$$I_{n-1} \to I_{n-4} \to I_{n-3} \to I_{n-2} \to I_n \ \ 1$$

and so forth. For the case of $n$ even, we first note that $f$ must have a periodic point of period 2. This follows from the above arguments provided we can guarantee that some $x_i$'s change sides under $f$ and some do not (use the facts that $I_{n-1} \leftarrow I_{n-2}$ and $I_{n-1} \to I_{n-2}$. If this is not the case, then all of the $x_i$'s must change sides and so $f[x_1, x_i] \supset [x_{i+1}, x_n]$ and $f[x_{i+1}, x_N] \supset [x_1, x_i]$. But then, our observation above produces a period 2 point in $[x_1, x_i]$.

The Theorem now will be proved for $n = 2^m$ as follows. Let $k = 2^\ell$ with $\ell < m$. Consider $g = f^{k/2}$. By assumption, $g$ has a periodic point of period $2^{m-\ell+1}$. Therefore, $g$ has a point which has period 2. This point has period

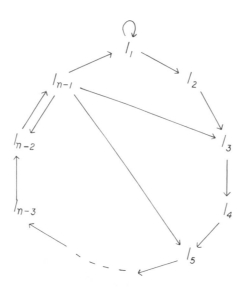

**Fig. 10.5**

$2^\ell$ for $f$. The final case is now $n = p \cdot 2^m$ where $p$ is odd. This case can be reduced to the previous two. We leave these reductions as Exercises.

q.e.d.

We now turn to the converse of Sarkovskii's Theorem. To produce a map with period 5 and no period 3, consider a map $f : [1, 5] \rightarrow [1, 5]$ which satisfies

$$f(1) = 3$$
$$f(3) = 4$$
$$f(4) = 2$$
$$f(2) = 5$$
$$f(5) = 1$$

so that 1 is periodic of period 5. Suppose that $f$ is linear between these integers, i.e. the graph is as shown in Fig. 10.6.

It is easy to check that

$$f^3[1, 2] = [2, 5]$$
$$f^3[2, 3] = [3, 5]$$
$$f^3[4, 5] = [1, 4]$$

so $f^3$ has no fixed points in any of these intervals. It is true that $f^3[3, 4] = [1, 5]$ so that $f^3$ has at least one fixed point in $[3, 4]$. But we claim that this

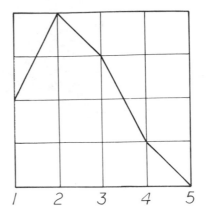

**Fig. 10.6**

point is unique, and therefore must be the fixed point for $f$, not the period 3 point. Indeed, $f:[3,4] \to [2,4]$ is monotonically decreasing, as is $f:[2,4] \to [2,5]$ and $f:[2,5] \to [1,5]$. Therefore $f^3$ is monotonically decreasing on $[3,4]$ and the fixed point is unique.

The graph, shown in Fig. 10.7, produces period 7 but not period 5.

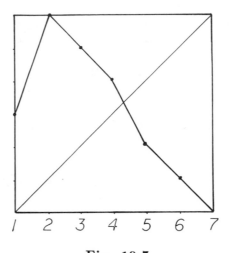

**Fig. 10.7**

This process is easily generalized to give the first portion of the Sarkovskii ordering. For the even periods, we will introduce a trick. Let $f:I \to I$ be continuous. We will construct a new function, the double of $f$, whose periodic points will have exactly twice the period of those of $f$, plus one additional fixed point. The procedure for producing $F$ is as follows. Divide the interval $I$ into thirds. Compress the graph of $f$ into the upper left corner of $I \times I$ as shown on Fig. 10.8.a. The rest of the graph is filled in as in Fig. 10.8.b.

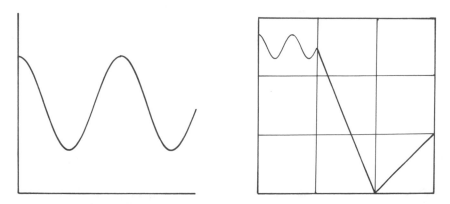

**Fig. 10.8.** Fig. 10.8.a. gives the graph of $f(x)$ while Fig. 10.8.b. gives the graph of its double, $F(x)$.

The map $F$ is piecewise linear on $[1/3, 2/3]$ and $[2/3, 1]$. Moreover, $F(\frac{2}{3}) = 0$, $F(1) = \frac{1}{3}$, and $F$ is continuous.

Note that $f$ maps $[0, \frac{1}{3}]$ into $[\frac{2}{3}, 1]$ and vice versa. Also note that if $x \in [\frac{1}{3}, \frac{2}{3}]$ and $x$ is not the fixed point, then there exists $n$ so that $f^n(x) \in [0, \frac{1}{3}] \cup [\frac{2}{3}, 1]$. This implies that there are no $F$-periodic points in $(\frac{1}{3}, \frac{2}{3})$. Exercise 7 shows that if $x$ is a periodic point of period $n$ for $f$, then $x/3$ is periodic of period $2n$ for $F$. On the other hand, if $y$ is $F$-periodic then either $y$ or $F(y)$ lies in $[0, \frac{1}{3}]$ and exercise 9 shows that $3y$ or $3F(y)$ is $f$-periodic. Thus to produce a map with period 10 but not period 8, we need only double the graph of a function with period 5 but not period 3.

As a final remark, we must emphasize that Sarkovskii's Theorem is very definitely only a one-dimensional result. There is no higher dimensional analogue of this result. In fact, the Theorem does not even hold on the circle. For example, the map which rotates all points on the circle by $120^o$ makes all points periodic with period three. There are no other periods whatsoever.

**Exercises**

**1.** Suppose $A_0, A_1, \ldots, A_n$ are closed intervals and $f(A_i) \supset A_{i+1}$ for $i = 0, \ldots, n-1$. Prove that there exists a point $x \in A_0$ such that $F^i(x) \in A_i$ for each $i$.

**2.** Prove that if $f$ has period $p \cdot 2^m$ with $p$ odd, then $f$ has period $q \cdot 2^m$ with $q$ odd, $q > p$.

**3.** Prove that if $f$ has period $p \cdot 2^m$ with $p$ odd, then $f$ has period $2^\ell$, $\ell \le m$.

**4.** Prove that if $f$ has period $p \cdot 2^m$ with $p$ odd, then $f$ has period $q \cdot 2^m$ with $q$ even.

**5.** Construct a piecewise linear map with period $2n + 1$.

**6.** Give a formula for $F(x)$ in terms of $f(x)$, where $F(x)$ is the double of $f(x)$.

**7.** Prove that $F(x)$, the double of $f(x)$, has a periodic point of period $2n$ at $x/3$ iff $x$ has $f$-period $n$.

**8.** Construct a map that has periodic points of period $2^j$ for $j < \ell$ but not period $2^\ell$.

**9.** Prove that if $F(x)$, the double of $f(x)$, has a periodic point $p$ that is not fixed, then either $p$ or $F(p)$ lies in $[0, \frac{1}{3}]$. Prove that, in this case, either $3p$ or $3F(p)$ is a periodic point for $f$.

## §1.11 THE SCHWARZIAN DERIVATIVE

In this section, we describe a tool first introduced into the study of one-dimensional dynamical systems by Singer in 1978. This is the Schwarzian derivative. Actually, the Schwarzian derivative plays an important role in complex analysis, where it is used as a criterion for a complex function to be a linear fractional transformation. In one-dimensional dynamics, the Schwarzian derivative is a valuable tool for a number of reasons. In this section, we will show how it may be used to establish an upper bound on the number of attracting periodic orbits that certain maps may have. We will also use it to prove that other maps have an entire interval on which the map is chaotic. Later, in §§ 17–19, the Schwarzian derivative will play an important role in our discussion of how families of maps like the quadratic family make the transition from simple to chaotic dynamics.

**Definition 11.1** The Schwarzian derivative of a function $f$ at $x$ is

$$ Sf(x) = \frac{f'''(x)}{f'(x)} - \frac{3}{2}\left(\frac{f''(x)}{f'(x)}\right)^2. $$

For example, if $F_\mu(x) = \mu x(1-x)$ is our quadratic model mapping, then $SF_\mu(x) = -6/(1-2x)^2$, so that $SF_\mu(x) < 0$ for all $x$ (even $x = 1/2$, the critical point, at which $SF_\mu(x) = -\infty$).

For us, functions with negative Schwarzian derivative will be most important. Besides the quadratic map, many other functions have negative Schwarzian derivatives. For example, $S(e^x) = -1/2$ and $S(\sin x) = -1 - \frac{3}{2}(\tan^2 x) < 0$. Many polynomials have this property, as the following proposition shows.

**Proposition 11.2.** *Let $P(x)$ be a polynomial. If all of the roots of $P'(x)$ are real and distinct, then $SP < 0$.*

*Proof.* Suppose

$$P'(x) = \prod_{i=1}^{N}(x - a_i)$$

with the $a_i$ distinct and real. Then we have

$$P''(x) = \sum_{j=1}^{N} \frac{P'(x)}{x - a_j} = \sum_{j=1}^{N} \frac{\prod_{i=1}^{N}(x - a_i)}{x - a_j}$$

$$P'''(x) = \sum_{j=1}^{N} \sum_{\substack{k=1 \\ k \neq j}}^{N} \frac{\prod_{i=1}^{N}(x - a_i)}{(x - a_j)(x - a_k)}.$$

Hence we have

$$SP(x) = \sum_{j \neq k} \frac{1}{(x - a_j)(x - a_k)} - \frac{3}{2}\left(\sum_{j=1}^{N} \frac{1}{x - a_j}\right)^2$$

$$= -\frac{1}{2}\sum_{j=1}^{N}\left(\frac{1}{x - a_j}\right)^2 - \left(\sum_{j=1}^{N} \frac{1}{x - a_j}\right)^2 < 0.$$

q.e.d.

One of the most important properties of functions which have negative Schwarzian derivative is the fact that this property is preserved under composition.

**Proposition 11.3.** *Suppose $Sf < 0$ and $Sg < 0$. Then $S(f \circ g) < 0$.*

*Proof.* Using the chain rule, one computes that

$$(f \circ g)''(x) = f''(g(x)) \cdot (g'(x))^2 + f'(g(x)) \cdot g''(x)$$

and

$$(f \circ g)'''(x) = f'''(g(x)) \cdot (g'(x))^3 + 3f''(g(x)) \cdot g''(x) \cdot g'(x)$$
$$+ f'(g(x)) \cdot g'''(x).$$

It follows that

$$S(f \circ g)(x) = Sf(g(x)) \cdot (g'(x))^2 + Sg(x)$$

so that $S(f \circ g)(x) < 0$.

q.e.d.

Of primary importance for us is the immediate consequence that, if $Sf < 0$, then $Sf^n < 0$ for all $n > 1$. The assumption that $Sf < 0$ has surprising implications for the dynamics of a one-dimensional map. One of the major results of this section is

**Theorem 11.4.** *Suppose $Sf < 0$ ( $Sf(x) = -\infty$ is allowed). Suppose $f$ has $n$ critical points. Then $f$ has at most $n + 2$ attracting periodic points.*

**Remarks.**

**1.** The quadratic function $F_\mu(x) = \mu x(1-x)$ has one critical point ($x = 1/2$). Hence, for each $\mu$ there exists *at most three* attracting periodic orbits. There may, of course, be none, as is the case for $\mu > 2 + \sqrt{5}$. Later we will see that the number of attracting periodic points can be reduced to at most one. Since, for large $\mu$, the map $F_\mu$ has infinitely many periodic points, it is indeed a surprise that at most one may be attracting.

**2.** This presents a computational dilemma. Suppose $F_\mu$ has an attracting periodic cycle of period three. By Sarkovskii's Theorem, $F_\mu$ must have periodic points of all other periods, but none of them can be attracting. On a computer, only attracting periodic points are "visible," so this raises the question: where are all of the other periodic points in this case? We will return to this question in §1.13.

**3.** The proofs below extend to non-hyperbolic periodic points as well. Consequently, the quadratic map $F_\mu$ has at most one periodic orbit which is *not* repelling.

To prove Theorem 11.4, we first need several lemmas.

**Lemma 11.5.** *If $Sf < 0$, then $f'(x)$ cannot have a positive local minimum or a negative local maximum.*

*Proof.* Suppose $x_0$ is a critical point of $f'(x)$, i.e., $f''(x_0) = 0$. Since $Sf(x_0) < 0$, we have $f'''(x_0)/f'(x_0) < 0$ so that $f'''(x_0)$ and $f'(x_0)$ have opposite signs.

It follows that, between any two successive critical points of $f'$, the graph of $f'(x)$ must cross the $x$-axis. In particular, there must be a critical point *for* $f$ between these two points.

q.e.d.

**Lemma 11.6.** *If $f(x)$ has finitely many critical points, then so does $f^m(x)$.*

*Proof.* For any $c, f^{-1}(c)$ is a finite set of points, since, between any two preimages of $c$, there must be at least one critical point of $f$. It follows easily that $f^{-m}(c) = \{x | f^m(x) = c\}$ is also a finite set.

Now suppose $(f^m)'(x) = 0$. By the Chain Rule, we have

$$(f^m)'(x) = \prod_{i=0}^{m-1} f'(f^i(x))$$

Hence for some $i, 0 \le i \le m-1$, $f^i(x)$ is a critical point of $f$. Therefore the set of critical points of $f^m$ is given by the union of inverse images of order less than $m$ of the critical point set of $f$ together with their orbits. By the above observation, this is a finite set of points.

q.e.d.

**Lemma 11.7.** *Suppose $f(x)$ has finitely many critical points and $Sf < 0$. Then $f$ has only finitely many periodic points of period $m$ for any integer $m$.*

*Proof.* Let $g = f^m$. Then periodic points of period $m$ for $f$ are fixed points for $g$. By Proposition 11.3, $Sg < 0$.

Suppose $g$ has infinitely many fixed points. By the Mean Value Theorem, there are infinitely many points at which $g'(x) = 1$. Between any three successive points for which $g'(x) = 1$, there must be a point for which $g' < 1$. Indeed, $g'(x)$ is not identically equal to one on an interval, for then $Sg = 0$ contradicting the fact that $Sg < 0$. Furthermore, by Lemma 11.5, $g'$ cannot have a positive local minimum between these three points. Hence there must be points for which $g' < 0$. Consequently, there are points for which $g' = 0$. But this implies that $g$ has infinitely many critical points. This contradicts Lemma 11.6 and completes the proof.

q.e.d.

We now complete the proof of Theorem 11.4. Let $p$ be an attracting periodic point of period $m$ for $f$. Let $W(p)$ be the maximal interval about $p$ in which all points tend asymptotically to $p$ under $f^m$, i.e., $W(p)$ is the connected component of $\{x | f^{mj}(x) \to p$ as $j \to \infty\}$ which contains $p$. Clearly, $W(p)$ is an open interval, and $f^m(W(p)) \subset W(p)$.

Let us suppose for the moment that $p$ is fixed. Since $f(W(p)) \subset W(p)$ and $W(p)$ is maximal, it follows that either $f$ preserves the endpoints of $W(p) = (\ell, r)$, or one or both of $\ell$ and $r$ are infinite. In the finite case, there are three possibilities

1. $f(\ell) = \ell$ and $f(r) = r$.
2. $f(\ell) = r$ and $f(r) = \ell$.
3. $f(\ell) = f(r)$.

If $f(\ell) = \ell$ and $f(r) = r$, then the graph of $f$ shows that there exists $a, b$ satisfying $\ell < a < p < b < r$ and $f'(a) = f'(b) = 1$. Since $f'(p) < 1$ and $f'$ cannot have a positive local minimum (by Lemma 11.5), it follows that there exists a critical point in the interval $(a, b)$. The second case follows similarly by considering $f^2$. In case 3, $f$ must have a minimum or maximum between $\ell$ and $r$, so that there is a critical point in $W(p)$ in this case as well. In the case of $\ell$ and/or $r$ infinite, this proof fails. However, these cases add at most two stable fixed points.

If $p$ is periodic, the same arguments produce a critical point for $f^m$ in $W(p)$. One point on the orbit of this critical point must in fact be a critical point for $f$ because of the chain rule.

q.e.d.

The above Theorem may be extended to the case of non-hyperbolic periodic points. Indeed, more can be said. If $f(x)$ has a fixed point $c$ with multiplier $\pm 1$, and $Sf < 0$, then $c$ must attract points from at least one side, and, as above, there must be a critical point in $W(c)$.

To see this, we first assume that $f'(c) = 1$ (otherwise consider $f^2$). By Lemma 11.7, $f$ has only finitely many fixed points. Hence there is an interval about $c$ in which $f$ has no other fixed points.

Suppose that $c$ is a "repelling" fixed point, i.e., for $x < c$ and near $c$, $f(x) < x$ and for $y > c$, $f(y) > y$. Clearly, $f'$ has a local minimum value of 1 in this case. This contradicts Lemma 11.5 and shows that either $f(x) > x$ for $a < x < c$ or $f(x) < x$ for $c < x < b$. Graphical analysis then shows that $c$ is attracting from at least one side. See Fig. 11.1.

The above proof shows that periodic points with bounded stable sets must attract a critical point. If the stable sets are unbounded, this need not be the case, as the following examples show.

**Example 11.8.** Let $A(x) = \lambda \arctan(x)$ with $\lambda > 1$.

$$SA(x) = -2/(1 + x^2)^2 < 0.$$

The graph of $A$ shows that there are two attracting fixed points with unbounded stable sets but no critical points. See Fig. 11.2.

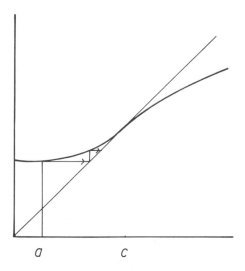

**Fig. 11.1.** $g(x) < x$ for $a < x < c$

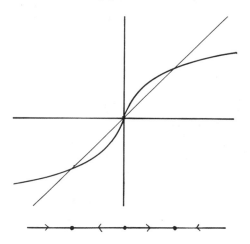

**Fig. 11.2.** The graph and phase portrait of
$A(x) = \lambda \arctan x$ with $\lambda > 1$

**Example 11.9.** Let $E(x) = e^{x-1}$. $E$ has a single fixed point at $p = 1$ which weakly attracts all points to the left but repels all points to the right. Again, $E$ has no critical points. See Fig. 11.3.

This example will play a prominent role in the next section, when we discuss bifurcation theory.

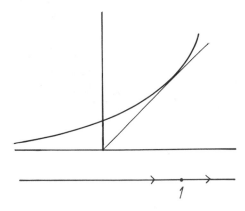

**Fig. 11.3.** The graph and phase portrait of
$$E(x) = e^{x-1}$$

For our quadratic map, the above theorem implies that there exists at most three attracting periodic orbits. However, since we know the behavior of these maps near $\infty$, we can say more.

**Corollary 11.10.** *Suppose* $F_\mu(x) = \mu x(1 - x)$. *Then there exists at most one attracting periodic orbit for each* $\mu$.

*Proof.* We have shown that $SF_\mu < 0$ and that, if $|x|$ is sufficiently large, then $|F_\mu^n(x)| \to \infty$. Hence there are no attracting periodic points with infinite stable sets.

q.e.d.

There may, of course, be no attracting periodic orbits, as in the case of $\mu > 2 + \sqrt{5}$ or $\mu = 4$.

Thus, when a map has negative Schwarzian derivative, the orbits of the critical points play an important role in determining the dynamics. Later, when we discuss the kneading theory, we will see that the orbits of critical points control *all* of the dynamics. For now, let us return to the quadratic map $F_4(x) = 4x(1 - x)$. Recall that, in Example 8.9, we proved that $F_4$ is chaotic on the unit interval. Since the critical point for this map is mapped onto a repelling fixed point, we know from the above considerations that $F_4$ cannot have an attracting periodic orbit. But the fact that $SF_4 < 0$ actually gives much more: using this fact, we may demonstrate that repelling periodic points are dense in $I$. Unlike the very special proof of this fact given in Example 8.9, this proof is considerably more general and applies to a wide variety of maps (see Exercises 1-2). We will use these ideas again much later when we discuss Julia sets of complex analytic dynamical systems.

Recall that $F_4$ has a repelling fixed point, a $p = 3/4$. Let $\hat{p} = 1/4$. Clearly, $F_4(\hat{p}) = p$. Let $J$ denote the half-open, half-closed interval $[\hat{p}, p)$. We will describe a "first return map" $R$ on the interior of $J - \{1/2\}$. Intuitively, $R(x) = F_4^n(x)$ where $n$ is the smallest integer for which $F_4^n(x) \in J$. To define $R$ precisely, we first note that $F_4$ maps $J$ onto the interval $[3/4, 1]$. That is, if $x \in J$, $F_4(x) \notin J$. Now $F_4$ maps $[3/4, 1]$ homeomorphically onto the interval $[0, 3/4]$, so certain points in $J$ are mapped back into $J$ by $F_4^2$. Indeed, a glance at Fig. 11.4 shows that there are two intervals $I_2$ and $\hat{I}_2$ in $J$ which are mapped homeomorphically onto $J$ by $F_4^2$. Note that both $I_2$ and $\hat{I}_2$ are half-open and half-closed intervals. We will define $R(x) = F_4^2(x)$ for $x \in I_2 \cup \hat{I}_2$.

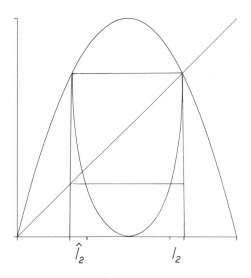

Fig. 11.4.

If $x \notin I_2 \cup \hat{I}_2$, then $F_4^2(x) \in [0, \frac{1}{4})$. Since $F_4$ is increasing on $[0, 1/4)$ and $F[0, 1/4) = [0, 3/4)$, it follows that as long as $x \neq 1/2$, the orbit of $x$ eventually returns to $J$. More precisely, if $x \in J - \{1/2\}$ there exists a least integer $n \geq 2$ for which $F_4^n(x) \in J$. Let $\phi(x)$ be this smallest integer; hence $\phi$ gives the "time" of first return to $J$. Note that $\phi = 2$ on $I_2 \cup \hat{I}_2$. More generally, we define

$$I_n = \{x \in (1/2, p) | \phi(x) = n\}$$
$$\hat{I}_n = \{x \in [\hat{p}, 1/2) | \phi(x) = n\}$$

It is easy to check that $I_n$ and $\hat{I}_n$ are half- open, half-closed intervals and that $F_4^n$ maps $I_n$ and $\hat{I}_n$ homeomorphically onto $J$. We therefore define $R\colon J - \{1/2\} \to J$ by

$$R(x) = F_4^{\phi(x)}(x).$$

Fig. 11.5 shows the graph of $R$. We emphasize that $R$ is not defined at $\frac{1}{2}$ and has infinitely many points of discontinuity, Nevertheless, a good understanding of the return map provides all of the information that we need about $F_4$.

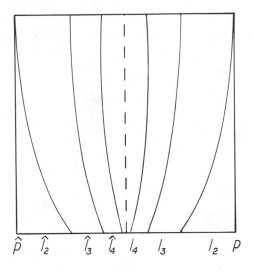

$$\hat{p} \quad \hat{I_2} \qquad \hat{I_3} \quad \hat{I_4} \quad I_4 \quad I_3 \qquad I_2 \quad p$$

**Fig. 11.5.**

The fact that each $F_4^n$ has negative Schwarzian derivative allows us to observe that if $K$ is any interval on which $(F_4^n)' \neq 0$, then the minimum value of $(F_4^n)'(x)$ occurs at one of the endpoints of $K$. This then allows us to prove the following fundamental result.

**Proposition 11.11.** $|R'(x)| > 1$ *for each* $x \in J$.

*Proof.* We work in the right hand intervals $I_n$; the result for $\hat{I}_n$ follows by symmetry. Let $I_k = [\ell_k, r_k)$. Let

$$W_k = \bigcup_{n>k}^{\infty} I_n.$$

$W_k$ is an open interval bounded by $1/2$ and $\ell_k$. We must show that $(F_4^k)'(x) > 1$ for $\ell_k \le x \le r_k$. By the above observation, since $(F_4^k)' \ne 0$ on $I_k$, it suffices to verify this condition at $\ell_k$ and $r_k$.

Now $F_4^k$ maps $I_k \cup W_k$ homeomorphically onto $(0, p)$ and $I_k$ onto $(\hat{p}, p)$. Since the length of $I_k$ is less than $1/4$, it follows that there exists $x_k \in I_k$ with $(F_4^k)'(x_k) > 1$. Now $F_4^k$ must map $W_k$ onto $(0, \hat{p})$. Since the length of $W_k$ is less than $1/4$ as well, it follows that there exists $x_k' \in W_k$ with $(F_4^k)'(x_k') > 1$. Since $(F_4^k)'$ cannot have a positive local minimum, it follows that $(F_4^k)'(\ell_k) > 1$, since $x_k' < \ell_k < x_k$.

To show that $(F_4^k)'(r_k) > 1$ as well, we note that

$$(F_4^k)'(r_k) = F_4'\big(F_4^{k-1}(r_k)\big) \cdot (F_4^{k-1})'(r_k)$$
$$= F_4'(\hat{p}) \cdot (F_4^{k-1})'(\ell_{k-1})$$
$$> 1$$

since both terms in this product are $> 1$. This completes the proof.

q.e.d.

To prove that repelling points are dense, let $U$ be any interval in $I$. We must produce a repelling periodic point in $U$. To do this we will find an $n > 0$ such that $F_4^n(U)$ is an interval containing $U$. The result then follows.

Since $|F_4'(x)| > 1$ if $x \notin J$, there is an $n > 0$ and a subinterval $U_0 \subset U$ with $V = F_4^n(U_0) \subset J$. Now $R$ expands the lengths of intervals in $J$ by Proposition 11.11. Hence there is a $k > 0$ and a subinterval $V_0 \subset V$ such that $R^k(V_0)$ contains a discontinuity point of $R$. Hence there is an $m > 0$ such that $p \in F_4^m(V_0)$. By graphical analysis, any neighborhood of $p$ is eventually expanded under iteration so that it covers $I$. In particular, there is $k > 0$ for which $F_4^{m+k}(V_0)$ covers $I$. The result then follows easily.

**Remarks.**

**1.** The above argument can be used to prove both sensitive dependence on initial conditions and topological transitivity as well. See Exercise 4.

**2.** To apply this method to other examples, we note that the crucial property that was used was that $F_4^k$ expanded both $W_k$ and $I_k$ over $(0, \hat{p})$ and $J$ respectively. This can often be verified by direct calculation or with a computer.

**Example 11.12.** Similar methods prove that $S(x) = 2\pi \sin x$ is chaotic on the interval $[0, 2\pi]$, since $\sin x$ has negative Schwarzian derivative.

**Example 11.13.** There exists a $c < 0$ for which the quadratic map $Q_c(x) = x^2 + c$ has the property that $Q_c^3(0)$ is a repelling fixed point $-p$. Numerically,

$c \approx -1.543689$ and $p \approx 0.839268 \ldots$. The graph of $Q_c$ is depicted in Fig. 11.6.

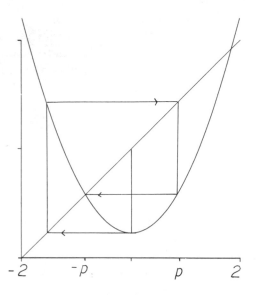

Fig. 11.6.

One may check that $Q_c$ has a repelling periodic point of period 2 at $q \approx .39039 \ldots$. The graph of $Q_c^2$ on the interval $[-p, p]$ is shown in Fig. 11.7. Note how $Q_c^2$ resembles $F_4$ on this interval and that $S(Q_c) < 0$. One may thus use the techniques of this section to prove that $Q_c$ is chaotic on $[-p, p]$. We will return to this example much later when we take up complex analytic dynamics.

As a remark, the techniques introduced in this section allow us to prove more than just the fact that the dynamics of a map are chaotic. Since repelling periodic points are dense, it follows that there can never be an interval which wanders around under iteration of the map and never reintersects itself. Such an interval is called a *wandering interval*. Thus we may exploit the fact that a map has negative Schwarzian derivative to rule out the possibility of wandering intervals in certain cases. As we shall see in §1.14, nontrivial wandering intervals may exist.

**Exercises.**

**1.** Prove that the map $S(x) = 2\pi \sin x$ is chaotic on the interval $[0, 2\pi]$.

**2.** Let $Q_c(x) = x^2 + c$ where $c \approx -1.543689$ as discussed in Example 11.13. Prove that there is an interval on which this map is chaotic.

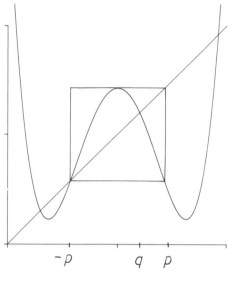

**Fig. 11.7**

**3.** Give an example of a polynomial with real coefficients which does *not* have negative Schwarzian derivative.

**4.** Use the return map $R$ to prove that $F_4(x) = 4x(1 - x)$ has sensitive dependence on initial conditions and is topologically transitive.

## §1.12 BIFURCATION THEORY

Bifurcation means a division in two, a splitting apart, a change. In dynamical systems, the object of bifurcation theory is to study the changes that maps undergo as parameters change. These changes often involve the periodic point structure, but may also involve other changes as well. In this section, we will consider one-parameter families of real-valued functions that depend smoothly on the parameter. More precisely, we will consider functions of two variables of the form

$$G(x, \lambda) = f_\lambda(x)$$

where, for fixed $\lambda, f_\lambda(x)$ is a $C^\infty$ function of the variable $x$. We will assume that $G$ depends smoothly on $\lambda$ as well. Examples include our friend, the quadratic family, $F_\mu(x) = \mu x(1 - x)$, as well as

1. $E_\lambda(x) = \lambda e^x$
2. $S_\lambda(x) = \lambda \sin(x)$
3. $Q_c(x) = x^2 + c$

and many others. For these and other families, our goal will be to understand how and when the periodic point structure of the family changes, i.e., the bifurcations that the family undergoes. We begin with some simple examples.

**Example 12.1** *(The Saddle-Node or Tangent Bifurcation.)* Consider the family $E_\lambda(x) = \lambda e^x$ where $\lambda > 0$. This family experiences a bifurcation when $\lambda = 1/e$. To see this we note that the graph of $f$ changes when $\lambda = 1/e$ as depicted in Fig. 12.1.

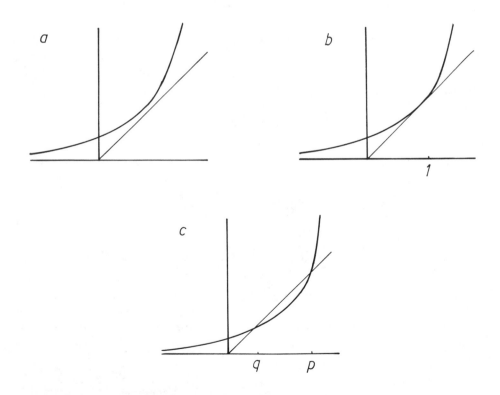

**Fig. 12.1.** The graphs of $E_\lambda(x) = \lambda e^x$ where a. $\lambda > 1/e$. b. $\lambda = 1/e$, and c. $0 < \lambda < 1/e$.

When $\lambda > 1/e$, the graph of $E_\lambda$ does not meet the diagonal, so $E_\lambda$ has no fixed points. When $\lambda = 1/e$, the graph meets the diagonal tangentially at $x = 1$, $y = 1$. For $\lambda < 1/e$, the graph meets the diagonal at two points, at $q$ with $E_\lambda'(q) < 1$ and at $p$ with $E_\lambda'(p) > 1$. Hence $E_\lambda$ has two fixed points for $\lambda < 1/e$. So, as the parameter decreases, two fixed points are born as $\lambda$ passes through $1/e$. Using graphical analysis, one can derive the entire phase portrait for each $E_\lambda$. We leave the following observations as an exercise.

1. When $\lambda > 1/e$, $E_\lambda^n(x) \to \infty$ for all $x$.
2. When $\lambda = 1/e$, $E_\lambda(1) = 1$. If $x < 1$, $E_\lambda^n(x) \to 1$. If $x > 1$, $E_\lambda^n(x) \to \infty$.
3. When $0 < \lambda < 1/e$, $E_\lambda(q) = q$ and $E_\lambda(p) = p$. If $x < p$, $E_\lambda^n(x) \to q$; if $x > p$, $E_\lambda^n(x) \to \infty$.

The phase portraits of $E_\lambda$ are sketched in Fig. 12.2. This is the typical change in the phase portrait which accompanies the saddle-node or tangent bifurcation. For later use, we note that, at the bifurcation ($\lambda = 1/e$, $x = 1$), we have $E_\lambda'(1) = 1$ and $E_\lambda''(1) = 1$.

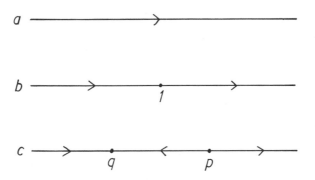

**Fig. 12.2.** The phase portraits of $E_\lambda(x) = \lambda e^x$ when
a. $\lambda > 1/e$, b. $\lambda = 1/e$, and c. $0 < \lambda < 1/e$.

This bifurcation can be described graphically in a *bifurcation diagram* in which we plot the location of fixed (or periodic) points versus the parameter. See Fig. 12.3. Each vertical slice of the bifurcation diagram gives the location of the fixed points of $E_\lambda$ on the real line.

**Example 12.2.** *(The Period-Doubling Bifurcation.)* We again consider the family $E_\lambda(x) = \lambda e^x$, this time with $\lambda < 0$. The graphs of $E_\lambda$ are given in Fig. 12.4 in three important cases. When $\lambda = -e$, $E_\lambda(-1) = -1$ and $E_\lambda'(-1) = -1$, so $-1$ is a non-hyperbolic fixed point for $E_\lambda$. When $\lambda > -e$,

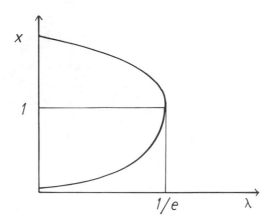

**Fig. 12.3.** The bifurcation diagram for $E_\lambda(x) = \lambda e^x$;
$x$ is plotted versus $\lambda$.

one may check (Exercise 3) that the fixed point for $E_\lambda$ is attracting; when
$\lambda < -e$, it is repelling. Hence the fixed point for $E_\lambda$ undergoes a change in
the nature of the nearby dynamics when $\lambda = -e$. This is not all that occurs,
however. Consider the graphs of $E_\lambda^2$. Using calculus, it is easy to see that
$E_\lambda^2$ is an increasing function that is concave up if $E_\lambda(x) < -1$, and concave
down if $E_\lambda(x) > -1$. See Fig. 12.5.

Thus $E_\lambda^2$ has 2 new fixed points at $q_1$ and $q_2$ when $\lambda$ decreases below $-e$.
These must in fact be periodic points of period 2, since we know that $E_\lambda$ has
a unique fixed point. Dynamically, this period-doubling bifurcation involves

1. a change from an attracting to a repelling fixed point, together with
2. the birth of a new period two orbit.

In the above example, we note that, as the fixed point loses its "attrac-
tiveness," the period two orbit acquires it. Also, for later use, note that, at
the bifurcation point ($\lambda = -e$, $x = -1$), we have $E_\lambda'(-1) = -1$, $E_\lambda''(-1) = 0$
and $E_\lambda'''(-1) \neq 0$.

Fig. 12.6 gives the bifurcation diagram for the period-doubling bifurcation. A similar period-doubling bifurcation occurs in the quadratic family.
See §1.4.

For maps of the real line, these two bifurcations turn out to be the most
common types of bifurcations – they are the ones that occur in a typical
family of maps. There are other, atypical bifurcations, however.

**Example 12.3.** Let $S_\lambda(x) = \lambda \sin x$. Note that $S_\lambda(0) = 0$ for all $\lambda$ and that

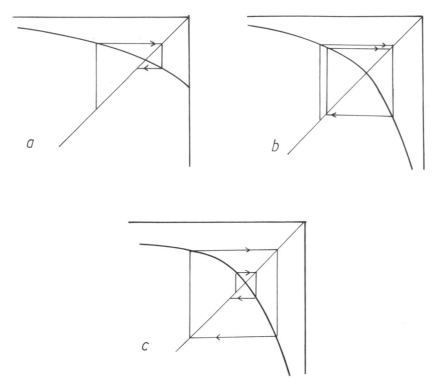

**Fig. 12.4.** The graphs of $E_\lambda(x) = \lambda e^x$ where a. $-e < \lambda < 0$, b. $\lambda = -e$, and c. $\lambda < -e$.

$S'_\lambda(0) = 1$ when $\lambda = 1$. Graphical analysis shows that the phase portrait of $S_\lambda$ is given as in Fig. 12.7 (Exercise 5) for $|x| < \pi$. The origin gives birth to two new fixed points in this case and changes from attracting to repelling as $\lambda$ increases through 1. The reason that this bifurcation is atypical is that, when $\lambda = 1$, we have both $S'_\lambda(0) = 1$ and $S''_\lambda(0) = 0$. Indeed, $S_\lambda$ is an odd function and hence it is symmetric near $x = 0$. Usually, a function has non-zero second derivative when its first derivative equals one and symmetry is not present.

**Example 12.4.** Consider again the quadratic family $F_\mu(x) = \mu x(1-x)$. When $\mu = 1$, $F_\mu$ has a unique fixed point, but for all other $\mu \neq 0$, there are two fixed points (one of which may be negative.) The phase portrait is given in Fig. 12.8. This bifurcation is also atypical since

$$\frac{d}{d\mu} f_\mu(0)\bigg|_{\mu=1} = 0.$$

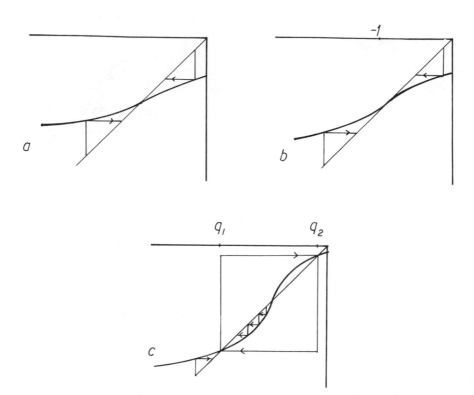

Fig. **12.5.** The graphs of $E_\lambda^2(x)$ when a. $-e < \lambda < 0$,
b. $\lambda = -e$, and c. $\lambda < -e$.

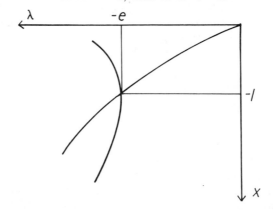

Fig. **12.6.** The bifurcation diagram for $E_\lambda(x) = \lambda e^x$;
$x$ is plotted versus $\lambda$.

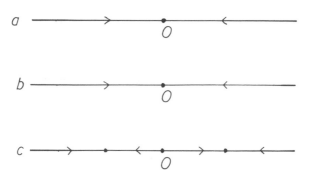

**Fig. 12.7.** The phase portraits for $S_\lambda(x) = \lambda \sin x$ when
a. $0 < \lambda < 1$, b. $\lambda = 1$, and c. $1 < \lambda < \pi/2$.

Usually we require that the bifurcation occur with non-zero "speed" in the
parameter variable.

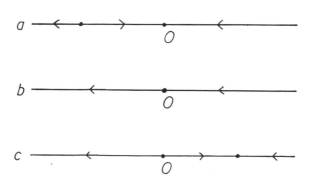

**Fig. 12.8.** The phase portraits for $F_\mu(x) = \mu x(1 - x)$ when
a. $\mu < 1$, b. $\mu = 1$, and c. $1 < \mu < 3$.

The above examples indicate that bifurcations occur near non-hyperbolic
fixed and periodic points. This is indeed the only place where bifurcations
of fixed points occur, as the following Theorem demonstrates.

**Theorem 12.5.** *Let $f_\lambda$ be a one-parameter family of functions and suppose
that $f_{\lambda_0}(x_0) = x_0$ and $f'_{\lambda_0}(x_0) \neq 1$. Then there are intervals $I$ about $x_0$
and $N$ about $\lambda_0$ and a smooth function $p: N \to I$ such that $p(\lambda_0) = x_0$ and
$f_\lambda(p(\lambda)) = p(\lambda)$. Moreover, $f_\lambda$ has no other fixed points in $I$.*

*Proof.* Consider the function defined by $G(x, \lambda) = f_\lambda(x) - x$. By hypothesis, $G(x_0, \lambda_0) = 0$ and

$$\frac{\partial G}{\partial x}(x_0, \lambda_0) = f'_{\lambda_0}(x_0) - 1 \neq 0.$$

By the Implicit Function Theorem, there are intervals $I$ about $x_0$ and $N$ about $\lambda_0$, and a smooth function $p: N \to I$ such that $p(\lambda_0) = x_0$ and $G(p(\lambda), \lambda) = 0$ for all $\lambda \in N$. Moreover, $G(x, \lambda) \neq 0$ unless $x = p(\lambda)$. This concludes the proof.

**Remarks.**

**1.** The content of the theorem is best understood by examining the graph of $f_\lambda$. Since $f_{\lambda_0}$ meets the line $y = x$ at an angle at $(x_0, x_0)$, nearby graphs $f_\lambda$ must have the same property. See Fig. 12.9. Hence there is one and only one fixed point near $x_0$ for $\lambda$ sufficiently near $\lambda_0$. The associated bifurcation diagram is shown in Fig. 12.10.

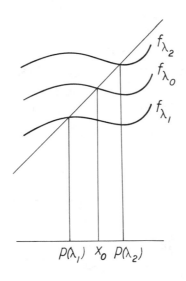

$$p(\lambda_1) \quad x_0 \quad p(\lambda_2)$$

**Fig. 12.9.**

**2.** For theoretical simplicity, it is often convenient to assume that the fixed point set of $f_\lambda$ is stationary as $\lambda$ is varied. The previous theorem allows us to make this assumption. Suppose that $f_\lambda$ is as in Theorem 12.5 and $f_\lambda(p(\lambda)) = p(\lambda)$ as in the theorem. Consider the new function

$$g_\lambda(z) = f_\lambda(z + p(\lambda)) - p(\lambda).$$

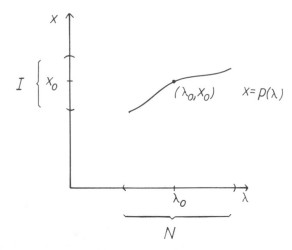

**Fig. 12.10.** The bifurcation diagram when $F'_{\lambda_0}(x_0) \neq 1$.

Clearly, $g_\lambda(0) = f_\lambda(p(\lambda)) - p(\lambda) = 0$ for all $\lambda$, so 0 is always fixed. Moreover, $g_\lambda$ is topologically conjugate to $f_\lambda$ via the simple map $h_\lambda(x) = x - p(\lambda)$. Hence the dynamics of $f_\lambda$ and $g_\lambda$ agree, but $g_\lambda$ is simpler to handle since its fixed point remains stationary at 0 as $\lambda$ varies.

**3.**   The above theorem (as well as all that follow) obviously hold for periodic points by replacing $f$ with $f^n$.

We now turn to the general setting of bifurcation theory.

**Theorem 12.6.** (The saddle-node bifurcation) *Suppose that*
1. $f_{\lambda_0}(0) = 0$
2. $f'_{\lambda_0}(0) = 1$
3. $f''_{\lambda_0}(0) \neq 0$
4. $\left.\dfrac{\partial f_\lambda}{\partial \lambda}\right|_{\lambda=\lambda_0} \neq 0.$

*Then there exists an interval $I$ about 0 and a smooth function $p: I \to \mathbf{R}$ such that*
$$f_{p(x)}(x) = x.$$
*Moreover, $p'(0) = 0$ and $p''(0) \neq 0$.*

**Remark.** The signs of $f''_\lambda(0)$ and
$$\left.\frac{\partial f_\lambda}{\partial \lambda}\right|_{\lambda=\lambda_0}$$

determine the "direction" of the bifurcation; if they have opposite signs, then the bifurcation diagram is as in Fig. 12.11.

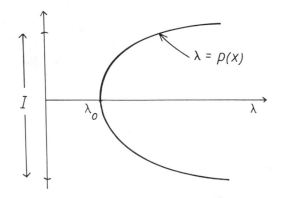

**Fig. 12.11.** The bifurcation diagram near a saddle-node bifurcation. The curve $\lambda = p(x)$ gives the fixed points for $f_\lambda$.

*Proof.* Define $G(x, \lambda) = f_\lambda(x) - x$. Note that $G(x, \lambda) = 0$ implies that $f_\lambda$ has a fixed point at $x$. We will apply the Implicit Function Theorem to $G$.

Note that $G(0, \lambda_0) = 0$ and that

$$\frac{\partial G}{\partial \lambda}(0, \lambda_0) = \frac{\partial f_\lambda}{\partial \lambda}\bigg|_{\lambda=\lambda_0}(0) \neq 0.$$

Hence there exists a smooth function $p(x)$ satisfying $G(x, p(x)) = 0$. From the chain rule, we have

$$\frac{\partial G}{\partial x} + \frac{\partial G}{\partial \lambda}p'(x) = 0.$$

Therefore

$$p'(x) = \frac{-\dfrac{\partial G}{\partial x}(x, p(x))}{\dfrac{\partial G}{\partial \lambda}(x, p(x))}.$$

Differentiating this expression and using the above gives

$$p''(0) = \frac{\dfrac{\partial^2 G}{\partial x^2}(0)\dfrac{\partial G}{\partial \lambda}\bigg|_{\lambda=\lambda_0}(0)}{\left(\dfrac{\partial G}{\partial \lambda}\right)^2}.$$

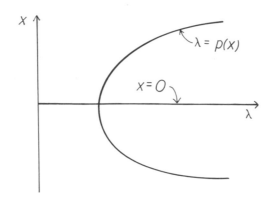

**Fig. 12.12.** The bifurcation diagram near a period-doubling bifurcation: $x = 0$ gives the fixed points, while $\lambda = p(x)$ gives the points of period two.

This completes the proof.

q.e.d.

**Theorem 12.7.** (Period-doubling bifurcation) *Suppose*

   *1. $f_\lambda(0) = 0$ for all $\lambda$ in an interval about $\lambda_0$.*
   *2. $f'_{\lambda_0}(0) = -1$*
   *3. $f'''_{\lambda_0}(0) \neq 0$*
   *4. $\dfrac{\partial(f_\lambda^2)'}{\partial \lambda}\bigg|_{\lambda=\lambda_0}(0) \neq 0$.*

*Then there is an interval $I$ about $0$ and a function $p: I \to \mathbf{R}$ such that*

$$f_{p(x)}(x) \neq x$$

*but*

$$f_{p(x)}^2(x) = x.$$

**Remarks.**

**1.** The condition $f'''_{\lambda_0}(0) \neq 0$ cannot be replaced by an assumption on $(f_{\lambda_0}^2)''(0)$, since $(f_{\lambda_0}^2)''(0) = 0$ is automatically satisfied.

**2.** The bifurcation diagram again depends on the signs of $f'''_{\lambda_0}(0)$ and

$$\frac{\partial f_\lambda^2}{\partial \lambda}\bigg|_{\lambda=\lambda_0}(0).$$

If these have opposite signs, the bifurcation diagram is as in Fig. 12.12.

*Proof.* For this proof we define $G(x, \lambda) = f_\lambda^2(x) - x$. Also set

$$H(x, \lambda) = \begin{cases} \dfrac{G(x, \lambda)}{x} & x \neq 0 \\ \dfrac{\partial G}{\partial x}(0, \lambda) & x = 0. \end{cases}$$

One checks easily that $H$ is smooth and satisfies

$$\frac{\partial H}{\partial x}(0, \lambda_0) = \frac{\partial^2 G}{\partial x^2}(0, \lambda_0)$$

$$\frac{\partial^2 H}{\partial x^2}(0, \lambda_0) = \frac{\partial^3 G}{\partial x^3}(0, \lambda_0).$$

We now apply the Implicit Function Theorem to $H$. Note that

$$\begin{aligned} H(0, \lambda_0) &= \frac{\partial G}{\partial x}(0, \lambda_0) \\ &= (f_{\lambda_0}^2)'(0) - 1 \\ &= f_{\lambda_0}'(0) \cdot f_\lambda'(0) - 1 \\ &= 0. \end{aligned}$$

We have by assumption that

$$\begin{aligned} \frac{\partial H}{\partial \lambda}(0, \lambda_0) &= \frac{\partial}{\partial \lambda}\Big|_{\lambda = \lambda_0} \left( (f_\lambda^2)'(0) - 1 \right) \\ &= \frac{\partial (f_\lambda^2)'}{\partial \lambda}(0) \\ &\neq 0. \end{aligned}$$

Hence there is a smooth function $p(x)$ defined on a neighborhood of 0 and satisfying $p(0) = \lambda_0$ and $H(x, p(x)) = 0$. In particular,

$$\frac{1}{x}G(x, p(x)) = 0$$

for $x \neq 0$ and it follows that $x$ is a period two point for $f_{p(x)}$.

As above, we compute

$$p'(0) = \frac{-\frac{\partial H}{\partial x}(0, \lambda_0)}{\frac{\partial H}{\partial \lambda}(0, \lambda_0)} = 0$$

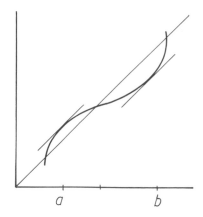

**Fig. 12.13.** The map $f_\lambda^2$ does not have negative Schwarzian derivative, since $(f_\lambda^2)'$ has a positive local minimum between $a$ and $b$.

since

$$(f_{\lambda_0}^2)''(0) = f_{\lambda_0}''(0) \cdot (f_{\lambda_0}'(0))^2 + f_{\lambda_0}''(0) \cdot f_{\lambda_0}'(0)$$

where we have used $f_{\lambda_0}'(0) = -1$.

Also

$$p''(0) = \frac{-\frac{\partial^2 H}{\partial x^2}(0, \lambda_0) \cdot \frac{\partial H}{\partial \lambda}(0, \lambda_0)}{\left(\frac{\partial H}{\partial \lambda}(0, \lambda_0)\right)^2}$$

$$= \frac{-f_{\lambda_0}'''(0)}{\frac{\partial}{\partial \lambda}\bigg|_{\lambda=\lambda_0}(f_\lambda^2)'(0)}.$$

This completes the proof.

q.e.d.

**Remark.** If we assume that $Sf_\lambda < 0$ for all $\lambda$ near $\lambda_0$, then the family $f_\lambda$ cannot have a "reverse" period-doubling bifurcation at $\lambda_0$. By this we mean a bifurcation of the following type. If $\lambda < \lambda_0$, $f_\lambda$ has a (locally) unique repelling fixed point. When $\lambda = \lambda_0$, $f_\lambda$ has a unique fixed point with multiplier $-1$. When $\lambda > \lambda_0$, $f_\lambda$ has a unique attracting fixed point with a repelling periodic point of period two. Figure 12.13 shows why this cannot happen, as the graph is impossible by Lemma 11.5.

**Exercises.**

**1.** Identify the bifurcations and discuss the phase portraits before and after the bifurcations which occur in the following families of maps at the indicated parameter values.

  a. $F_\mu(x) = \mu x(1 - x)$, $\mu = 3$.
  b. $F_\lambda(x) = \lambda x - x^3$, $\lambda = 1$.
  c. $F_\lambda(x) = \lambda x - x^3$, $\lambda = -1$.
  d. $Q_c(x) = x^2 + c$, $c = -\frac{3}{4}$.
  e. $A_\lambda(x) = \lambda \arctan(x)$, $\lambda = -1$.
  f. $H_\lambda(x) = \lambda \sinh(x)$, $\lambda = 1$.

**2.** Let $\lambda > 0$. Prove that the phase portraits for $E_\lambda$ are as given in Fig. 12.2.

**3.** Let $\lambda < 0$. Prove that $E_\lambda(x)$ has a unique fixed point that is attracting if $\lambda > -e$ and repelling if $\lambda < -e$.

**4.** Prove that $E_\lambda^2$ is concave up for $x$ such that $E_\lambda'(x) > -1$, and concave down if $E_\lambda'(x) < -1$.

**5.** Discuss the phase portrait and the bifurcation diagram for $S_\lambda(x) = \lambda \sin x$ for $|x| < \pi$, $0 < \lambda \le \pi/2$.

**6.** Consider the quadratic family $Q_\lambda(x) = x^2 + \lambda$. Where does this family have a saddle-node bifurcation? A period-doubling bifurcation? Describe the phase portraits and the bifurcation diagrams nearby in each case.

**7.** Using Sarkovskii's Theorem and graphical analysis, describe the phase portraits of $F_\mu(x) = \mu x(1 - x)$ as $\mu$ increases through 3. Using graphical analysis, explain why $F_\mu$ must undergo a series of successive period-doubling bifurcations.

## §1.13 ANOTHER VIEW OF PERIOD THREE

   In this section, we return to the consideration of the quadratic mapping $F_\mu(x) = \mu x(1 - x)$. This time we will consider the specific parameter value $\mu = 3.839$, following some ideas of Smale and Williams. We will drop the subscript $\mu$ for the remainder of this section and let $F(x) = 3.839x(1 - x)$. Our goal is to sharpen the results of §1.10 regarding the implications of the existence of a periodic point of period three. With a calculator, one may check easily that there is an attracting orbit of period three for $f$ given to six decimals by

$$a_1 = .149888$$
$$a_2 = .489172$$
$$a_3 = .959299$$

Moreover, $(F^3)'(a_i) \approx -.78$ approximately. The existence of such a periodic point can be proved rigorously by hand computation: one simply finds a small interval about $a_1$ which is mapped inside itself by $F^3$ with derivative everywhere less than one. We will relegate the tedious details to the exercises, but it is important to realize that the computations involved can be done by hand and with complete accuracy.

By Sarkovskii's Theorem, $F$ has periodic points of all periods. By the results of §1.11, none besides $a_i$ can be attracting. Hence, for all practical purposes, these points are invisible to the computer. This brings up the question: where are all of these other peridic points and exactly how many of them are there? We answer these questions below by introducing a more general concept from symbolic dynamics, *the subshift of finite type.*

We first define the shift on $N$ symbols. Let $\Sigma_N$ denote the set of all possible sequences of natural numbers between 1 and $N$, i.e.,

$$\Sigma_N = \{(\mathbf{s}) = (s_0 s_1 s_2 \ldots) | s_j \in \mathbf{Z}, 1 \le s_j \le N\}.$$

Note that, unlike the case of $\Sigma_2$ introduced before, we will not allow 0 as an entry in a sequence $\Sigma_N$; we will use instead the symbols 1 through $N$. This will help with the "bookkeeping" later. As in §1.6, there is a natural metric or distance function on $\Sigma_N$ defined by

$$d_N[\mathbf{s}, \mathbf{t}] = \sum_{i=0}^{\infty} \frac{|s_i - t_i|}{N^i}$$

where $\mathbf{s} = (s_0 s_1 s_2 \ldots)$ and $\mathbf{t} = (t_0 t_1 t_2 \ldots)$. The proof of the following Proposition is similar to that of Proposition 6.2 and is therefore left as an exercise.

**Proposition 13.1.**
1. $d_N$ is a metric on $\Sigma_N$.
2. If $s_i = t_i$ for $i = 0, \ldots, k$, then $d_N[\mathbf{s}, \mathbf{t}] \le 1/N^k$.
3. If $d_N[\mathbf{s}, \mathbf{t}] < 1/N^k$ then $s_i = t_i$ for $i \le k$.

Just as in the case of $\Sigma_2$, we have the shift map given by $\sigma(s_0 s_1 s_2 \ldots) = (s_1 s_2 s_3 \ldots)$. Proposition 6.3 transfers over verbatim to this case to show that $\sigma$ is continuous.

Our goal is to describe certain subsets of $\Sigma_N$ which arise naturally and which provide a more general setting for symbolic dynamics. Let $A$ be an $N \times N$ matrix whose entry in the $i^{th}$ row and $j^{th}$ column, which we denote by $a_{ij}$, is either 0 or 1. That is, $A$ is an $N \times N$ square array of 0's and 1's. $A$ is called the *transition matrix* for the system. We will use $A$ to describe which sequences in $\Sigma_N$ lie inside a subset which we denote by $\Sigma_A$. A sequence $\mathbf{s} = (s_0 s_1 s_2 \ldots)$ lies in $\Sigma_A$ if it obeys the following rule. Each adjacent pair of entries in the sequence $\mathbf{s}$ determines a location in the matrix $A$, the $a_{s_i s_{i+1}}$ entry. The sequence lies in $\Sigma_A$ if and only if every such entry is 1. More concisely,

$$\Sigma_A = \left\{ (\mathbf{s}) = (s_0 s_1 s_2 \ldots) \middle| a_{s_i s_{i+1}} = 1 \text{ for all } i \right\}.$$

That is, we use the transition matrix to rule out certain pairs of entries in a sequence which lies in $\Sigma_A$.

**Example 13.2.** Let

$$A = \begin{pmatrix} 1 & 0 \\ 0 & 1 \end{pmatrix}.$$

Since $a_{12} = a_{21} = 0$, it follows that 1 and 2 can never be adjacent in a sequence in $\Sigma_A$. Consequently, there are only two allowable sequences in $\Sigma_A$, the constant sequences $(111 \ldots)$ and $(222 \ldots)$.

**Example 13.3.** Let

$$A = \begin{pmatrix} 1 & 1 \\ 0 & 1 \end{pmatrix}.$$

In this example, 2 may follow 1, but not vice-versa. Thus, $\Sigma_A$ consists of the constant sequences plus any sequence of the form $(11 \ldots 11\ 222 \ldots)$ where there are arbitrarily many 1's followed by infinitely many 2's.

**Example 13.4.** Let

$$A = \begin{pmatrix} 1 & 1 \\ 1 & 0 \end{pmatrix}.$$

Any combination of 1's and 2's are allowed in a sequence in $\Sigma_A$, except a pair of adjacent 2's.

We denote by $\sigma_A$ the restriction of $\sigma$ to the set $\Sigma_A$. The following proposition guarantees that this makes sense.

**Proposition 13.5.** $\Sigma_A$ *is a closed subset of* $\Sigma_N$ *which is invariant under* $\sigma_A$.

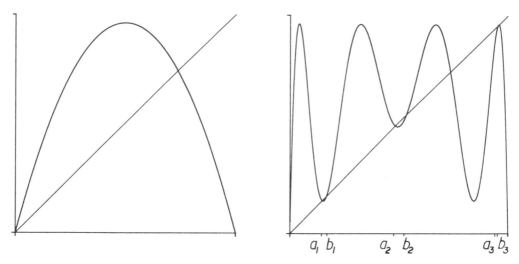

**Fig. 13.1.** The graphs of $F(x)$ and $F^3(x)$.

*Proof.* Invariance is clear. To prove that $\Sigma_A$ is closed, we suppose that $s_i$ is a sequence of elements of $\Sigma_A$, i.e., a sequence of sequences, which converge to $\mathbf{t}$. If $\mathbf{t} \notin \Sigma_A$, there is an integer $\alpha$ for which $a_{t_\alpha t_{\alpha+1}} = 0$.

Since the $s_i$ converge to $\mathbf{t}$, there is another integer $K$ such that, if $i > K$, then $d_N[s_i, \mathbf{t}] < 1/N^{\alpha+1}$. By Proposition 13.1, this forces $t_0, t_1, \ldots, t_{\alpha+1}$ to agree with the corresponding entries of $s_i$ for $i \geq K$. In particular, we must have $A_{t_\alpha t_{\alpha+1}} = 1$, since $s_i \in \Sigma_A$. This contradiction establishes the result.

q.e.d.

We call $\sigma_A$ a *subshift of finite type*, since it is determined by the finitely many conditions imposed by the transition matrix $A$. There are subshifts which are not of finite type, but we will not discuss them in this book.

We now return to the quadratic map $F(x) = 3.839x(1 - x)$. Recall that there is an attracting periodic orbit in the vicinity of $a_1 = .149888$, $a_2 = .489172$, and $a_3 = .959299$. The graphs of $F$ and $F^3$ may be sketched as in Fig. 13.1.

There is a second periodic orbit of period 3 for $F$ which we denote by $b_1, b_2, b_3$. These points are given approximately by

$$b_1 = .169040$$
$$b_2 = .539247$$
$$b_3 = .953837$$

with $F(b_1) \approx b_2, F(b_2) \approx b_3$. One may calculate that $(F^3)'(b_i) \approx 2.66$. Again,

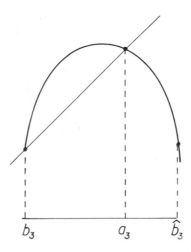

**Fig. 13.2.** A portion of the graph of $F^3(x)$ depicting $\hat{b}_3$ and $W(a_3)$.

the existence of this periodic orbit can be proved by computing both $F^3$ and $(F^3)'$ on a small interval about $b_i$ and noting that $F^3$ expands this interval over itself.

Recall from §1.11 that there is a maximal open interval about each $a_i$ which consists of points which tend to $a_i$ under iteration of $F^3$. We denote this interval by $W(a_i)$. From the proof of Theorem 11.4, we see that one of the endpoints of each $W(a_i)$ is fixed by $F^3$. Hence $b_i$ is one of the endpoints of $W(a_i)$. Let us denote by $\hat{b}_i$ the point on the opposite side of $a_i$ from $b_i$ which is mapped to $b_i$ by $F^3$. See Fig. 13.2.

Let $A_1 = (\hat{b}_1, b_1)$, $A_2 = (\hat{b}_2, b_2)$, and $A_3 = (b_3, \hat{b}_3)$. Note that $F$ maps $A_1$ and $A_3$ monotonically onto $A_2$ and $A_1$ respectively, but $F$ has a critical point at $1/2 \in A_2$ so $F$ is not monotonic on this interval. Since, however, the maximum value of $F$ is .95975, it follows that $F(A_2)$ is contained in $A_3$. Also note that $F(b_2) = F(\hat{b}_2) = b_3$.

As in §1.5, it is easy to prove that if $x < 0$ or $x > 1$ then $F^n(x) \to -\infty$. Moreover, if $x \in A_i$ for some $i$, then $F^n(x)$ tends to the orbit of $a_i$. Hence all of the other periodic points must lie in the complement of the $A_i$ in $I$. There are four closed intervals in the complement of the $A_i$ in $I$; we call these intervals $I_0, I_1, I_2$, and $I_3$, from left to right. Since we know the behavior of the $b_i$ under iteration of $F$, we know how these four intervals are mapped by $F$. Their images are depicted in Fig. 13.3.

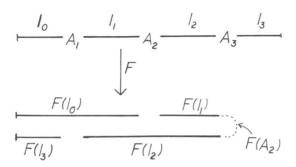

**Fig. 13.3.** The images of $I_0, I_1, I_2$, and $I_3$.

Since there are no periodic points for $F$ in the $W(a_i)$, it follows that all of the infinitely many periodic points must lie in the $I_j$. In fact, we can say more.

**Proposition 13.6.** *All periodic points of $F$ lie in $I_1 \cup I_2$ with the exception of the fixed point at 0 and the periodic point of period 3: $a_1, a_2$, and $a_3$.*

*Proof.* Note that $F$ is monotonic on each of the intervals $I_j$. From Figure 13.3, we see that $F$ maps $I_0$ across both $I_0$ and $I_1$, $I_1$ onto $I_2$, $I_2$ across both $I_1$ and $I_2$, and finally $I_3$ onto $I_0$. From this it follows that if a periodic point $x \in I_1 \cup I_2$, then $F(x) \in I_1 \cup I_2$. Thus, if $x \in I_1 \cup I_2$ lies on a periodic orbit, then the entire orbit of $x$ lies in $I_1 \cup I_2$.

Now if $x \in I_0$ and $x \neq 0$, we note that $F(x) > x$. Hence there is an $n$ for which $F^n(x) \notin I_0$. Either $F^n(x) \in A_1$, in which case $x$ is not periodic, or $F^n(x) \in I_1$. In the latter case, the forward orbit of $F^n(x)$ can never leave $I_1$ or $I_2$ to return to $x$, so, again, $x$ is not periodic. Finally, if $x \in I_3, F(x) \in I_0$, so $x$ is again not periodic.

q.e.d.

Consequently, all of the remaining periodic points for $F$ lie in $I_1 \cup I_2$. Let us denote by $\Lambda$ the set of points whose entire orbit is contained in these two intervals. To understand the dynamics of $F$ on $\Lambda$, we again invoke symbolic dynamics. We define the sequence associated to $x$ by the rule

$$S(x) = (s_0 s_1 s_2 \dots)$$

where $s_j = 1$ if $F^j(x) \in I_1$ and $s_j = 2$ if $F^j(x) \in I_2$. Since $F(I_1) = I_2$, it follows that a 1 can only be followed by a 2, i.e., that $S$ takes its values in $\Sigma_A$ where

$$A = \begin{pmatrix} 0 & 1 \\ 1 & 1 \end{pmatrix}.$$

We remark that we encountered a similar phenomenon in the proof of Sarkovskii's Theorem. Indeed, the existence of a point of period 3 forces the existence of a pair of intervals that behave like $I_1$ and $I_2$ under iteration of $F$. In the present case, however, we can say much more.

**Theorem 13.7.** *The restriction of $F$ to $\Lambda$ is topologically conjugate to the subshift of finite type given by $\sigma_A$ on $\Sigma_A$.*

*Proof.* The proof of surjectivity and continuity of $S$ proceeds exactly as in §1.7 so we omit the details. The only difference from §1.7 arises in the proof that $S$ is one-to-one, so we concentrate on this fact. The difficulty here is that $|F'(x)|$ is not everywhere greater than one on $I_1 \cup I_2$. However, we can say that $|F'(x)| > \nu = F'(\hat{b}_2) \approx .3$, since $F'' < 0$ and the interval $(\hat{b}_2, b_2)$ containing the critical point has been removed. Hence $|F'|$ is bounded from below.

We claim that there exists $\lambda > 1$ such that, if $x \in \Lambda$, then $|(F^3)'(x)| > \lambda$. To prove this, we note that there are three closed intervals in $I_1 \cup I_2$ in which $|(F^3)'(x)| \leq 1$. Two of them, $B_1$ and $B_2$, are symmetrically located with respect to $1/2$, while the third, $B_3$, lies in $I_2$. We note that the $F^3$-image of $B_3$ is contained in $(\hat{b}_1, b_1)$ and so $B_3 \cap \Lambda = \emptyset$. See Fig. 13.1.

We claim that $B_1$ and $B_2$ are mapped by $F^3$ into $(b_3, \hat{b}_3)$. Indeed, one may check easily that $B_2$ is contained in the interval $.661 < x < .683$ and that $(F^3)'(.661) > 1$, $(F^3)'(.683) < -1$. Furthermore, $F^3$ maps this interval inside $(b_3, \hat{b}_3)$ as required. By symmetry, $F^3(B_1) \subset (b_3, \hat{b}_3)$ as well. Hence $B_1 \cap \Lambda = \emptyset$ and $B_2 \cap \Lambda = \emptyset$.

We now prove that $\Lambda$ is a hyperbolic set. Choose $K$ such that $\nu^2 \lambda^K > 1$. Let $N = 3K + 2$. If $n > N$, then we may write $n = 3(K + \alpha) + i$ where $\alpha > 0$ and $0 \leq i \leq 2$ are integers. Hence, if $x \in \Lambda$, using the Chain Rule, we have

$$|(F^n)'(x)| = |(F^i)'(F^{3(K+\alpha)}(x))| \cdot |(F^{3(K+\alpha)})'(x)| > \nu^2 \lambda^{K+\alpha} > 1.$$

This proves that $\Lambda$ is a repelling hyperbolic set.

With hyperbolicity established, the remainder of the proof is similar to that of Theorem 7.2. This completes the proof.

q.e.d.

As a remark, we emphasize that although this proof is most easily carried out with a calculator in hand, the numbers are not so horrendous that actual hand computation is ruled out. This is a real proof!

**Remark.** The techniques used in this proof may be used to prove that the quadratic map $F_\mu(x) = \mu x(1 - x)$ admits a hyperbolic set when $4 < \mu \leq 2 + \sqrt{5}$, the cases we omitted in §1.5. See Exercise 7.

We now see that there are periodic points for $F$ of all periods, as guaranteed by Sarkovskii's Theorem. Indeed, to produce a periodic point of period $k$ in $\Sigma_A$, we need only list a string of $k - 1$ 2's followed by a 1, and then repeat this sequence. These are *exactly* the orbits produced in the proof of Sarkovskii's Theorem. Of course there are many other allowable repeating sequences in $\Sigma_A$, so this raises the question of exactly how many periodic points of period $k$ $F$ has.

To answer this, we first need a definition.

**Definition 13.8.** Let $A = (a_{ij})$ be an $N \times N$ matrix. The trace of $A$ is given by

$$\text{Tr}(A) = \sum_{i=1}^{N} a_{ii} \; ,$$

i.e., by the sum of the diagonal entries of $A$.

The trace is an important invariant of the conjugacy class of a matrix which is studied in advanced linear algebra. For us, its purpose is quite different: the trace of the powers of $A$ gives an accurate count of the periodic points in $\Sigma_A$. Recall that if $A = (a_{ij})$ and $B = (b_{ij})$ are $N \times N$ matrices, then the product of $A$ and $B$ is the $N \times N$ matrix $A \cdot B = (c_{ij})$ where

$$c_{ij} = \sum_{k=1}^{N} a_{ik} b_{kj}.$$

In particular, if

$$A = \begin{pmatrix} 0 & 1 \\ 1 & 1 \end{pmatrix}$$

then

$$A^2 = A \cdot A = \begin{pmatrix} 1 & 1 \\ 1 & 2 \end{pmatrix}$$

$$A^3 = A \cdot A^2 = \begin{pmatrix} 1 & 2 \\ 2 & 3 \end{pmatrix}$$

$$A^4 = A \cdot A^3 = \begin{pmatrix} 2 & 3 \\ 3 & 5 \end{pmatrix}.$$

**Proposition 13.9.** *Let $A$ be an $N \times N$ transition matrix. Then*

$$\text{card } \text{Per}_K \sigma_A = \text{Tr} \left( A^K \right).$$

*Proof.* Recall that a sequence s in $\Sigma_A$ is fixed by $\sigma^K$ iff s is a repeating sequence of the form $(i_0 i_1 \ldots i_{K-1} \, i_0 i_1 \ldots i_{K-1} \ldots)$. Such a sequence lies in $\Sigma_A$ iff $a_{i_0 i_1} = a_{i_1 i_2} = \ldots = a_{i_{K-1} i_0} = 1$ or, equivalently $a_{i_0 i_1} a_{i_1 i_2} \cdot \ldots \cdot$ $a_{i_{K-1} i_0} = 1$. Thus, the products $a_{i_0 i_1} a_{i_1 i_2} \cdot : \ldots \cdot a_{i_{K-1} i} = 1$ if and only if the string $i_0 i_1 \ldots i_{K-1} i_0$ is an allowed piece of a sequence in $\Sigma_A$, and equal to zero otherwise. Consequently,

$$\sum_{i_0 i_1, \ldots, i_{K-1}} a_{i_0 i_1} a_{i_1 i_2} \cdot \ldots \cdot a_{i_{K-1} i_0}$$

gives the cardinality of $\mathrm{Per}_K \sigma_A$. On the other hand, it is easily checked that this sum is $\mathrm{Tr}(A^K)$.

$$\text{q.e.d.}$$

Note that we may therefore compute easily that, for

$$A = \begin{pmatrix} 0 & 1 \\ 1 & 1 \end{pmatrix}$$

we have

$$\mathrm{Tr}(A) = 1$$
$$\mathrm{Tr}(A^2) = 3$$
$$\mathrm{Tr}(A^3) = 4$$
$$\mathrm{Tr}(A^4) = 7$$
$$\mathrm{Tr}(A^5) = 11.$$

In general, for $K > 2$,

$$\mathrm{Tr}(A^K) = \mathrm{Tr}(A^{K-1}) + \mathrm{Tr}(A^{K-2}).$$

This recursion is curiously the same as the well-known Fibonacci recursion relation

$$p_k = p_{k-1} + p_{k-2}$$

with $k > 2$. The usual Fibonacci sequence begins with $p_1 = p_2 = 1$; here we have $p_1 = 1$ but $p_2 = 3$.

### Exercises

**1.** Let $\mu = 3.839$ and $a_1 = .149888$. Prove that there is a small interval about $a_1$ which is mapped inside itself by $F_\mu^3$.

**2.** Show that there is a smaller interval about $a_1$ which is mapped inside itself and on which $(F_\mu^3)' < 1$. This proves that there is a unique attracting periodic point with period 3 near $a_1$.

3.  Define a metric on $\Sigma_N$ by

$$d_n[\mathbf{s},\mathbf{t}] = \sum_{k=0}^{\infty} \frac{|s_i - t_i|}{n^k}.$$

a. Prove that $d_n$ is a metric.

b. Prove the analogue of Proposition 6.2, i.e., if $s_i = t_i$ for $i = 0,\ldots,k$ then $d_n[\mathbf{s},\mathbf{t}] \leq 1/n^k$. Similarly, if $d_n[\mathbf{s},\mathbf{t}] < 1/n^k$, then $s_i = t_i$ for $i \leq k$.

4.  Let $A = \begin{pmatrix} 0 & 1 \\ 1 & 1 \end{pmatrix}$. Prove that $\mathrm{Tr}(A^K) = \mathrm{Tr}(A^{K-2}) + \mathrm{Tr}(A^{K-1})$.

5.  Using the intervals $I_0$ and $I_3$ as well as $I_1$ and $I_2$, show that the set of points whose orbits remain for all time in these four intervals also determines a subshift of finite type. What is the transition matrix?

6.  Let

$$A = \begin{pmatrix} 1 & 1 & 0 \\ 1 & 1 & 0 \\ 0 & 0 & 1 \end{pmatrix}.$$

Find a formula for the trace of $A^K$.

7.  Prove that $F_\mu(x) = \mu x(1 - x)$ admits a hyperbolic set in $[0,1]$ when $4 < \mu \leq 2 + \sqrt{5}$.

## §1.14. MAPS OF THE CIRCLE

In this section, we specialize some of our previous results to the case of maps of the circle. The dynamics of these maps are somewhat different from maps of $\mathbf{R}$ since the circle is bounded. In particular, diffeomorphisms of $S^1$ are more interesting, since the circle permits non-trivial recurrence while the real line does not. Diffeomorphisms of $S^1$ may have periodic points of any given period, but diffeomorphisms of $\mathbf{R}$ may have only fixed or period two points.

For simplicity, we will restrict attention in this section to orientation-preserving diffeomorphisms of $S^1$, i.e., diffeomorphisms $f\colon S^1 \to S^1$ which preserve the order of points on the circle. Orientation-reversal does not add significant difficulties, so we deal with this case in the exercises.

To study the dynamics of a circle map, it is helpful to *lift* the map to $\mathbf{R}$. That is, we define the map $\pi\colon \mathbf{R} \to S^1$ by

$$\pi(x) = \exp(2\pi i x) = \cos(2\pi x) + i\sin(2\pi x).$$

The map $\pi$ is an example of a *covering map*, since it wraps $\mathbf{R}$ around $S^1$ without doubling back (i.e., without critical points).

**Definition 14.1.** $F: \mathbf{R} \to \mathbf{R}$ is a lift of $f: S^1 \to S^1$ if

$$\pi \circ F = f \circ \pi.$$

We remark that $\pi$ is not a topological conjugacy between $F$ and $f$ since it is many-to-one.

**Example 14.2.** Let $\tau_\omega(\theta) = \theta + 2\pi\omega$ be translation by angle $2\pi\omega$. For each $k \in \mathbf{Z}$, the map $T_{\omega,k}(x) = x + \omega + k$ is a lift of $\tau_\omega$. Similarly, if $f(\theta) = \theta + \epsilon \sin(\theta)$, then $F_{\epsilon,k}(x) = x + \frac{\epsilon}{2\pi} \sin(2\pi x) + k$ is a lift of $f$.

**Remarks.**

**1.** There are always infinitely many different lifts for a given map $f: S^1 \to S^1$. Indeed, one may easily prove that any two lifts of $f$ differ by an integer (see Exercise 3).

**2.** In the above examples, there is a similarity between the formulas for the map on the circle and its lift. However, it is important to realize that these maps are defined on different spaces and thus one should expect that they have very different dynamics. Indeed, if $\omega$ is rational, then all points of $S^1$ are periodic under $\tau_\omega$, but no points of $\mathbf{R}$ are periodic under $T_\omega$ (unless $\omega = 0$).

**3.** If $F$ is a lift of $f$ then we must have $F'(x) > 0$ so that $F$ is increasing. Furthermore, we must have $F(x+1) = F(x) + 1$, and, more generally, $F(x + k) = F(x) + k$ for any integer $k$. We stress that these facts hold since $f$ is an orientation preserving diffeomorphism of the circle. For other types of maps, one may also define the lift to the real line, but this last equality need not hold. Consequently,

$$F(x + 1) - (x + 1) = F(x) - x,$$

so that $F - id$ is a periodic function with period 1, where $id(x) = x$ is the identity function. Similarly, $F^n - id$ is periodic with period 1 as well, since $F^n$ is a lift of $f^n$. Using this, one may check easily that, if $|x - y| < 1$, then we must also have $|F^n(x) - F^n(y)| < 1$.

The most important invariant associated to a circle map is its *rotation number*. This number, between 0 and 1, essentially measures the average amount points are rotated by an iteration of the map. Before defining the actual rotation number, however, we introduce a preliminary concept.

Let $f: S^1 \to S^1$ be an orientation-preserving diffeomorphism and choose any lift $F$ of $f$. Define

$$\rho_0(F) = \lim_{n \to \infty} \frac{|F^n(x)|}{n}.$$

Note that this limit, if it exists, does not depend upon the choice of $x$. Indeed, since $F^n - $ id is periodic, we have

$$|F^n(x) - F^n(y)| \leq |(F^n(x) - x) - (F^n(y) - y)| + |x - y|$$
$$\leq 1 + |x - y|$$

where this second inequality follows from Remark 3 above. Therefore,

$$\lim_{n \to \infty} \frac{|F^n(x) - F^n(y)|}{n} = 0$$

so that $\rho_0$ is independent of $x$. However, $\rho_0$ does depend on the choice of the lift.

**Example 14.3.** Let $\tau_\omega(\theta) = \theta + 2\pi\omega$ be translation and consider the lift $T_k(x) = x + \omega + k$. We have

$$\rho_0(T_k) = \lim_{n \to \infty} \frac{x + n\omega + nk}{n} = \omega + k$$

so that different lifts produce different values for $\rho_0$. Note, however, that they all differ by integers.

**Example 14.4.** Suppose $f: S^1 \to S^1$ has a fixed point at $\theta = 0$ (we may always arrange this by conjugation with a translation). Suppose $F$ is a lift of $f$. Then $F(0)$ is an integer, say $F(0) = k$. It follows that $F^n(0) = nk$ so that $\rho_0(F) = k$. Hence we also have

$$\lim_{n \to \infty} \frac{F^n(x)}{n} = k$$

for all $x \in \mathbf{R}$.

As before, we note that any two lifts produce an integer difference in $\rho_0$. This is a general fact. Let $F_1$ and $F_2$ be lifts of $f$. By Exercise 3, there is an integer $k$ for which $F_2(x) = F_1(x) + k$. It follows easily that $F_2^n(x) = F_1^n(x) + nk$ so that $\rho_0(F_2) = \rho_0(F_1) + k$. We may thus remove the dependence of $\rho_0$ on the lift by eliminating the integer part of $\rho_0$.

**Definition 14.5.** The rotation number of $f$, $\rho(f)$, is the fractional part of $\rho_0(F)$ for any lift $F$ of $f$. That is, $\rho(f)$ is the unique number in $[0, 1)$ such that $\rho_0(F) - \rho$ is an integer.

We remark that nothing in our definitions of $\rho_0(F)$ or $\rho(f)$ requires any differentiability; rotation numbers are equally well-defined for maps which are only homeomorphisms.

We have not yet verified that the limit $\rho_0(F)$ actually exists. This is easy if $f$ has a periodic point. Suppose that $f^m(\theta) = \theta$ and $\pi(x) = \theta$. Then $F^m(x) = x + k$ for some integer $k$. Hence $F^{jm}(x) = x + jk$, and we have

$$\lim_{j \to \infty} \frac{|F^{jm}(x)|}{jm} = \lim_{j \to \infty} \left( \frac{x}{jm} + \frac{k}{m} \right) = \frac{k}{m}.$$

More generally, we may write any integer $n$ in the form $n = jm + r$ where $0 \le r < m$. Note that there is a constant $M$ such that

$$|F^r(y) - y| \le M$$

for all $y \in \mathbf{R}$ and $0 \le r < m$. Thus we have

$$\frac{|F^n(x) - F^{jm}(x)|}{n} = \frac{|F^r(F^{jm}(x)) - F^{jm}(x)|}{n}$$

$$\le \frac{M}{n}.$$

Consequently

$$\lim_{n \to \infty} \frac{|F^n(x)|}{n} = \lim_{j \to \infty} \frac{|F^{mj}(x)|}{mj} = \frac{k}{m}.$$

This shows that the rotation number $\rho(f)$ exists whenever $f$ has a periodic point. Moreover, $\rho(f)$ is rational in this case.

In case $f$ has no periodic points, we need a slightly more complicated argument. Since $F^n(x) - x$ is never an integer if $n \ne 0$, there is an integer $k_n$ such that

$$k_n < F^n(x) - x < k_n + 1$$

for every $x \in \mathbf{R}$. Applying this inequality repeatedly to the cases $x = 0$, $F^n(0), F^{2n}(0), \ldots$, we find

$$k_n < F^n(0) < k_n + 1$$

$$k_n < F^{2n}(0) - F^n(0) < k_n + 1$$

$$\vdots$$

$$k_n < F^{mn}(0) - F^{(m-1)n}(0) < k_n + 1.$$

Adding each of these inequalities, we have

$$mk_n < F^{mn}(0) < m(k_n + 1).$$

Hence

$$\frac{k_n}{n} < \frac{F^{mn}(0)}{mn} < \frac{(k_n + 1)}{n}.$$

The original inequality gives immediately

$$\frac{k_n}{n} < \frac{F^n(0)}{n} < \frac{k_n + 1}{n}.$$

Thus, combining these two expressions, we find

$$\left| \frac{F^{mn}(0)}{mn} - \frac{F^n(0)}{n} \right| < \frac{1}{n}.$$

Now this entire argument may be repeated with $n$ and $m$ interchanged, yielding

$$\left| \frac{F^{mn}(0)}{mn} - \frac{F^m(0)}{m} \right| < \frac{1}{m}.$$

It follows that

$$\left| \frac{F^n(0)}{n} - \frac{F^m(0)}{m} \right| < \frac{1}{n} + \frac{1}{m}.$$

This means that the sequence $\{F^n(0)/n\}$ is an example of a Cauchy sequence. Such a sequence in **R** is easily seen to converge. See Exercise 2. Hence we have proved

**Theorem 14.6.** *Let $f: S^1 \to S^1$ be an orientation-preserving diffeomorphism with lift $F$. Then*

$$\rho_0(F) = \lim_{n \to \infty} \frac{|F^n(x)|}{n}$$

*exists and is independent of $x$. Consequently, the rotation number $\rho(f)$ is well-defined.*

The proof of this Theorem also allows us to establish the result that $\rho(f)$ depends continuously on $f$.

**Corollary 14.7.** *Suppose $f: S^1 \to S^1$ is an orientation-preserving diffeomorphism. Let $\epsilon > 0$. There exists $\delta > 0$ such that if $g: S^1 \to S^1$ is also a diffeomorphism which is $C^0$-$\delta$ close to $f$, then $|\rho(f) - \rho(g)| < \epsilon$.*

*Proof.* Choose $n$ such that $2/n < \epsilon$. We may choose a lift $F$ of $f$ with $r - 1 < F^n(0) < r + 1$ for some integer $r$. We may also choose $\delta > 0$ small enough so that there is a lift $G$ of $g$ with $r - 1 < G^n(0) < r + 1$ as well. As in the previous proof, we have

$$m(r - 1) < F^{nm}(0) < m(r + 1)$$

$$m(r - 1) < G^{nm}(0) < m(r + 1).$$

Consequently,

$$\left| \frac{F^{nm}(0)}{nm} - \frac{G^{nm}(0)}{nm} \right| < \frac{2}{n} < \epsilon$$

for all $m$. Since

$$\lim_{m \to \infty} \frac{F^{nm}(0)}{nm} = \rho_0(F),$$

we are done.

q.e.d.

As we remarked above, the rotation number measures the average rotation that a diffeomorphism induces on $S^1$. For example, $\rho(\tau_\omega) = \omega$ where $\tau_\omega(\theta) = \theta + 2\pi\omega$ is rotation by angle $2\pi\omega$. For the map $f(\theta) = \theta + \sin^2(\theta/2)$, we have $\rho(f) = 0$. Indeed, $f$ fixes the point $\theta = 0$, but all other points are advanced slightly by $f$. One may check easily that $f^n(\theta) \to 0$ as $n \to \pm\infty$, so that points never make one complete circuit of $S^1$ under iteration of $f$.

An important property of $\rho(f)$ is its invariance under topological conjugacy. If $f$ and $g$ are both orientation-preserving diffeomorphisms of $S^1$, then it is easy to check that $\rho(f) = \rho(g^{-1}fg)$. See Exercise 4.

We have shown above that if $f$ has a periodic point, then $\rho(f)$ is rational. When $\omega$ is irrational, the translation $\tau_\omega$ has no periodic points by Jacobi's Theorem. Thus we are led to suspect that $\rho(f)$ is irrational if $f$ has no periodic points. This is indeed the case, as we now show.

**Proposition 14.8.** $\rho(f)$ *is irrational if and only if $f$ has no periodic points.*

*Proof.* Given our previous results, it suffices to show that if $f$ has no periodic points, then $\rho(f)$ is irrational. Let us assume that $\rho(f)$ is rational and derive a contradiction. As one may easily check, $\rho_0(F^m) = m\rho_0(F)$ for any lift $F$. Thus we may assume at the outset that $\rho(f) = 0$, but $f$ has no fixed points.

The lift $F$ also has no fixed points, so we may assume that $F(x) > x$ for all $x \in \mathbf{R}$ (the other case is handled similarly). Then either $F^n(0) < 1$ for all $n$ or else there exists $k > 0$ for which $F^k(0) > 1$. In the latter case, we find $F^{mk}(0) > m$. Hence $\rho_0(F) > 1/k$, which gives a contradiction.

In the former case, the sequence $F^n(0)$ is monotonically increasing in $[0, 1]$ and therefore converges. If $p$ is the limit point of this sequence, we have

$$F(p) = F\big(\lim_{n \to \infty} F^n(0)\big)$$
$$= \lim_{n \to \infty} F^{n+1}(0) = p$$

so that $p$ is a fixed point for $F$. This again is a contradiction and establishes the result.

q.e.d.

Let us now discuss the case of a map with irrational rotation number in more detail. From the previous result, we know that such a map cannot have any periodic points. One such map is the rotation map $\tau_\omega(\theta) = \theta + 2\pi\omega$ where $\omega$ is irrational. One might be hard pressed to think of another example of a homeomorphism with irrational rotation number, but there are many examples of such maps which are not topologically conjugate to an irrational rotation. Recall that, by Jacobi's Theorem, all orbits of an irrational rotation are dense in $S^1$. This property must be shared by any map topologically conjugate to an irrational rotation. So, to produce a different map of $S^1$, we need only find a map which has an orbit that is not dense. The following example, due to Denjoy, shows how to manufacture such a map.

**Example 14.9.** *(A Denjoy map.)* We will perform "surgery" on $\tau_\omega$ where $\omega$ is irrational. Take any point $\theta_0 \in S^1$. We cut out each point on the orbit of $\theta_0$ and replace it with a small "interval." That is, at the point $\tau_\omega^n(\theta_0)$ we cut apart the circle and glue in a small interval $I_n$ in its place. Provided we take $I_n$ small enough so that the lengths $\ell(I_n)$ are positive and satisfy

$$\sum_{n=-\infty}^{\infty} \ell(I_n) < \infty \;,$$

the result of this "operation" is still a "circle"—a little larger than before but still a simple closed curve. See Fig. 14.1.

We may now extend the map to the union of the $I_n$'s by choosing any orientation-preserving diffeomorphism $h_n$ taking $I_n$ to $I_{n+1}$. This extends our original map to be a homeomorphism of the new circle. Note that the new map has no periodic points, so its rotation number is irrational. Moreover,

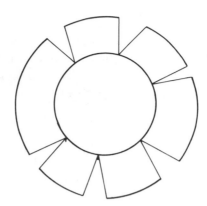

**Fig. 14.1** Opening up the circle to construct
a Denjoy homeomorphism.

no point in the interior of $I_n$ ever returns to $I_n$ under iteration of the map,
so the orbits of these points are certainly not dense.

**Remarks.**

**1.** The Denjoy example is clearly a *homeomorphism* of the circle, but its construction probably casts doubt in the reader's mind as to whether it can be
made a *diffeomorphism*. Actually, one may choose the $h_n$'s carefully enough
so that the resulting map is a $C^1$ diffeomorphism. See Exercise 5. This
construction however, cannot yield a $C^2$ diffeomorphism. It is known that
a $C^2$ diffeomorphism with irrational rotation number is always topologically
conjugate to $\tau_\omega$ for appropriate $\omega$. Thus, there are surprisingly complicated
differences between the dynamics of $C^1$ and $C^2$ diffeomorphisms of the circle.

**2.** The Denjoy example gives an example of a wandering domain for a map
of the circle. Eliminating this type of behavior for maps of **R** necessitated
special assumptions such as negative Schwarzian derivative and detailed computations. See §1.11.

At this point, it is useful to see how the previous discussion allows us
to analyze fairly completely the bifurcation structure of an important two-
parameter family of circle diffeomorphisms. This is the family of maps known
as the *standard* or *canonical family*.

**Example 14.10.** Consider the two-parameter family of maps of $S^1$ given
by
$$f_{\omega,\epsilon}(\theta) = \theta + 2\pi\omega + \epsilon\sin\theta.$$

This map has the lift

$$F_{\omega,\epsilon}(x) = x + \omega + \frac{\epsilon}{2\pi}\sin(2\pi x)$$

which we will deal with exclusively when discussing $f$.

When $\epsilon = 0$, this map reduces to the rotation map $T_\omega$. For $0 \leq \epsilon < 1$, $f_{\omega,\epsilon}$ is a diffeomorphism of $S^1$; when $\epsilon = 1$, the map is only a homeomorphism. When $\epsilon > 1$, the map is no longer one-to-one.

Observe that, if $\omega_1 > \omega_2$, then

$$F_{\omega_1,\epsilon}(x) > F_{\omega_2,\epsilon}(x)$$

for all $x \in \mathbf{R}$. It follows that

$$F_{\omega_1,\epsilon}^n(x) > F_{\omega_2,\epsilon}^n(x)$$

so that $\rho_0(F_{\omega_1,\epsilon}) \geq \rho_0(F_{\omega_2,\epsilon})$. Hence $\rho$ is a non-decreasing function of $\omega$ for each fixed $\epsilon$. Moreover, by Corollary 14.7, $\rho$ varies continuously with $\omega$. Let us fix $\epsilon \neq 0$ and consider $f_\omega = f_{\omega,\epsilon}$.

Suppose that $\rho(f_{\omega_0}) = p/q$ is rational. It follows that $f_{\omega_0}$ has a periodic point of period $q$. Hence there exists $x_0 \in \mathbf{R}$ such that

$$F_{\omega_0}^q(x_0) = x_0 + k$$

for some integer $k$. Actually, $k = p$. We claim that there is an interval of $\omega$-values for which the rotation number of $f_{\omega_0}$ is $p/q$. To see this, consider the graph of $F_{\omega_0}^q$. This graph meets the straight line $y = x + k$ at the point $(x_0, x_0 + k)$. If $(F_{\omega_0}^q)'(x_0) \neq 1$, then it follows immediately from the Implicit Function Theorem that there is an open interval about $\omega_0$ for which the graph of each $F_\omega^q$ also pierces the line $y = x + k$. Hence $\rho(f_\omega) = p/q$ for all of these values. If, on the other hand, $(F_{\omega_0}^q)'(x_0) = 1$, the argument is more complicated. To sketch the proof in this case, we observe that, since $F_{\omega_0}$ is analytic, there is an integer $j$ for which the $j^{th}$ derivative $(F_{\omega_0}^q)^{[j]}(x_0) \neq 0$. Otherwise, $F_{\omega_0}^q(x)$ would be identically equal to $x + k$. If $j$ is odd, it follows immediately that the graphs of nearby $F_\omega^q$ must pierce the line $y = x + k$. If $j$ is even, then either $F_{\omega_0}^q$ is concave up or concave down at $x_0$. In either event, the graphs of nearby $F_\omega^q$ for $\omega < \omega_0$ or $\omega > \omega_0$ must cross the line $y = x + k$. We leave the details to the reader.

Thus we see that, for each rational number $p/q$, there is an interval with non-empty interior on which $\rho(f_\omega) = p/q$. On the other hand, there is a unique $\omega$ for which $\rho(f_\omega)$ is a given irrational number. This is a fairly

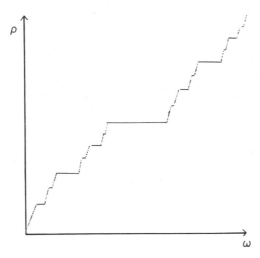

**Fig. 14.2.** The graph of $\rho(f_\omega)$ is a Cantor function.

deep result whose proof we will omit. The graph of $\rho(f_\omega)$ is an example of a *Cantor function*; it is constant on intervals corresponding to rational rotation numbers, yet everywhere continuous. This graph has also been called a "devil's staircase." See Fig. 14.2.

The bifurcation diagram for the standard family is most interesting. Let us plot the regions in the $\epsilon - \omega$ plane where $\rho(f_{\omega,\epsilon})$ is a fixed rational number. These regions are necessarily "tongues" which flare out from each point of the form $\epsilon = 0$, $\omega = p/q$. None of these tongues can overlap when $\epsilon < 1$ and all have non-empty interior. See Fig. 14.3.

The bifurcation diagram for $\epsilon > 1$, when $f_{\omega,\epsilon}$ is no longer a homeomorphism, is also interesting. See Exercise 6.

The dynamics of the maps $f_\omega$ for fixed $\epsilon > 0$ are easy to describe, given our previous work on bifurcation theory. Note that $f_\omega$ has a fixed point provided

$$\sin\theta = \frac{-2\pi\omega}{\epsilon}.$$

From the graph of $\sin(\theta)$ in the interval $0 \le \theta \le 2\pi$, we therefore read off that this equation has two solutions if $|2\pi\omega| < \epsilon$, one solution if $|2\pi\omega| = \epsilon$, and no solutions if $|2\pi\omega| > \epsilon$. One may easily compute that $|f'_\omega(\theta_i)| \ne 1$ at the fixed points $\theta_i$, $i = 1,2$, which occur when $|2\pi\omega| < \epsilon$, whereas $f'_\omega(\theta) = 1$ at the unique fixed point for the map with $|2\pi\omega| = \epsilon$. We may interpret this dynamically as follows. A fixed point for $f_\omega$ is born in a saddle node bifurcation at $\theta = \pi/2$ when $\epsilon = -2\pi\omega$. This fixed point separates into two

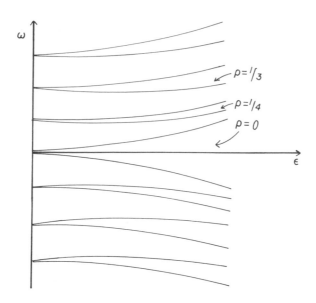

**Fig. 14.3.** The bifurcation diagram for the standard family.

fixed points which race around the unit circle in opposite directions as $\omega$ increases. Finally, the two fixed points meet and coalesce in another saddle node bifurcation at $\theta = 3\pi/2$, and then disappear. A similar phenomenon occurs in the other tongues.

**Exercises**

**1.** Prove that an orientation-reversing diffeomorphism of $S^1$ must have two fixed points.

**2.** A sequence of real numbers $\{a_n\}$ is called a Cauchy sequence if, for any $\epsilon > 0$, there is an integer $N$ such that, for all $n, m > N$, $|a_n - a_m| < \epsilon$. Prove that any Cauchy sequence in $\mathbf{R}$ converges.

**3.** Prove that any two different lifts of a circle map must differ by an integer. Conversely, prove that if $F(x)$ is a lift of $f$, then so too is $F(x) + k$ where $k \in \mathbf{Z}$.

**4.** Suppose $f$ and $g$ are orientation-preserving diffeomorphisms of $S^1$. Prove that $\rho(f) = \rho(g^{-1}fg)$.

**5.** *The Denjoy example revisited.* In this series of exercises, we show that the Denjoy homeomorphism constructed in this section may actually be made $C^1$.

a. For each integer $n$, let

$$\ell_n = \frac{1}{(|n|+1)(|n|+2)}.$$

Show that

$$\sum_{n=-\infty}^{\infty} \ell_n < \infty$$

$$\lim_{|n| \to \infty} \frac{\ell_{n+1}}{\ell_n} = 1$$

b. For each $n$, let $I_n = [a_n, b_n]$ be an interval with length $\ell_n$. Define a map $f$ on $[a_n, b_n]$ by

$$f(x) = a_{n+1} + \int_{a_n}^{x} 1 + \frac{6(\ell_{n+1} - \ell_n)}{\ell_n^3}(b_n - t)(t - a_n) \, dt.$$

Show that $f'(a_n) = f'(b_n) = 1$.
c. Prove that $f$ is a diffeomorphism mapping $[a_n, b_n]$ onto $[a_{n+1}, b_{n+1}]$.
d. Prove that

$$f''\left(\frac{a_n + b_n}{2}\right) = 0$$

but

$$\lim_{x \to a_n} |f''(x)| \to \infty.$$

Conclude that $f$ is not $C^2$.
e. Define the variation of $f'$ on $I_n$ to be

$$V_n = |f'(a_n) - f'(\frac{a_n + b_n}{2})|.$$

Prove that

$$\sum_{n=1}^{\infty} V_n$$

is unbounded so that $f'(x)$ is not of bounded variation.

If we now paste in the intervals $I_n$ in place of an orbit of the irrational rotation map, exactly as we did in this section, and define $f$ on the $I_n$ as above, then the resulting circle map is $C^1$.

**6.**   The rotation number may also be defined even if $f: S^1 \to S^1$ is not a homeomorphism. In this case, however, the rotation number depends on $x$.

For the standard family when $\epsilon > 1$, prove that the set of rotation numbers which occur form an interval in $\mathbf{R}$.

## §1.15 MORSE-SMALE DIFFEOMORPHISMS

This section continues the discussion of orientation preserving diffeomorphisms of the circle begun in the last section. Here we take a somewhat different point of view: our goal is to understand as completely as possible the large class of circle diffeomorphisms known as Morse-Smale diffeomorphisms. This class of maps has two important properties. First, each element of this class is structurally stable and has a phase portrait which is easy to describe. Unlike the quadratic family, Morse-Smale diffeomorphisms exhibit no chaotic behavior whatsoever. Second, any circle diffeomorphism may be approximated as closely as we desire by a Morse-Smale diffeomorphism. Thus, the "generic" diffeomorphism of $S^1$ is Morse-Smale and structurally stable.

**Definition 15.1.** An orientation preserving diffeomorphism of $S^1$ is Morse-Smale if it has rational rotation number and all of its periodic points are hyperbolic.

**Example 15.2.** The diffeomorphism $f(\theta) = \theta + \frac{\pi}{n} + \epsilon \sin(2n\theta)$ is Morse-Smale when $\epsilon > 0$ is small. There are two periodic orbits of period $2n$, an attracting periodic point whose orbit contains $\theta = \frac{\pi}{2n}$, and a repelling periodic point at $\theta = 0$. The points on these orbits alternate around the circle.

If $f$ is a Morse-Smale diffeomorphism with $\rho(f) = p/q$, then, as in the above example, all of the periodic points of $f$ have period $q$. Thus $f^q$ has only fixed points. Since each of these fixed points is hyperbolic, they must be alternately sinks and sources around the circle. So the phase portrait of an iterate of a Morse-Smale diffeomorphism is quite simple, as depicted in Fig. 15.1. Thus, the fact that these maps are structurally stable comes as no surprise.

**Theorem 15.3.** *A Morse-Smale diffeomorphism of $S^1$ is $C^1$-structurally stable.*

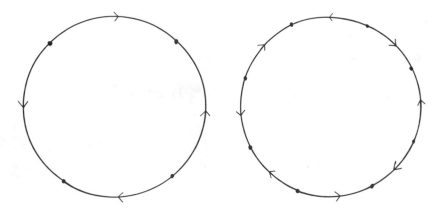

**Fig. 15.1.** The phase portraits of some Morse-Smale diffeomorphisms.

*Proof.* We will prove this in the case where $f$ is orientation-preserving and satisfies $\rho(f) = 0$, so that $f$ has only fixed points. The other cases are left as exercises.

Let $F$ be the lift of $f$ that has fixed points; only one lift of $f$ has this property. We will show that, if $G$ is $C^1$-$\epsilon$ close to $F$ on $\mathbf{R}$, then $G$ is topologically conjugate to $F$. The result on the circle follows immediately.

Since $f$ is Morse-Smale, $F$ has only finitely many fixed points in $[0, 1]$. Let $p_1, \ldots, p_n$ be the fixed points for $F$. We may choose disjoint neighborhoods $U_j = (\alpha_j, \beta_j)$ of $p_j$ with $F'(x) \neq 1$ on $U_j$. There exists $\epsilon_j > 0$ such that $|F'(x) - 1| > \epsilon_j$ for each $x \in U_j$ and, moreover, $|F(\alpha_j) - \alpha_j| > \epsilon_j$, $|F(\beta_j) - \beta_j| > \epsilon_j$. Hence if a diffeomorphism $G$ is $C^1$-$\epsilon_j$ close to $F$ on $U_j$, it follows that $G'(x) \neq 1$ on $U_j$ and that $G$ has a unique fixed point in $U_j$. See Fig. 15.2.

Now $F$ has no fixed points in the complement of the $U_j$. Hence there exists $\epsilon_0 > 0$ such that $|F(x) - x| > \epsilon_0$ for all $x \in I - (\cup_{j=1}^n U_j)$. If $G$ is $C^0$-$\epsilon_0$ close to $F$ on these intervals, then $G$ has no fixed points in these regions as well.

If we choose $\epsilon < \min \epsilon_j$ for $j = 0, \ldots, n$, then any diffeomorphism $G$ which is $C^1 - \epsilon$ close to $F$ has the same phase portrait as $F$. If $G$ is a lift of a diffeomorphism $g$ of $S^1$, one may apply the above remarks over all of $\mathbf{R}$. Consequently, $g$ is Morse-Smale and has the same phase portrait of $f$. One may then produce the conjugacy between $f$ and $g$ via fundamental domain arguments as in §1.9. This completes the proof.

q.e.d.

We now turn to the question of the size of the set of structurally sta-

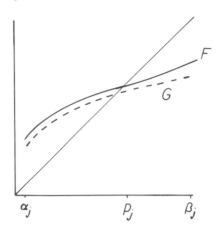

**Fig. 15.2.** Perturbations of a Morse-Smale diffeomorphism.

ble diffeomorphisms of $S^1$. Circle maps come in two basic varieties: those with rational and those with irrational rotation numbers. We show in Theorem 15.4 that diffeomorphisms with irrational rotation numbers can never be structurally stable. Moreover, such diffeomorphisms can be approximated as closely as we wish (with respect to the $C^r$-distance ) by a map with periodic points. This result, known as the Closing Lemma, has straightforward proof in the case of the circle but is much more difficult in higher dimensions. Indeed, it is only known in the $C^1$ case when the dimension is greater than one. We remark that the bifurcation diagram for the standard family foreshadows this result, since an open and dense subset of the $\omega$-$\epsilon$ plane gives a map $f_{\omega,\epsilon}$ with rational rotation number.

Before we prove this result, we recall the definition of a recurrent point (Exercise 7.3). A point $\theta \in S^1$ is recurrent under $f$ if, for any neighborhood $U$ of $\theta$, there exists $n > 0$ such that $f^n(\theta) \in U$. That is, the images of a recurrent point must pile up on itself. If $f$ has irrational rotation number, then there must be at least one recurrent point for $f$. Indeed, since $f$ has no periodic points, if there were no recurrent points as well, there would be a $\delta > 0$ such that $|f^n(\theta) - \theta| > \delta$ for all $n > 0$ and all $\theta \in S^1$. It follows that all of the images of each $\theta \in S^1$ must be separated by an arc of length $\delta$. Since $\theta$ is not periodic, there must be infinitely many distinct images of $\theta$, and this yields a contradiction.

The Closing Lemma applies to just this situation. It allows us to "close up" a recurrent orbit and make it periodic by an arbitrarily $C^r$-small perturbation.

**Theorem 15.4.** (*The Closing Lemma*). *Suppose $f$ is a diffeomorphism of*

$S^1$ *with an irrational rotation number. Then, for any $\epsilon > 0$, there exists a diffeomorphism $g\colon S^1 \to S^1$ which is $C^r - \epsilon$ close to $f$ and which has rational rotation number.*

*Proof.* Let $\theta_0$ be a recurrent point for $f$ and let $U = (\theta_0 - \delta, \theta_0 + \delta)$ be an arc of the circle about $\theta_0$. Since $\theta_0$ is recurrent, there exists a sequence of integers $n_i \to \infty$ for which $f^{n_i}(\theta_0) \in U$. Let us assume that all of the $f^{n_i}(\theta_0)$ belong to the arc $(\theta_0 - \delta, \theta_0)$. We will modify $f$ slightly on $U$ so that $\theta_0$ becomes periodic for the perturbed map.

Let $V \subset U$ be a neighborhood of $\theta_0$. Let $\phi$ be a bump function on $V$ as in Exercise 2.8. That is, $\phi(\theta) = 1$ if $\theta \in V$ but $\phi(\theta) = 0$ if $\theta \in S^1 - U$. Now consider the map $f_\epsilon(\theta) = f(\theta) + \epsilon\phi(\theta)$. For $\epsilon > 0$ sufficiently small, $f_\epsilon$ is also a diffeomorphism of $S^1$. The $C^r$-distance between $f_\epsilon$ and $f$ is given by

$$\sup_{\theta \in U} \epsilon |\phi^{[r]}(\theta)|$$

where $\phi^{[r]}(\theta)$ denotes the $r^{th}$ derivative of $\phi$ at $\theta$. Thus, by choosing $\epsilon$ small, we may make $f_\epsilon$ as close as we please to $f$. Intuitively, $f_\epsilon$ behaves exactly as $f$ does off of the neighborhood $U$, while $f_\epsilon$ advances points by $\epsilon\phi(\theta)$ each time $f$ maps a point to $\theta \in U$. In particular, each time an orbit meets $V$ it is advanced by at least $\epsilon$ units.

We emphasize that $f_\epsilon$ only advances points as they land in $U$. Since $f$ and $f_\epsilon$ are order-preserving on $S^1$, we never "lose ground" when applying the perturbation (this is what makes the Closing Lemma so easy on $S^1$.) Since the $f^{n_i}(\theta_0)$ accumulate on $\theta_0$, there is a smallest $n_i$ for which $f_\epsilon^{n_i}(\theta_0) \geq \theta_0$. By decreasing $\epsilon$ to 0, there must then be an $\epsilon_0$ for which $f_{\epsilon_0}^{n_i}(\theta_0) = \theta_0$. Thus we have a periodic point for $f_{\epsilon_0}$. We remark that this uses the continuous dependence of the family $f_\epsilon$ on $\epsilon$.

q.e.d.

We now turn our attention to the approximation of circle diffeomorphisms by Morse-Smale maps. We will prove the following theorem, which is a special case of a result known as the Kupka-Smale Theorem.

**Theorem 15.5.** *Let $f$ be an orientation-preserving diffeomorphism of $S^1$. For any $\epsilon > 0$, there is a $C^1$- Morse-Smale diffeomorphism $g$ which is $C^1 - \epsilon$ close to $f$.*

We begin the proof with several preliminary reductions. By the Closing Lemma, we may assume at the outset that $\rho(f)$ is rational. We will in fact assume that $\rho(f) = 0$, so that $f$ has only fixed points. The proof in the more general periodic point case is analogous.

We will divide the proof of this theorem into a sequence of three steps. First, we will show how to perturb $f$ so that it has no intervals of periodic points. Second, we will show that any diffeomorphism map may be approximated by one with isolated periodic points. Finally, we will perturb $f$ again so that all of the isolated periodic points become hyperbolic.

For a diffeomorphism $f$ of $S^1$ with $\rho(f) = 0$, we always choose the lift $F$ of $f$ which has fixed points: there is only one such lift for which this happens. The next proposition shows that we may always break up intervals of fixed points by a small perturbation.

**Proposition 15.6.** *Let $f: S^1 \to S^1$ be an orientation diffeomorphism which satisfies $f(\theta) = \theta$ for all $\theta$ in the interval $|\theta - \theta_0| \leq 2\pi\delta$. For any $\epsilon > 0$, there exists a diffeomorphism $g, C^r$-$\epsilon$ close to $f$ which satisfies*

1. *$g(\theta) = f(\theta)$ if $|\theta - \theta_0| \geq 2\pi\delta$.*
2. *$g(\theta_0) = \theta_0$.*
3. *$g(\theta) \neq \theta$ if $0 < |\theta - \theta_0| < 2\pi\delta$.*

*Proof.* Suppose first that the lift $F$ satisfies $F(x) = x$ for all $x$ in the interval $J$ given by $|x - x_0| \leq \delta$. We will perturb $F$ to a new map $\hat{F}$ which is $C^r - \epsilon$ close to $F$ and which has $x_0$ as the unique fixed point in the interior of $J$. To define $\hat{F}$, we take a bump function $\phi$ on $J$ that satisfies $\phi(x) \neq 0$ for $|x - x_0| < \delta$ and $\phi(x_0) = 1$. Then, given any $\epsilon > 0$, we set

$$\hat{F}(x) = F(x) + \nu\phi(x)\sin\left(\frac{\pi(x - x_0)}{\delta}\right).$$

Then, provided $\nu$ is chosen small enough, it follows easily that $\hat{F}$ is $C^r$-$\epsilon$ close to $F$ and that $\hat{F}$ has the desired properties.

q.e.d.

The next step in the proof of the Kupka-Smale Theorem is to show that any diffeomorphism may be approximated by one with isolated periodic points. The previous proposition shows how intervals of periodic points may be eliminated; the next shows how accumulation points of periodic points may be destroyed by a small perturbation. As before, we will deal only with fixed points.

**Proposition 15.7.** *Suppose $f$ is an orientation-preserving diffeomorphism of the circle with $\rho(f) = 0$. Then there is a $C^1$ diffeomorphism $g$ which is arbitrarily close to $f$ with respect to the $C^1$-distance and which has only isolated fixed points.*

*Proof.* Let us assume that $x_0$ is an accumulation point of fixed points for the lift $F$ of $f$. Hence we must have $F'(x_0) = 1$ and $F''(x_0) = 0$. Let $J = [a, b]$ be an interval with $a \le x_0 \le b$ and on which

$$|F'(x) - 1| < \epsilon/4$$

$$|F(x) - x| < \epsilon/4.$$

Let us assume also that $F'(a) = F'(b) = 1$. We will replace $F$ by a $C^1$ function on this interval which has at most one fixed point in $(a, b)$.

Let $G(x) = F(x) - x$. Note that $|G(x) - G(y)| \le \epsilon/2$ for any $x, y \in J$. We assume that $G(b) - G(a) > 0$. The cases $G(b) - G(a) = 0$ and $G(b) - G(a) < 0$ are handled in an analogous fashion. By the Mean Value Theorem, we have

$$\frac{G(b) - G(a)}{b - a} \le \max |G'(x)| < \frac{\epsilon}{4}$$

for $x \in J$. Let $\phi(x)$ be a bump function on $J$ which satisfies

1. $\phi(x) \le \epsilon/2$ for all $x \in J$
2. $\int_a^b \phi(x)dx = G(b) - G(a)$.

Since we have

$$G(b) - G(a) \le \frac{\epsilon}{4}(b - a),$$

it follows that it is possible to select a $\phi$ which satisfies both 1 and 2. Now define

$$h(x) = G(a) + \int_a^x \phi(t)dt.$$

By the Fundamental Theorem of Calculus, $h$ is a $C^1$ function on $J$. We have

$$|h(x) - G(x)| \le |G(a) - G(x)| + \left|\int_a^b \phi(t)dt\right| \le \epsilon$$

and

$$|h'(x) - G'(x)| = |\phi(x) - G'(x)| \le \epsilon.$$

Hence $G$ and $h$ are $C^1 - \epsilon$ close on $J$. Therefore, $F$ and $h(x) + x$ are also $C^1$-$\epsilon$ close on $J$. Since $h'(x) = \phi(x) > 0$ on $J$, it follows that $h(x) + x$ has at most one fixed point in $J$.

Thus we define a new map $\hat{F}$ by

$$\hat{F}(x) = \begin{cases} F(x) & x \notin J \\ h(x) + x & x \in J. \end{cases}$$

Since $h'(a) = h'(b) = 0$, it follows that the derivatives of $\hat{F}$ match at $a$ and $b$, and so $\hat{F}$ is a $C^1$- function which has the desired properties.

q.e.d.

Intuitively, in the proof of the preceding proposition, we have cut all of the "wiggles" out of the graph of $F$ on $J$ and replaced them with a smooth graph which meets the diagonal at most once. Applying this technique at each point which is an accumulation of fixed points yields a perturbed map with isolated and hence finitely many fixed points.

The final step in the proof of the Kupka-Smale Theorem is to perturb $f$ so that all of its periodic points are hyperbolic. The previous proposition guarantees that all periodic points are isolated, and hence a small local "push" is all that is needed to make each periodic point hyperbolic.

**Proposition 15.8.** *Let $f$ be an orientation-preserving diffeomorphism of the circle which has isolated periodic points. There is a diffeomorphism $g$ which is $C^r$-$\epsilon$ close to $f$ and which has only hyperbolic periodic points.*

*Proof.* We consider only the case where the lift has an isolated fixed point at $x_0$ with $F'(x_0) = 1$. There are three cases: $x_0$ is a weak attractor, a weak repeller, or the hybrid case where $x_0$ attracts from one side and repels from the other. Several examples of phase portraits of such maps were depicted in Fig. 4.5.

We discuss only the first case; the others are left as exercises. Since $x_0$ is an attracting fixed point, there is $\delta > 0$ such that $|F(x) - F(x_0)| < |x - x_0|$ as long as $|x - x_0| \leq \delta$. We will perturb $F$ in this interval so that the new map $\hat{F}$ has a unique hyperbolic attracting fixed point in this interval. Let $\phi(x)$ be a bump function on $|x - x_0| \leq \delta$ with $\phi(x_0) = 1$. Define

$$\hat{F}(x) = F(x) - \epsilon\phi(x)\sin\left(\frac{\pi(x - x_0)}{\delta}\right).$$

We have $\hat{F}(x_0) = x_0$ but $|\hat{F}(x) - x_0| < |x - x_0|$ if $x \neq x_0$ provided $\epsilon > 0$ is small enough. Moreover,

$$\hat{F}'(x_0) = \frac{\pi}{\delta}(1 - \epsilon),$$

so $x_0$ is hyperbolic. Clearly, if $\epsilon$ is chosen small enough $\hat{F}$ is $C^r$-$\epsilon$ close to $F$.

q.e.d.

This completes the proof of the Kupka-Smale Theorem. The content of this last proposition is depicted in Fig. 15.3.

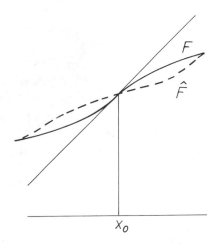

**Fig. 15.3.** Perturbing a non-hyperbolic fixed point.

We remark that the Kupka-Smale Theorem is also valid for $C^r$ small perturbations. The only place where we restricted our attention to $C^1$ diffeomorphisms was in the proof of Proposition 15.7, but this restriction may be removed. See Exercise 6.

**Exercises**

**1.** Extend the proof of structural stability of Morse-Smale diffeomorphisms to the case of periodic points.

**2.** Prove that a Morse-Smale diffeomorphism which reverses orientation is structurally stable.

**3.** Prove that the $C^r$ distance between diffeomorphisms of $S^1$ actually gives a metric on the set of all $C^r$ diffeomorphisms of $S^1$.

**4.** Prove Proposition 15.5 in case $f$ has periodic rather than fixed points.

**5.** Suppose $F: \mathbf{R} \to \mathbf{R}$ satisfies $F(x_0) = x_0$, $F'(x_0) = 1$, and $x_0$ is attracting from one side and repelling from the other. For $\epsilon$ arbitrarily small, construct a $C^r$-$\epsilon$ small perturbation of $F$ that has no fixed points in a neighborhood of $x_0$.

**6.** Construct a $C^\infty$ diffeomorphism $g$ which is $C^r$-$\epsilon$ close to $f$ and which satisfies the conclusion of Proposition 15.7.

**7.** Construct explicitly a Morse-Smale diffeomorphism of $S^1$ which has exactly $k$ repelling and $k$ attracting periodic orbits of period $n$.

**8.** Construct explicitly a Morse-Smale diffeomorphism of the circle which is orientation-reversing and which has 2 fixed points and $k$ periodic orbits of period 2.

## §1.16 HOMOCLINIC POINTS AND BIFURCATIONS

In this section, we return to the study of the bifurcations that occur in one-dimensional dynamical systems. Here we will investigate the profound effect that the presence of a single homoclinic point has on a dynamical system. We will show that the existence of a homoclinic point often implies the existence of a hyperbolic invariant set on which the map is chaotic. Moreover, as a family of dynamical systems develops a homoclinic point, the family undergoes a remarkably complicated sequence of bifurcations known collectively as a homoclinic bifurcation.

Let $p$ be a repelling fixed point. For simplicity, we will assume throughout this section that $f'(p) > 1$ (otherwise replace $f$ by $f^2$). We remark that everything below applies equally well to repelling periodic points with only minor modifications. Recall that if $p$ is a repelling fixed point, then there is an open interval about $p$ on which $f$ is one-to-one and satisfies the expansion property

$$|f(x) - p| > |x - p|.$$

We define the *local unstable set* at $p$ to be the maximal such open interval about $p$. We denote this set by $W^u_{\text{loc}}(p)$.

**Example 16.1** Let $F_\mu(x) = \mu x(1 - x)$ be the quadratic map with $\mu > 4$. $F_\mu$ has a repelling fixed point at 0. One may easily check that $W^u_{\text{loc}}(0) = (-\infty, \frac{1}{2})$.

**Definition 16.2.** *Let $f(p) = p$ and $f'(p) > 1$. A point $q$ is called homoclinic to $p$ if $q \in W^u_{\text{loc}}(p)$ and there exists $n > 0$ such that $f^n(q) = p$. The point $q$ is heteroclinic if $q \in W^u_{\text{loc}}(p)$ and there exists $n > 0$ such that $f^n(q)$ lies on a different periodic orbit.*

We remark that if $p$ has a homoclinic point, then $p$ is sometimes called a "snap-back repellor." Since a homoclinic point $q$ lies by definition in the local unstable set about $p$, we can define a sequence of preimages of $q$, each of which lies closer to $p$ in the local unstable set. These preimages are uniquely

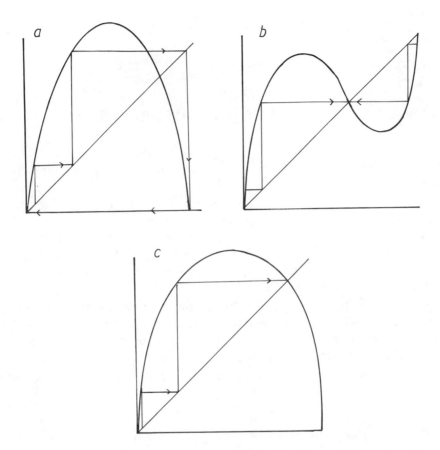

**Fig. 16.1.** In Fig 16.1.a. $f$ admits a homoclinic point,
while in b. and c.. $f$ admits heteroclinic points.

defined since $f$ is one-to-one on the local unstable set. A homoclinic point,
together with its backward orbit defined by the above preimages, and its
(finite) forward orbit, is called a *homoclinic orbit.* Thus a homoclinic orbit is
one which tends to a fixed point under backward iteration and which lands
on the same fixed point under forward iteration.

**Example 16.3.** If $F_\mu(x) = \mu x(1 - x)$ with $\mu > 4$. there are two fixed
points, 0 and $p_\mu$, both of which admit infinitely many homoclinic as well as
heteroclinic points.

Fig. 16.1 illustrates several homoclinic and heteroclinic points.

**Definition 16.4.** A homoclinic orbit is called nondegenerate if $f'(x) \neq 0$ for all points $x$ on the orbit. Otherwise, the orbit is degenerate.

Note that, for the quadratic map above, all homoclinic points to both 0 and $p_\mu$ are nondegenerate when $\mu > 4$. Indeed, from §1.5, we know that the only critical point tends to infinity. If $\mu = 4$, $1/2$ lies on a degenerate homoclinic orbit to 0. We will see below that nondegenerate homoclinic orbits lead to chaotic behavior, at least on some invariant subset, while degenerate homoclinic orbits often lead to complicated bifurcations when a parameter is varied.

**Theorem 16.5.** *Suppose $q$ lies on a nondegenerate homoclinic orbit to a fixed point $p$. Then for each neighborhood $U$ of $p$, there is an integer $n \geq 0$ such that $f^n$ has a hyperbolic invariant subset in $U$ on which $f^n$ is topologically conjugate to the shift automorphism.*

*Proof.* Let $W$ be a neighborhood of $p$ contained in $U$ and, on which, $f'(x) > \delta > 1$ for all $x$. By taking preimages if necessary, we may assume that $q \in W$. There is an $n > 0$ such that $f^n(q) = p$. Since there are only a finite number of points on the orbit of $q$ which do not lie in $W$, and since $f'(f^i(q)) \neq 0$ for all $i$, we may find a neighborhood $V$ of $q$ in $W$ and an $\epsilon > 0$ such that $|(f^n)'(x)| > \epsilon$ for all $x \in V$. Since $(f^n)'(x) \neq 0$ for all $x \in V$, $f^n$ maps $V$ diffeomorphically onto an interval $f^n(V)$ which contains $p$ in its interior.

Now choose $j$ so that $\delta^j \epsilon > 1$. By choosing $V$ smaller if necessary, we may assume that $f^{n+i}(V) \subset W$ for $i = 1, 2, \ldots, j$, but $f^{n+i}(V) \cap V = \emptyset$. That is, each map $f^{n+i}$ expands the interval $f^n(V)$ about $p$. Since $V$ itself belongs to the local unstable set, there is an integer $k \geq j$ such that $f^{n+k}(V)$ covers $V$ and $|(f^{n+k})'(x)| > 1$ for $x \in V$. More precisely, $f^{n+k}: V \to f^{n+k}(V)$ is a diffeomorphism onto its image which contains both $p$ and $V$.

To introduce symbolic dynamics, let us choose another neighborhood $V'$ of $p$ which is contained in $W$ and which satisfies

1. $f^{n+k}$ is one-to one on $V'$ and $f^{n+k}(V') \subset W$.
2. $|(f^{n+k})'(x)| > 1$ for $x \in V'$.
3. $f^{n+k}(V') \supset V$.
4. $V \cap V' = \emptyset$. Clearly, $f^{n+k}(V')$ also covers $V'$.

Hence the map $f^{n+k}$ expands both $V$ and $V'$ and their images contain both $V$ and $V'$. Using the techniques of §1.7, it is now easy to construct a topological conjugacy between $f^{n+k}$ and the shift. This completes the proof.

q.e.d.

**Corollary 16.6.** *Suppose $f$ admits a nondegenerate homoclinic point to $p$. Then, in every neighborhood of $p$, there are infinitely many distinct periodic points.*

The orbits of these periodic points do not, of course, lie in a neighborhood of $p$. Rather, the orbits of the periodic points move far away, roughly following the homoclinic orbit.

By Sarkovskii's Theorem, $f$ must have periodic points of all periods of the form $2^k$, since $f^{n+k}$ has infinitely many periodic orbits with distinct periods.

**Remarks.**

**1.** Note that the above procedure gives a conjugacy between $f^{n+k}$ and the shift. It is easy to modify the above construction to find a conjugacy between $f^i$ and the shift for any $i > k + n$.

**2.** By varying the number of intervals chosen, one may also find various subshifts of finite type near fixed points which admit nondegenerate homoclinic orbits.

Thus, nondegenerate homoclinic points lead to the existence of a chaotic regime for a map. A natural question is how do these types of homoclinic points arise. Generally, the answer is they are spawned by a degenerate homoclinic orbit as a parameter is varied. This gives another example of how structural stability can fail as well as a different and much more complex type of bifurcation than those considered in §1.12.

Observe that homoclinic orbits are preserved by topological conjugacy. Indeed, one may also show that homoclinic points are nonwandering but not recurrent (see Exercises 7.2 and 7.3).

Degenerate homoclinic orbits do not occur in structurally stable systems. This is illustrated by the quadratic map $F_4(x) = 4x(1-x)$. The point $1/2$ clearly lies on a degenerate homoclinic orbit. When $\mu < 4$, the map $F_\mu(x) = \mu x(1-x)$ has maximum value $\mu/4 < 1$. Hence there are no points homoclinic to 0 for these $\mu$-values. On the other hand, when $\mu > 4$, there are infinitely many distinct homoclinic orbits (Exercise 3). Consequently, $F_4$ is not structurally stable. More generally, if $f$ admits a degenerate homoclinic orbit, a $C^1$-small change in $f$ can change the number of homoclinic orbits. (See Fig. 16.2).

To describe the bifurcations which occur when a degenerate homoclinic point is created, let us again use the quadratic map as a model. We will show that, in every neighborhood of the critical parameter value $\mu = 4$, there are $\mu$-values for which the corresponding maps have bifurcations of either saddle

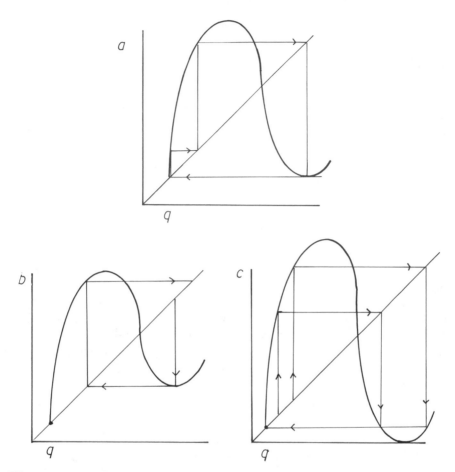

**Fig. 16.2.** In Fig. 16.2.a, there is a degenerate homoclinic point to $q$.
Small perturbations yield either no homoclinic points as in b,
or infinitely many homoclinic points as in c.

node or period-doubling type. Thus, these bifurcations are accumulation points of simple bifurcations. We use the term homoclinic bifurcation to describe this phenomenon.

The remainder of this section is somewhat technical, due in part to the fact that we are dealing with a specific family of maps. Nevertheless, the ideas presented below are quite general. First, we need a lemma.

**Lemma 16.7.** Let $F_\mu(x) = \mu x(1 - x)$.

    a. $\frac{d}{d\mu} F_\mu^2\left(\frac{1}{2}\right) < 0$ if $\mu > \frac{8}{3}$.

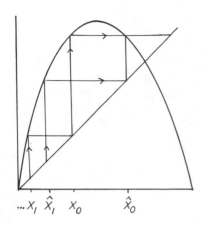

$$\ldots X_1 \; \hat{X}_1 \; X_0 \qquad \hat{X}_0$$

Fig. 16.3                              Fig. 16.4

*b.* $(F_\mu^2)''(\tfrac{1}{2}) = \mu^2(\mu - 2)$.

In *b*, the differentiation is with respect to $x$, not $\mu$. Both parts of this lemma are proved by elementary calculus. We leave the details to the reader.

Now fix $\mu_0$ with $\tfrac{8}{3} < \mu_0 < 4$. We will show that there are infinitely many values of $\mu$ in the interval $[\mu_0, 4]$ for which $F_\mu$ has a saddle node bifurcation. Since $\mu > 2$, we may find a neighborhood $J$ of $\tfrac{1}{2}$ such that $(F_\mu^2)''(x) > 0$ for all $x \in J$ and all $\mu \in [\mu_0, 4]$. We may assume that $J$ is symmetric about $\tfrac{1}{2}$, say $J = [x_0, \hat{x}_0]$. Since $\mu > 2$, $F_\mu(\tfrac{1}{2}) > \tfrac{1}{2}$   We may also assume that $F_\mu(\tfrac{1}{2}) > \hat{x}_0$. Since $F_\mu$ is an increasing sequence on $[0, \tfrac{1}{2})$ and therefore on $[0, x_0]$, there is a well-defined double sequence

$$0 < \ldots < x_2 < \hat{x}_2 < x_1 < \hat{x}_1 < x_0 < \hat{x}_0$$

where $F_\mu(\hat{x}_j) = \hat{x}_{j-1}$ and $F_\mu(x_j) = x_{j-1}$. Let $I_j$ be the interval $[x_j, \hat{x}_j]$. See Fig. 16.3.

Clearly, $F_\mu$ maps $I_j$ monotonically onto $I_{j\,1}$. Hence it follows that the graph of $F_\mu^{j+2} = F_\mu^2 \circ F_\mu^j$ on $I_j$ resembles that of $F_\mu^2$ on $I_0$. This is made more precise in the following

**Proposition 16.8.**

1. *For each $j \geq 0$, $F_\mu^{j+2}$ has a unique critical point $c_j(\mu)$ in $I_j$, and $F_\mu^{j+2}$ has a minimum at this point. Also, $F_\mu^{j+2}(c_j(\mu)) = F_\mu^2(\tfrac{1}{2})$.*

2. *$F_\mu^{j+2}(x_j) = F_\mu^{j+2}(\hat{x}_j) = F_\mu^2(x_0)$.*

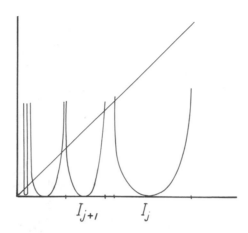

**Fig. 16.5.** The graph of $F_4^{j+2}$ on $I_j$.

*Proof.* For part 1, we note that on $I_j$, we have $F_\mu^{j+2} = F_\mu^2 \circ F_\mu^j$. Hence

$$(F_\mu^{j+2})'(x) = (F_\mu^2)'(F_\mu^j(x)) \cdot (F_\mu^j)'(x).$$

Since $F_\mu^j \colon I_j \to I_0$ is a diffemorphism, $(F_\mu^j)'(x) > 0$ for $x \in I_j$. Thus the only critical points occur where $(F_\mu^2)'$ vanishes, namely at $F_\mu^j = \frac{1}{2}$. The result now follows immediately. Part 2 follows from the fact that $F_\mu(x_0) = F_\mu(\hat{x}_0)$ by symmetry. The content of this proposition is illustrated in Fig. 16.4.

q.e.d.

When $\mu = 4$, the situation is simpler to describe. The minimum value of $F_4^{j+2}$ on each $I_j$ is 0. Since

$$F_4^{j+2}(x_j) = F_4^{j+2}(\hat{x}_j) = F_4^2(x_0) > 0,$$

it follows that for $j$ sufficiently large, $F_4^{j+2}$ has precisely two fixed points on $I_j$. (See Figure 16.5).

To show that there are infinitely many bifurcations, we simply observe that as $\mu$ increases, the graph of $F_\mu^{j+2}$ on $I_j$ "descends," eventually crossing the diagonal and creating a bifurcation. Let us make this more precise. For each $\mu$, the intervals $I_j$ converge to 0. We claim that there exists an integer $N = N(\mu)$ such that $F_\mu^{j+2}(x) > x$ for all $x \in I_j$ and $j > N$. Indeed, by Proposition 16.8, the minimum value of $F_\mu^{j+2}$ on $I_j$ is $F_\mu^2(\frac{1}{2}) > 0$. So if $N(\mu)$ is chosen so that $I_N \subset [0, F_\mu^2(\frac{1}{2})]$, the result then follows immediately.

Now we may show that there are infinitely many maps among the $F_\mu$, $\mu \in [\mu_0, 4]$ which undergo bifurcations. Choose $N(\mu_0)$ as above. For each $j > N(\mu_0)$ we have $F_\mu^{j+2}(x) > x$ for all $x \in I_j$. Consequently, $F_\mu^{j+2}$ has no fixed points in $I_j$. Now let $\mu$ increase. When $\mu = 4$, $F_4^{j+2}$ has two fixed points in $I_j$. Hence there must be at least one $\mu$ in $[\mu_0, 4]$ for which $F_\mu^{j+2}$ suffers a bifurcation of fixed points.

**Remarks.**

**1.**   One may show that there must be a saddle node bifurcation in $I_j$ (which is possibly degenerate) as $\mu$ increases.

**2.**   There must also be a period-doubling bifurcation as $\mu$ increases, since one of the fixed points for $F_4^{j+2}$ in $I_j$ has negative derivative.

**3.**   Indeed, one may find an interval on which $F_\mu^{j+2}$ may be "renormalized" as we shall describe in the next section. In particular, it follows that there are infinitely many parameter values in $[\mu_0, 4]$ for which $F_\mu$ admits a degenerate homoclinic point.

**Exercises**

**1.**   Prove that homoclinic orbits are preserved by topological conjugacy.

**2.**   Prove that homoclinic orbits are nonwandering but not recurrent (See Exercises 7.2 and 7.3).

**3.**   Let $F_\mu(x) = \mu x(1 - x)$. Prove that when $\mu > 4$, $F_\mu$ has infinitely many distinct orbits homoclinic to 0, all of which are nondegenerate. Prove that $F_4$ has infinitely many degenerate homoclinic orbits.

**4.**   Prove Lemma 16.7.

**5.**   Let $p_1$ and $p_2$ be repelling fixed point and suppose both points admit nondegenerate heteroclinic orbits connecting each other. That is, suppose there exists $q_i \in W_{\text{loc}}^u(p_i)$ and integers $n_1$ and $n_2$ such that

$$f^{n_1}(q_1) = p_2, \quad f^{n_2}(q_2) = p_1$$

Prove that $f$ admits a hyperbolic invariant set on which the map is chaotic.

**6.**   Prove that the results of Exercise 5 apply to the family of maps $g_\lambda(x) = x^3 - \lambda x$ for $\lambda$ sufficiently large.

## §1.17 THE PERIOD-DOUBLING ROUTE TO CHAOS

As we have seen in §1.5, the quadratic map $F_\mu(x) = \mu x(1-x)$ is simple dynamically for $0 \leq \mu \leq 3$ but chaotic when $\mu \geq 4$. The natural question is: how does $F_\mu$ become chaotic as $\mu$ increases? Where do the infinitely many periodic points which are present for large $\mu$ come from? In this section, we will give a geometric and intuitive answer to this question. In subsequent sections, we will provide a more rigorous approach. This will necessitate the introduction of a new, more powerful version of symbolic dynamics, the kneading theory.

Sarkovskii's Theorem provides a partial answer to the question of how infinitely many periodic points arise as the parameter is varied. Before $F_\mu$ can possibly have infinitely many periodic points with distinct periods, it must have periodic points with all periods of the form $2^j$. The local bifurcation theory provides two "typical" ways that these periodic points can arise: in saddle node bifurcations and via period-doublings. The question then becomes which type of bifurcations occur as $F_\mu$ becomes more chaotic.

As we shall see, the usual scenario for $F_\mu$ to become chaotic is for $F_\mu$ to undergo a series of period-doubling bifurcations. This is not always the case, but it is a typical route to chaos. We remark that, although we deal here with the quadratic map, the ideas below apply to a much wider class of maps, namely the *unimodal maps* which we will describe in the next section.

Recall that the graphs of $F_\mu$ for various values of $\mu$ are as depicted in Fig. 17.1. For $1 < \mu < 3$, $F_\mu$ has a unique attracting fixed point at $p_\mu = (\mu-1)/\mu$ so that $0 < p_\mu < 1$. Note that, as long as $F_\mu'(p_\mu) < 0$, there exists a "partner" $\hat{p}_\mu$ for $p_\mu$ in the sense that $F_\mu(\hat{p}_\mu) = p_\mu$ and $\hat{p}_\mu < p_\mu$.

Using graphical analysis of $F_\mu$, we may also sketch the graphs of $F_\mu^2$ for various $\mu$-values. These are depicted in Fig. 17.2. Note in particular the portion of the graph of $F_\mu^2$ in the interval $[\hat{p}_\mu, p_\mu]$. We have enclosed this portion of the graph inside a box. Let us make three observations about this graph.

    a. The graph of $F_\mu^2$, although "upside-down," resembles the graph of the original quadratic map (for a different $\mu$-value ) in a sense to be made precise later.

    b. Indeed, inside the box, $F_\mu^2$ has one fixed point at an endpoint of the

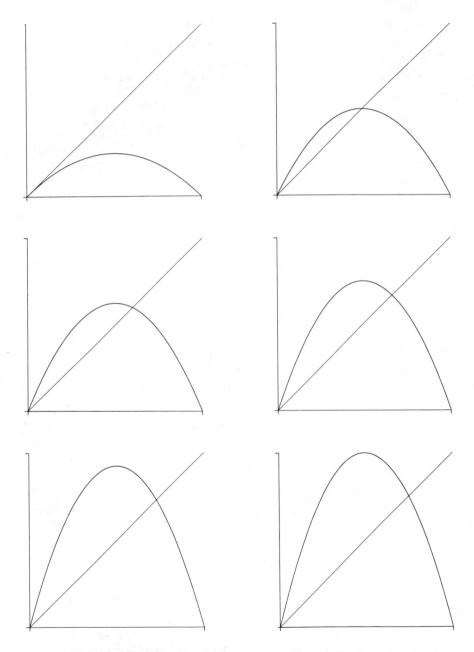

**Fig. 17.1.** The graphs of $F_\mu(x) = \mu x(1-x)$ for $\mu = 1$, $\mu = 2$, $\mu = 2.5$, $\mu = 3$, $\mu = 3.5$, $\mu = 4$ from left to right.

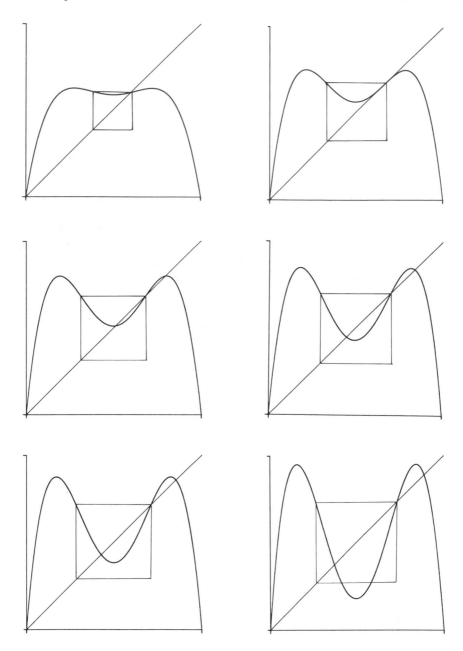

**Fig. 17.2** The graphs of $F_\mu^2(x)$ for $\mu = 2.5$, $\mu = 3$, $\mu = 3.2$, $\mu = 3.4$, $\mu = 3.5$, $\mu = 3.8$ from left to right.

interval $[\hat{p}_\mu, p_\mu]$ and a unique critical point within this interval.

c. As $\mu$ increases, the "hump" in this quadratic-like map grows until it eventually protrudes through the bottom of the box.

That is, the behavior of $F_\mu^2$ on the interval $[\hat{p}_\mu, p_\mu]$ is very similar to that of $F_\mu$ on its original domain $[0, 1]$. In particular, as $\mu$ increases, we first expect a new fixed point in $[\hat{p}_\mu, p_\mu]$ for $F_\mu^2$ (i.e., a period 2 point for $F_\mu$ ) to be born. Eventually, this "fixed point" will itself period-double, just as $p_\mu$ did for $F_\mu$, producing a period 4 point. Continuing this procedure, we may find a small box in which the graphs of $F_\mu^4, F_\mu^8$, etc. resemble the original quadratic function. Thus we are led to expect that $F_\mu$ undergoes a series of period-doublings as $\mu$ increases.

To make these ideas precise, we introduce the *renormalization* operator $R$. $R$ is a function of functions: $R$ converts certain given functions on $I$ to new functions on $I$. To define $R$, we first suppose that $\mu$ is large enough so that $\hat{p}_\mu$ is defined and $\hat{p}_\mu < p_\mu$. For $F_\mu, \mu > 2$ suffices. Let $L_\mu$ denote the linear map which takes $p_\mu$ to 0 and $\hat{p}_\mu$ to 1. That is,

$$L_\mu(x) = \frac{1}{\hat{p}_\mu - p_\mu}(x - p_\mu).$$

One may check easily that the inverse of $L_\mu$ is given by

$$L_\mu^{-1}(x) = (\hat{p}_\mu - p_\mu)x + p_\mu.$$

Note that $L_\mu$ expands the small interval $[\hat{p}_\mu, p_\mu]$ onto $[0, 1]$ with a change of orientation.

We now define the renormalization of $F_\mu$ by

$$(RF_\mu)(x) = L_\mu \circ F_\mu^2 \circ L_\mu^{-1}(x).$$

The renormalized function $RF_\mu$ is defined on $I$ and shares many of the features of $F_\mu$. We single these out in a Proposition.

**Proposition 17.1.**

*1. $(RF_\mu)(0) = 0$ and $RF_\mu(1) = 0$.*

*2. $(RF_\mu)'(\frac{1}{2}) = 0$ and $\frac{1}{2}$ is the only critical point for $RF_\mu$.*

*3. $S(RF_\mu) < 0$, where $S$ is the Schwarzian derivative.*

The proof of each statement is straightforward and is left to the reader. We observe that the renormalization of $F_\mu$ converts periodic points of period

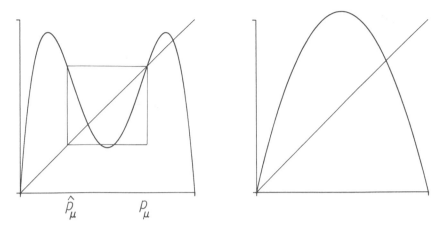

**Fig. 17.3.** The graphs of $F_\mu^2$, left, and $RF_\mu$ for $\mu = 3.7$

2 for $F_\mu$ into fixed points for $RF_\mu$. Also, before $\mu$ reaches 4, the peak of the graph of $RF_\mu$ already protrudes through the top of the unit square. See Fig. 17.3.

Thus, as we noted above, we expect $RF_\mu$ to undergo a period-doubling bifurcation as $\mu$ increases. In fact, as long as $RF_\mu$ admits a fixed point with negative derivative at some point $p_1(\mu)$, then we may find $\hat{p}_1(\mu)$ as before and define a second renormalization. The linear map in this case takes $p_1(\mu)$ to 0 and $\hat{p}_1(\mu)$ to 1 and thus is a different linear map. Hence we see that the entire picture repeats itself and we get another period-doubling bifurcation, this time for $F_\mu^2$. Continuing this process leads to a succession of period-doubling bifurcations as $\mu$ increases. Hence we expect that the bifurcation diagram for $F_\mu$ will include at least the complication shown in Fig. 17.4.

While this section has been for the most part heuristic, we have managed to introduce the important notion of renormalization. The operator $R$ allows us to examine the second iterate of a given map on the same scale as the original map. $R$ acts like a microscope, allowing us to view phenomena that occur for $f^2$ in the same detail as for $f$. One might naturally ask what happens in the limit when $R$ is applied over and over again to a given map. This is, in fact, the ultimate goal of renormalization group analysis from physics and leads to the important concept of universality. These ideas are beyond the scope of this text; we will, however, discuss this operator from the point of view of symbolic dynamics. This necessitates the introduction of a new and more powerful version of symbolic dynamics, the kneading theory, which is the topic of the next section.

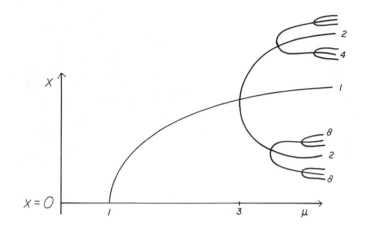

**Fig. 17.4.** The bifurcation diagram for $F_\mu$ showing the repeated period doublings. The integers represent the periods.

### Exercises.

*The Adding Machine* (Misiurewicz). The purpose of this set of exercises is to construct and analyze a continuous map of $I$ which has exactly one periodic point of period $2^j$ for each $j$ and no other periodic points. The construction of the map relies on the notion of the double of a map, a topic discussed in §1.10. Start with $f_0(x) = 1/3$. Let $f_1(x)$ denote the double of $f_0$, i.e., $f_1(x)$ is obtained from $f_0$ by the following procedure:

1. $f_1(x) = \frac{1}{3}f_0(3x) + \frac{2}{3}$ if $0 \le x \le 1/3$.
2. $f_1(2/3) = 0; f_1(1) = 1/3$.
3. $f_1$ is continuous and linear on the intervals $1/3 \le x \le 2/3$ and $2/3 \le x \le 1$.

That is, the graph of $f_1$ is obtained from $f_0$ as shown in Fig. 17.5. Inductively, we define $f_{n+1}(x)$ to be the double of $f_n(x)$. See Fig. 17.6. Finally, let $F(x) = \lim_{n \to \infty} f_n(x)$.

**1.** Prove that $f_{n+\alpha}(x) = f_n(x)$ for all $\alpha > 0$ and $x \ge 1/3^n$. Conclude that if we define $F(0) = 1$, then $F(x)$ is a well-defined continuous map of $I$.

**2.** Prove that $f_n(x)$ has a unique periodic orbit $2^j$ for each $j \le n$. Prove that each of these periodic points is repelling. if $j < n$.

**3.** Prove that $f_n(x)$ has no other periodic points.

**4.** Prove that $F(x)$ has a unique periodic orbit of period $2^j$ for each $j$ and no other periodic points. Show that this periodic orbit is repelling.

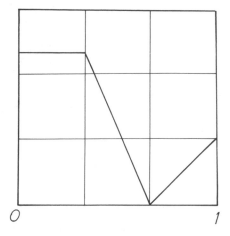

**Fig. 17.5.** The double of $f_0$.

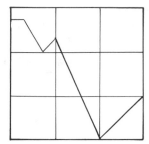
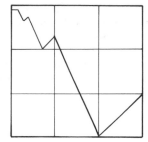

**Fig. 17.6.** The graph of $f_2$ and $f_3$.

Recall from Example 5.5 the construction of the Cantor Middle-Thirds set. Let $A_0 = (\frac{1}{3}, \frac{2}{3})$ be the middle third of the unit interval $I$. Let $I_0 = I - A_0$. Let $A_1 = (\frac{1}{9}, \frac{2}{9}) \cup (\frac{7}{9}, \frac{8}{9})$ be the middle third of the two intervals in $I_0$. Let $I_1 = I_0 - A_1$. Inductively, let $A_n$ denote the middle third of the intervals in $I_{n-1}$ and let $I_n = I_{n-1} - A_n$. Finally, let

$$I_\infty = \bigcap_{n \geq \infty} I_n.$$

$I_\infty$ is the classical middle-thirds Cantor set.

5.  Show that the periodic point of period $2^j$ for $F$ lies in union of intervals which comprise $A_j$.

6.  Prove that $F(I_n) = I_n$.

**7.** Prove that if $x \in A_n$ and $x$ is not periodic, then there exists $k > 0$ such that $F^k(x) \in I_n$.

**8.** Prove that $I_\infty$ is invariant under $F$.

**9.** Prove that, if $x \notin I_\infty$, then the orbit of $x$ tends to $I_\infty$ or eventually lies in $I_\infty$.

Thus all of the non-periodic points for $F$ are attracted to the set $I_\infty$. Thus, to understand the dynamics of $F$, it suffices to understand the dynamics of $F$ on $I_\infty$. For each point $p \in I_\infty$, we attach an infinite sequence of 0's and 1's, $S(p) = (s_0 s_1 s_2 \ldots)$, according to the rule: $s_0 = 1$ if $p$ belongs to the left component of $I_0$; $s_0 = 0$ if $p$ belongs to the right component. Note that this is slightly different from our coding for the quadratic map! Now $p$ belongs to some component of $I_{n-1}$, and $I_n$ is obtained by removing the middle third of this interval. Therefore we may set $s_n = 1$ if $p$ belongs to the left hand interval in $I_n$ and $s_n = 0$ otherwise.

Let $\Sigma_2$ be the set of all sequences of 0's and 1's. Define the adding machine $A \colon \Sigma_2 \to \Sigma_2$ by $A(s_0 s_1 s_2 \ldots) = (s_0 s_1 s_2 \ldots) + (100 \ldots) \bmod 2$, i.e., $A$ is obtained by adding 1 mod 2 to $s_0$ and carrying the result. For example, $A(110\,\overline{110}\ldots) = (001\,110\,\overline{110}\ldots)$ and $A(11\overline{1}\ldots) = (00\overline{0}\ldots)$.

**10.** Let $d$ be the usual distance on $\Sigma_2$ (see Proposition 6.1). Prove that $S \colon I_\infty \to \Sigma_2$ is a topological conjugacy between $F$ on $I_\infty$ and $A$ on $\Sigma_2$.

**11.** Prove that $A$ has no periodic points.

**12.** Prove that every orbit of $A$ is dense in $\Sigma_2$.

Since $\Sigma_2$ has no proper closed invariant subsets under $A$, $\Sigma_2$ is an example of a *minimal set*.

## §1.18 THE KNEADING THEORY

In previous sections, we have shown how symbolic dynamics may be used to understand completely the dynamics of certain quadratic maps. When $\mu$ is sufficiently large or when $\mu = 3.839$, we have seen that all of the interesting dynamics of $F_\mu(x) = \mu x(1 - x)$ occurs on a Cantor set. The map on this set is equivalent to the shift map or a subshift of finite type. For other values of the parameter, the situation is more complicated. For example, the maps $F_4(x) = 4x(1 - x)$ and $Q_c(x) = x^2 + c$ where $c \approx -1.543689$ both have intervals on which the map is chaotic (see Examples 8.9 and 11.13).

One difference between these two pairs of examples is the behavior of the critical point under iteration. In the case of $F_\mu$ when $\mu > 4$, the orbit of the critical point tends to $-\infty$, whereas it tends to an attracting periodic orbit when $\mu = 3.839$. In the latter two examples, the critical point eventually lands on a repelling fixed point. Thus, in some sense, the orbit of the critical point determines the dynamics of the map. Our goal in this section is to make this statement precise. We will introduce a more elaborate version of symbolic dynamics, the kneading theory, which keeps track of the orbit of the critical point and thereby allows us to handle many of the additional complications. The kneading theory also enables us to understand on a symbolic level the transition from simple to chaotic dynamics which was described heuristically in the previous section.

**Definition 18.1.** Let $f: I \to I$. The map is unimodal if
1. $f(0) = f(1) = 0$.
2. $f$ has a unique critical point $c$ with $0 < c < 1$.

Clearly, unimodal maps are increasing on the interval $[0, c)$ and decreasing on $(c, 1]$. The quadratic map $F_\mu(x) = \mu x(1-x)$ is unimodal for $0 < \mu \le 4$, as is $S_\lambda(x) = \lambda \sin(\pi x)$ for $0 < \lambda < 1$. See Fig. 18.1. For the remainder of this section, we will work with a fixed unimodal map $f$.

Note that, for a unimodal map, the orbit of the critical point is trapped in the unit interval. It cannot escape to $-\infty$ as in the case of $F_\mu(x) = \mu x(1-x)$ with $\mu > 4$. But there are many other possible fates for this orbit. To highlight the role of the critical point, we will extend slightly our definition of the itinerary of a point by adding a third symbol "C".

**Definition 18.2.** Let $x \in I$. The itinerary of $x$ under $f$ is the infinite sequence $S(x) = (s_0 s_1 s_2 \ldots)$ where

$$s_j = \begin{cases} 0 & \text{if } f^j(x) < c \\ 1 & \text{if } f^j(x) > c \\ C & \text{if } f^j(x) = c. \end{cases}$$

Most important for us will be the itinerary of the critical point.

**Definition 18.3.** The kneading sequence $K(f)$ of $f$ is the itinerary of $f(c)$, i.e., $K(f) = S(f(c))$.

**Example 18.4.** If $f(x) = F_4(x) = 4x(1 - x)$, then $c = 1/2$, $f(c) = 1$, and $f^j(c) = 0$ for all $j > 1$. Hence

$$K(f) = (100\bar{0}\ldots).$$

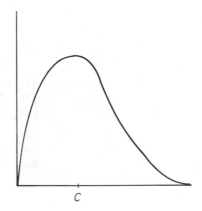

**Fig. 18.1** The graph of a unimodal map.

If $f(x) = F_2(x) = 2x(1-x)$, then $c = 1/2$ and $f^j(c) = c$ for all $j$. Hence

$$K(f) = (CC\overline{C}\ldots).$$

Here the overbar means that the given symbol or symbols is repeated ad infinitum.

There are many possible itineraries for a unimodal map, but there are some restrictions. For example, if $s_j = C$, then we must have $s_{j+k} = \alpha_k$, where

$$K_f = (\alpha_1 \alpha_2 \alpha_3 \ldots).$$

We call a sequence *regular* if $s_j = 0$ or $1$ for all $j$, i.e., $s_j \neq C$ for any $j$. Unimodal maps which feature any given regular sequence may be constructed, but not all sequences need occur in a given map. In fact, a unimodal map may have very few itineraries.

**Example 18.5.** Let $f(x) = F_\mu(x) = \mu x(1-x)$ where $1 < \mu < 2$. From graphical analysis (see §1.5 ), we see that the only itineraries for this map are

$$(00\overline{0}\ldots)$$

$$(100\overline{0}\ldots)$$

$$(C00\overline{0}\ldots).$$

**Example 18.6.** When $2 < \mu < 3$ in the above example, there are more possible itineraries:

$$(C11\overline{1}\ldots)$$

$$(00\overline{0}\ldots)$$

$$(11\overline{1}\ldots)$$

$$(0\ldots0\,11\overline{1})$$

$$(0\ldots0C11\overline{1})$$

as well as all of the above preceded by a 1. In the last two sequences, there is an arbitrary but finite number of zeroes.

One may also easily enumerate the possible sequences for $\mu = 2, \mu = 3$, and the case of the attracting period two orbit when $\mu$ is slightly larger than 3. See Exercise 1.

We note that, although $F_\mu$ has a single attracting fixed point for all $\mu$ with $1 < \mu < 3$, the number of possible sequences has changed. The change occurs at $\mu = 2$ where the critical point is itself periodic, i.e., the kneading sequence is $(CCC\ldots)$. The case of a periodic kneading sequence will add complications below for precisely this reason.

We now define an ordering $\prec$ on the set of itineraries. Let $s = (s_0 s_1 s_2 \ldots)$ and $t = (t_0 t_1 t_2 \ldots)$. We say that s and t have *discrepancy* $n$ if $s_i = t_i$ for $0 \le i < n$ but $s_n \ne t_n$. Let $\tau_n(s)$ denote the number of 1's among $s_0, s_1, \ldots, s_n$. This number is important to us for the following reason. The sign of the derivative of $f^n$ at $x$ governs the local dynamics near $x$. Note that $f'(x)$ is negative whenever $x \in (c, 1]$. Hence, by the Chain Rule, the number of 1's in the itinerary of $x$ governs the sign of $(f^n)'(x)$ (provided $f'(f^i(x)) \ne 0$ for all $i$).

We will define the ordering on sequences inductively. To begin, we set $0 < C < 1$.

**Definition 18.7.** Suppose s and t have discrepancy $n$. We say $s \prec t$ if either

    a. $\tau_{n-1}(s)$ is even and $s_n < t_n$

    b. $\tau_{n-1}(s)$ is odd and $s_n > t_n$.

**Example 18.8.** The above definition implies

$$(0101\ldots) \prec (010C\ldots) \prec (0100\ldots)$$

$$(110\ldots) \prec (11C\ldots) \prec (111\ldots).$$

This ordering, while somewhat cumbersome, is nevertheless reflected on the real line. More precisely, if $x, y \in I$ and $x < y$, then $S(x) \preceq S(y)$.

Conversely, if $S(x) \prec S(y)$, then $x < y$. Before proving this, let us check this ordering in a simple example.

**Example 18.9.** Let $f(x) = F_\mu(x)$ where $2 < \mu < 3$. Using Example 18.6, we see that

$$(00\bar{0}\ldots) \prec (C11\bar{1}\ldots) \prec (11\bar{1}\ldots) \prec (1C11\bar{1}\ldots) \prec (100\bar{0}\ldots).$$

Between $(00\bar{0}\ldots)$ and $(C11\bar{1}\ldots)$, there are infinitely many itineraries of the form:

$$\ell_n = (\underbrace{0\ldots0}_{n\ 0's}\ 11\bar{1}\ldots)$$

$$\ell'_n = (\underbrace{0\ldots0}_{n\ 0's}\ C11\bar{1}\ldots).$$

One may check easily that

$$(00\bar{0}\ldots) \prec \ldots \prec \ell'_2 \prec \ell_2 \prec \ell'_1 \prec \ell_1 \prec (C11\bar{1}\ldots).$$

Preceding these orbits by a 1 reverses the ordering:

$$(1C11\bar{1}\ldots) \prec 1\ell_1 \prec 1\ell'_1 \prec 1\ell_2 \prec 1\ell'_2 \prec \ldots \prec (100\bar{0}\ldots).$$

We now show that the ordering on itineraries is the same as that on the real line.

**Theorem 18.10.** *Let $x, y \in I$.*
  *1. If $S(x) \prec S(y)$, then $x < y$.*
  *2. If $x < y$, then $S(x) \preceq S(y)$.*

**Remark.** The equality in part two cannot be removed, as the existence of an attracting periodic point usually implies the existence of an interval of points in the real line with the same itinerary.

*Proof.* We prove part 1. Part 2 then follows immediately. Let $S(x) = (s_0 s_1 s_2 \ldots)$ and $S(y) = (t_0 t_1 t_2 \ldots)$. We use induction on $n$, where $n$ is the discrepancy of $S(x)$ and $S(y)$. If $n = 0$, the result is clear, since $0 < C < 1$, which is precisely the order on the real line. So we assume that the result is true for sequences with discrepancy $n - 1$ and prove it for discrepancy $n$.

We first apply $f$ to $x$ and $y$. Using the shift, we have

$$S(f(x)) = (s_1 s_2 s_3 \ldots)$$

$$S(f(y)) = (t_1 t_2 t_3 \ldots).$$

There are three cases, $s_0 = 0, C$, and 1. If $s_0 = 0$, then $S(f(x)) \prec S(f(y))$ since we have not changed the number of 1's before the discrepancy. By induction, we have $f(x) < f(y)$. But since $f$ is increasing on $[0, c)$, it follows that $x < y$ as well. If $s_0 = 1$, then $S(f(x)) \succ S(f(y))$ since there is one less 1 among $s_1, \ldots, s_n$. Hence $f(x) > f(y)$ by induction. But then $x < y$ since $f$ decreases on $(c, 1]$. Finally, if $s_0 = C$, it follows that $x = y = c$.

q.e.d.

We say that a sequence s is *admissible* for $f$ if there exists $x \in I$ with $S(x) = s$. Let $\Sigma_f$ denote the set of all possible $f$-admissible sequences. Our goal is to find a method whereby we may determine all sequences in $\Sigma_f$. The key to this is the kneading sequence. The kneading sequence gives one necessary condition which must be satisfied by any sequence in $\Sigma_f$. Since $f(c)$ is the maximum of $f$, it must be true that $f^n(x) \le f(c)$ for all $x \in I$ and all $n \ge 1$. Consequently, if $s \in \Sigma_f$, then $\sigma^n(s) \preceq K(f)$ for all $n \ge 1$. This condition is not quite sufficient, as the following example shows.

**Example 18.11.** Let $f(x) = F_4(x) = 4x(1 - x)$. Note that $f(1) = 0$ so that $K(f) = S(1) = (100\bar{0}\ldots)$. The only preimage of 1 is $c$, so that the only admissible sequence which is a preimage of $(100\bar{0}\ldots)$ is $(C100\bar{0}\ldots)$. Hence sequences of the form $t = (0 \ldots 0100\bar{0}\ldots)$ with $n$ initial 0's are not admissible. However, $\sigma^i(t) \prec K(f)$ for $i \ne n$ and $\sigma^n(t) = K(f)$.

The kneading sequence may be used to give a sufficient condition which holds in certain cases, at least.

**Theorem 18.12.** *Suppose $f$ is unimodal and $c$ is not periodic. If t is a sequence which satisfies $\sigma^n(t) \prec K(f)$ for all $n \ge 1$, then there exists $x \in I$ with $S(x) = t$. That is, $t \in \Sigma_f$.*

*Proof.* If $t = (00\bar{0}\ldots)$ or $(100\bar{0}\ldots)$, then we are done, since $S(0) = (00\bar{0}\ldots)$ and $S(1) = (100\bar{0}\ldots)$. So we may assume that $t \ne (00\bar{0}\ldots)$ or $(100\bar{0}\ldots)$. Define

$$L_t = \{x \in I | S(x) < t\}$$
$$R_t = \{x \in I | S(x) > t\}.$$

We will show below that both $L_t$ and $R_t$ are open in $I$. Since $0 \in L_t$ and $1 \in R_t$ (recall: $t \ne (00\bar{0}\ldots)$ or $(100\bar{0}\ldots)$ ), it follows that both $L_t$ and $R_t$ are non-empty. Since $L_t \cap R_t = \emptyset$, it thus follows that there is a non-empty closed set in $I$ with itinerary $t$. This completes the proof, except for the openness result.

We will only show that $L_t$ is open; the proof for $R_t$ is similar. We begin with the following observation. Let $s = (s_0 s_1 s_2 \ldots)$ and suppose $s_i \neq C$ for $i = 0, \ldots, n$. Then $\{x \in I | s_i(x) = s_i \text{ for } i = 0, \ldots, n\}$ is open in $I$. Indeed, if $S(y) = s$, there are open neighborhoods $W_i$ of $y$ such that $f^i(W_i)$ lies on the same side of $c$ as $f^i(y)$. The intersection of these neighborhoods yields a neighborhood of $y$ with the desired property.

Now let $z \in L_t$ and suppose $S(z) = s = (s_0 s_1 s_2 \ldots) < t$. Since $s \neq t$, $s$ and $t$ have discrepancy $n$, for some $n \geq 0$, i.e., $s_n \neq t_n$. There are then two cases: $t_n = C$ and $t_n \neq C$. If $t_n = C$, then it follows that $\sigma^{n+1}(t) = K(f)$ which contradicts our assumption. Hence we must have $t_n \neq C$. Let us assume that $t_n = 1$; the case $t_n = 0$ is similar. If $s_n = 0$, then we again invoke our preliminary observation to conclude that $L_t$ is open.

Thus the only remaining possibility is that $s_n = C$; that is, $K(f) = (s_{n+1}, s_{n+2}, s_{n+3}, \ldots)$. In this case, there exists $\alpha > 0$ such that $s_{n+\alpha} \neq t_{n+\alpha}$, for otherwise we would have $\sigma^{n+1}(t) = (s_{n+1}, s_{n+2}, s_{n+3}, \ldots) = K(f)$. Since $c$ is not periodic, we must have $s_{n+i} \neq C$ for all $i > 0$. Let $W$ be the neighborhood of $z$ such that, if $x \in W$, then

$$S(x) = (s_0 \ldots s_{n-1} * s_{n+1} \ldots s_{n+2} \ldots)$$

where $*$ may be $0, 1$, or $C$. That is, $W$ consists of all points whose itineraries agree with that of $z$ up to the $(n + \alpha)$-entry, except possibly in the $n^{th}$ slot. Hence $W$ is open. Clearly, we have

$$(s_0 \ldots s_{n-1} * \ldots) \preceq (s_0 \ldots s_{n-1} t_n \ldots) = (t_0 \ldots t_{n-1} t_n \ldots)$$

since $s \prec t$. Consequently, if $x \in W$, then $S(x) \prec t$ and we are done.

q.e.d.

**Remarks.**

**1.**   The condition that $\sigma^i(s) \prec K(f)$ for $i \geq 1$ cannot be weakened. Obviously, $(100\overline{0} \ldots) \succ t$ for any sequence $t$, and $S(1) = (1000 \ldots)$ so that this sequence is admissible.

**2.**   The assumption that $c$ not be periodic can be eliminated. To accomplish this, however, we must exclude one additional sequence. Suppose $K(f) = (\alpha_1 \ldots \alpha_n C \alpha_1 \ldots \alpha_n C \ldots)$. If there is an even number of 1's among the $\alpha_i$'s then the sequence $(\alpha_1 \ldots \alpha_n \, 0 \, \alpha_1 \ldots \alpha_n 0 \ldots)$ is less than $K(f)$. However,

$$\{x | S(x) = \alpha_1 \ldots \alpha_n \, 0 \, \alpha_1 \ldots \alpha_n 0 \ldots\}$$

is not closed in $I$. This contradicts the results in the proof of Theorem 18.12. As an example, note that if $f(x) = F_2(x) = 2x(1-x)$, then $K(f) = (CCC\ldots)$. Every point in $[0, c)$ has itinerary $(000\ldots)$, so that

$$\{x|S(x) = (000\ldots)\}$$

is not closed.

Thus we must modify the hypotheses of Theorem 18.12 by assuming that $\sigma^n(t) < (\alpha_1 \ldots \alpha_n \ 0 \ \alpha_1 \ldots \alpha_n \ 0 \ldots)$ for all $n \geq 1$. In case the number of 1's among the $\alpha_i$'s is even, then we must assume that $\sigma^n(t) < (\alpha_1 \ldots \alpha_n \ 1 \ \alpha_1 \ldots \alpha_n \ 1 \ldots)$. With this proviso, Theorem 18.12 then holds in case $c$ is periodic. We leave the details to the reader (see Exercise 2).

**Exercises**

**1.** Let $F_\mu(x) = \mu x(1-x)$. List all possible itineraries for $F_\mu$ when $\mu = 2$ and $\mu = 3$.

**2.** Prove Theorem 18.12 in the case where $c$ is periodic, incorporating the exclusions mentioned in Remark 2 above.

**3.** *Renormalization and the Kneading Theory.* Recall that in the last section we defined the renormalization $Rf$ of a unimodal map $f$. $Rf$ was defined if $f$ admitted a fixed point $p$ with $f'(p) < 0$. Let $K(f) = (\alpha_1 \alpha_2 \alpha_3 \ldots)$.
   a. Show that if $Rf$ is defined and is a unimodal map, then $K(f) = (1\alpha_2 \ 1\alpha_4 \ 1\alpha_6 \ldots)$, i.e., $\alpha_{2n+1} = 1$ for all $n$.
   b. Define $\hat{\alpha}_j$ to be 0 if $\alpha_j$ is 1 or 1 if $\alpha_j$ is 0. So $\hat{\alpha}_j \neq \alpha_j$. Show that $K(Rf) = (\hat{\alpha}_2 \hat{\alpha}_4 \hat{\alpha}_6 \ldots)$.

Thus we may define a renormalization operator $R$ on regular sequences by defining $R(\alpha_1 \alpha_2 \alpha_3 \ldots) = (\hat{\alpha}_2 \hat{\alpha}_4 \hat{\alpha}_6 \ldots)$. Intuitively, this operator eliminates every odd entry in the sequence and changes every even entry.
   c. Assuming that both $Rf$ and $R^2 f$ are unimodal maps, show that $\alpha_2 = \alpha_6 = \alpha_{10} = \ldots = 0$, i.e., that $\alpha_{4n+2} = 0$.
   d. Assuming that $R^i f$ is a unimodal map for $i \leq n$, prove that the entries $\alpha_j$ are determined, where $j = 2^N k + 2^{N-1}$.
   e. Conclude that all entries of $K(f)$ are determined if $R^i f$ is a unimodal map for all $i$. What is $K(f)$?

We now assume that $R^i f$ is a unimodal map for all $i$. For a repeating sequence $(s_0 \ldots s_n \ \overline{s_0 \ldots s_n} \ldots)$ we will adopt the shorthand notation $\overline{(s_0 \ldots s_n)}$. We introduce some special repeating sequences

$$\tau_0 = (\overline{1})$$

$$\tau_1 = (\overline{10})$$

$$\tau_2 = (\overline{1011})$$

$$\tau_3 = (\overline{10111010}).$$

Inductively, $\tau_{j+1}$ is obtained from $\tau_j$ by doubling $\tau_j$ and changing the last entry

f. Prove that $\tau_j$ has period $2^j$.

g. Prove that $R(\tau_{j+1}) = \tau_j$ if $j \geq 0$.

h. Let $\tau_\infty = \lim\limits_{j\to\infty} \tau_j$. $\tau_\infty$ is not a repeating sequence. Prove that $R(\tau_\infty) = \tau_\infty$, so that $\tau_\infty$ is a "fixed point" for the renormalization operator on sequences.

i. Prove that if $R^i f$ is a unimodal map for all $i$, then $K(f) = \tau_\infty$.

Thus the set of unimodal maps which can be renormalized infinitely often all share the same kneading sequence $\tau_\infty$. We will meet this sequence as well as the $\tau_j$'s again when we discuss the genealogy of periodic points in the next section.

# §1.19 GENEALOGY OF PERIODIC POINTS OF UNIMODAL MAPS

The kneading theory provides a powerful tool for studying the dynamics of a unimodal map. In this section, we will investigate the ramifications of this theory on the structure of the set of periodic points of a unimodal map. When such a map also has negative Schwarzian derivative, we will see that there are restrictions on the number and type of periodic points that can arise. In particular, we will give an almost complete answer to the question we asked earlier: how do maps like the quadratic map $F_\mu(x) = \mu x(1 - x)$ make the transition from simple to chaotic dynamics.

Unlike the quadratic map $F_\mu$ with $\mu$ large where there is a unique periodic point corresponding to any repeating itinerary, the situation for general unimodal maps is quite different. There may be more than one periodic point which shares the same itinerary. For example, when $1 < \mu < 2$, $F_\mu$ has two fixed points, both of which have itinerary $(000\ldots)$. Moreover, whenever $f$ has an attracting periodic orbit, there is generally an entire interval of points which share the same itinerary. However, as the following Theorem shows, there is always at least one periodic point which corresponds to a given repeating itinerary.

**Theorem 19.1.** *Let* $s = (s_0 \ldots s_{n-1} \, \overline{s_0 \ldots s_{n-1}} \ldots)$ *be an $f$-admissible repeating sequence which satisfies $\sigma^i(s) \prec K(f)$ for all $i$. Then there exists a periodic point $p$ of period either $n$ or $2n$ with $S(p) = s$.*

*Proof.* Let us assume that $K(f)$ is not repeating. As in the proof of Theorem 18.11, special arguments are necessary in this case for one of the admissible sequences. We leave the details to the reader (see Exercise 1). Since $\sigma^i(s) \prec K(f)$ for all $i$, Theorem 18.11 shows that

$$J = \{z \in I | S(z) = s\}$$

is a non-empty closed interval in $I$. Observe that $f^n(J) \subset J$ since all points in $f^n(J)$ have itinerary s as well.

Now if $J$ is a single point, it follows that this point is periodic and has the desired itinerary. If $J = [a, b]$ with $a \neq b$, then we argue as follows. If $x \in [a, b]$, then $f^i(x) \neq c$ for any $i$ since $S(f^{i+1}(x)) = \sigma^{i+1}(s) \prec K(f)$. Consequently, $(f^n)'(x) \neq 0$ for any $x \in [a, b]$. Thus $f^n$ either increases or decreases on $J$. Moreover, $f^n$ preserves the endpoints of $J$. This follows from the fact that, if $f^n(a)$ belongs to the interior of $(J)$, then there exists an open interval $N$ about $a$ having the following properties. If $x \in N$ then

    1. $f^i(x) \neq c$ for any $i < n$
    2. $f^n(x) \in J$.

Consequently, all points in $N$ have itinerary s and so $J$ is larger than $[a, b]$, contrary to our assumption.

If $f^n$ is increasing on $J$, it therefore follows that both $a$ and $b$ are periodic with period $n$ and itinerary s. If $f^n$ is decreasing, it is clear that $a$ and $b$ have period $2n$ with $f^n(a) = b$ and $f^n(b) = a$. By the Intermediate Value Theorem, there exists $z$ between $a$ and $b$ with $f^n(z) = z$. This completes the proof.

q.e.d.

We remark that this result is still true under the slightly weaker hypothesis that $\sigma^i(s) \preceq K(f)$. In this case $\{z \in I | S(z) = s\}$ need no longer be a closed set. We also note that the case of a periodic point with period $2n$ but whose itinerary repeats with period $n$ can actually happen. Indeed, in the quadratic family, just after the first period doubling, both period two points lie close to the fixed point which spawned them. Hence their itineraries are both $(11\overline{1}\ldots)$.

If we assume in addition that a unimodal map has negative Schwarzian derivative, then this puts strong restrictions on the number of periodic points which can share the same itinerary.

**Corollary 19.2.** *Suppose $Sf < 0$. Let s be a non-zero repeating sequence with period $n$ which satisfies $\sigma^i(\mathrm{s}) \leq K(f)$ for all $i$. Then there exists at most two periodic orbits with itinerary s.*

*Proof.* By the previous Theorem, any periodic point with itinerary s is fixed by $f^{2n}$. Hence we suppose that there are three distinct periodic orbits with itinerary $n$. For simplicity, let us assume that each of these orbits actually has period $n$. The more general case is handled similarly (see Exercise 2).

Let $x_1 < x_2 < x_3$ be three consecutive points fixed by $f^n$ and with the same itinerary. There cannot be a critical point for $f^n$ in $[x_1, x_3]$, for all points in this interval must have itinerary s by our ordering. By Proposition 11.3, $Sf^n < 0$. If two of the $x_i$ are attracting (even weakly attracting from one side), then the proof of Theorem 11.4 shows that they must either attract a critical point of $f$ or else have infinite basin of attraction. This latter possibility cannot occur, since $0 < x_i < 1$ and neither 0 nor 1 lie in the basin of attraction. But then both of the points must attract a critical point of $f$, which is again impossible since $f$ is unimodal.

The only other possibility is that one of the $x_i$ is attracting (from both sides). Clearly, this point must be $x_2$. Then, however, $(f^n)'$ has a positive local minimum between $x_1$ and $x_3$. This contradicts Lemma 11.5 and establishes the result.

q.e.d.

**Corollary 19.3.** *Suppose $Sf < 0$. Let $\mathrm{s} = (s_0 \ldots s_{n-1}\overline{s_0 \ldots s_{n-1}} \ldots)$ be a regular repeating sequence with*

$$I_{n-1}(\mathrm{s}) = \sum_{i=0}^{n-1} s_i$$

*an odd number and $\sigma^i(\mathrm{s}) \preceq K(f)$ for all $i$. Then*

    *1. There exists a unique periodic point $z_{\mathrm{s}}$ for $f$ of period $n$ and with itinerary s.*

    *2. If, in addition, $(f^n)'(z_{\mathrm{s}}) < -1$ and $K(f) = \sigma^i(\mathrm{s})$ for some $i$, then there exists a pair of periodic points of period $2n$ for $f$ which has itinerary s as well.*

*Proof.* For part 1, we suppose that $x$ is a periodic point of period $n$ with itinerary s. Since $\sum_{i=0}^{n-1} s_i$ is odd, it follows that $(f^n)'(x) < 0$. Consequently, if $f$ admits two periodic points with itinerary s, not both can have period $n$. This completes the proof of part 1; the second part is left to the reader (see Exercise 3).

q.e.d.

A good illustration of this Corollary is provided by the quadratic map $F_\mu$. For $1 < \mu < 2$, this map has 2 fixed points, both with itinerary $(00\overline{0}\ldots)$. At $\mu = 2$, one of these fixed points becomes the critical point. Thereafter, for $2 < \mu < 3$, this fixed point has negative derivative and a different itinerary $(11\overline{1}\ldots)$. As $\mu$ increases through 3, a period-doubling bifurcation occurs. As we noted above, just after $\mu = 3$, the new period two orbit shares the itinerary $(11\overline{1}\ldots)$.

Now let us turn to the question of how a unimodal map progresses from finitely many to infinitely many distinct periodic points. To answer this, we combine our previous results on bifurcation theory (§1.12) and negative Schwarzian (§1.11) with the kneading theory. The result is a nearly complete topological or qualitative picture of the transition from simple to complicated dynamics.

For the moment, we will deal with the periodic point structure of a fixed unimodal map $f$. Later we will turn to families of such maps. As we will deal with regular repeating itineraries in this section, we will drop the "tail" of the itineraries and work with a finite sequence of 0's and 1's instead. That is, $(s_0 s_1 \ldots s_n)$ will denote the infinite repeating itinerary $(s_0 s_1 \ldots s_n \overline{s_0 \ldots s_n} \ldots)$.

**Definition 19.4.** Let s be a repeating itinerary. Let $M(\mathrm{s})$ denote the maximal sequence in the orbit of s, i.e., $M(\mathrm{s}) = \sigma^j(\mathrm{s})$ where $\sigma^j(\mathrm{s}) \succeq \sigma^i(\mathrm{s})$ for all $i$.

We need some notation. Let $\mathrm{s} = (s_0 \ldots s_n)$ and $\mathrm{t} = (t_0 \ldots t_k)$ be repeating sequences. We denote the concatenation of these two sequences by $\mathrm{s} \cdot \mathrm{t} = (s_0 \ldots s_n\, t_0 \ldots t_k)$. We also write $\hat{\mathrm{s}} = (s_0 \ldots s_{n-1} \hat{s}_n)$ where $\hat{s}_n = 1$ if $s_n = 0$ or $\hat{s}_n = 0$ if $s_n = 1$. That is, $\hat{\mathrm{s}}$ is the same sequence as s, except that the last entry has been changed.

We will consider some special repeating sequences. Define

$$\tau_0 = (1)$$
$$\tau_1 = (10)$$
$$\tau_2 = (1011)$$
$$\tau_3 = (1011\ 1010)$$

and, inductively,

$$\tau_{j+1} = \tau_j \cdot \hat{\tau}_j.$$

Finally, we set

$$\tau_\infty = \lim_{n \to \infty} \tau_n$$
$$= (1011\ 1010\ 1011\ 1011\ldots).$$

Note that $\tau_\infty$ is a non-repeating sequence.

**Proposition 19.5.**

    *1. $\tau_j$ has prime period $2^j$.*
    *2. $\tau_j$ has an odd number of 1's*
    *3. $\tau_0 \prec \tau_1 \prec \tau_2 \prec \ldots$*

*Proof.* The proofs of 1 and 2 are straightforward. See Exercise 18.3. For 3, we write

$$\tau_j = (s_0 \ldots s_\alpha \nu).$$

If $\nu = 1$, then by 2, there is an even number of 1's among $(s_0, \ldots, s_\alpha)$. Consequently

$$\tau_j \succ \hat{\tau}_j = \tau_{j-1} \cdot \tau_{j-1} = \tau_{j-1}.$$

The argument is similar if $\nu = 0$.

q.e.d.

    The $\tau_j$'s play a special role in the transition to chaos: they are always the first periodic orbits to appear in any family of unimodal maps.

**Proposition 19.6.** $M(\tau_j) = \tau_j.$

*Proof.* We use induction on $j$. The cases $j = 0$ and $j = 1$ are clear. Let us assume that $M(\tau_{j-1}) = \tau_{j-1}$. We observe that

$$\hat{\tau}_j = \tau_{j-1} \cdot \tau_{j-1} = M(\tau_{j-1}) = \tau_{j-1} \prec \tau_j$$

by the previous proposition. Now suppose that $M(\tau_j) = \sigma^i(\tau_j)$ for some $i \neq 0$. If $1 \leq i < 2^{j-1}$, then we have

$$\sigma^i(\tau_j) = \sigma^i(\tau_{j-1}) \cdot \sigma^i(\hat{\tau}_{j-1}) \prec \tau_{j-1} \cdot \hat{\tau}_{j-1} = \tau_j.$$

Similarly, if $2^{j-1} + 1 \leq i < 2^j$, we may write $\ell = i - 2^{j-1}$ and we have

$$\sigma^i(\tau_j) = \sigma^\ell(\hat{\tau}_{j-1}) \cdot \sigma^\ell(\tau_{j-1}) \prec \tau_{j-1} \cdot \hat{\tau}_{j-1} = \tau_j$$

since

$$M(\hat{\tau}_{j-1}) = M(\tau_{j-2}) = \tau_{j-2} \prec \tau_{j-1}.$$

Finally, if $i = 2^{j-1}$, we have

$$\sigma^i(\tau_j) = \hat{\tau}_{j-1} \cdot \tau_{j-1} \prec \tau_{j-1} \cdot \hat{\tau}_{j-1}$$

since $\hat{\tau}_{j-1} \prec \tau_{j-1}$ as we observed above.

q.e.d.

The next proposition shows that the periodic points with itinerary $\tau_j$ occur before periodic points with other itineraries for a unimodal map.

**Proposition 19.7.** *Let* **t** *be any regular repeating sequence with* **t** $\neq (0)$ *or* $\tau_j$ *for any* $j$. *Then* $M(\mathbf{t}) \succ \tau_j$.

*Proof.* Since $\mathbf{t} \neq (0)$ or $(1) = \tau_0$, it follows that there exists $i \geq 0$ such that

$$\sigma^i(\mathbf{t}) = (10\ldots) \succ (1) = \tau_0.$$

Consequently, $M(\mathbf{t}) \succ \tau_0$.
  Now suppose that

$$\tau_{j-1} \prec M(\mathbf{t}) \prec \tau_j.$$

We have

$$\tau_{j-1} = \tau_{j-1} \cdot \tau_{j-1} \prec M(\mathbf{t}) \prec \tau_{j-1} \cdot \hat{\tau}_{j-1}.$$

Since the only discrepancy in the above sequences occurs in the $2^j$-th slot, it follows that $M(\mathbf{t}) = \tau_{j-1}$ or $M(\mathbf{t}) = \tau_j$. This contradicts our assumption and completes the proof.

q.e.d.

We turn now to the consideration of families of unimodal maps $f_\lambda$, where the maps depend smoothly on the parameter $\lambda$. That is, the function $G(x, \lambda) = f_\lambda(x)$ is $C^\infty$ in both variables.

**Definition 19.8.** Let $f_\lambda$ be a family of unimodal maps with $\lambda_0 \leq \lambda \leq \lambda_1$. $f_\lambda$ is called a transition family if
  1. $f_{\lambda_0}(x) \equiv 0$ for all $x \in I$.
  2. When $\lambda = \lambda_1$, $K(f_\lambda) = (100\bar{0}\ldots)$.
  3. $Sf_\lambda < 0$ for all $\lambda$.

**Remarks.**

**1.**  Transition families are called full families by some authors.

**2.**  Condition 1 may be relaxed; all we really need is that $K(f_\lambda) = (00\bar{0}\ldots)$.

**Example 19.9.** The quadratic family $F_\mu(x) = \mu x(1-x)$ forms a transition family for $0 \le \mu \le 4$. Also, $S_\lambda(x) = \lambda \sin(\pi x)$ forms a transition family for $0 \le \lambda \le 1$.

Conditions 1 and 2 above guarantee that a transition family becomes dynamically complex as $\lambda$ increases: there are no dynamics at all when $\lambda = \lambda_0$, while $f_{\lambda_1}$ has at least one periodic point corresponding to any regular repeating itinerary. This last statement follows immediately from Theorem 19.1 and the kneading theory of §1.17.

Our previous results allow us to say more. For each $j$, as long as $K(f_\lambda) \succeq \tau_j$, there exists a unique periodic point in $[0,1]$ of period $2^j$ and with itinerary $\tau_j$. Let us denote this point by $\gamma_j(\lambda)$. Note that, since

$$(f^{2^j})'(\gamma_j(\lambda)) < 0,$$

the bifurcation theory of §1.12 guarantees that the $\gamma_j(\lambda)$ depend continuously on $\lambda$. If we plot the bifurcation diagram for the family $f_\lambda$, it follows that the $\gamma_j(\lambda)$ must lie on a continuous curve in the $x$-$\lambda$ plane.

The domain of definition of $\gamma_j$ can be extended somewhat. If $K(f_\lambda) = \tau_j$ and

$$(f^{2^j})'(\gamma_j) < -1,$$

then there exists a unique periodic orbit of period $2^{j+1}$ for $f$ which shares the itinerary $\tau_j$. See Corollary 19.3. We denote the largest of these points by $\gamma_{j+1}(\lambda)$. Thus $\gamma_{j+1}(\lambda)$ is defined for all $\lambda$ for which $K(f_\lambda) \ge \tau_j$ and $(f_\lambda^{2^j})'(\gamma_j) < -1$. Since $(f_\lambda^{2^{j+1}})'(\gamma_{j+1}) \ne 1$ for all $\lambda$ (Exercise 4), it follows that $\gamma_{j+1}(\lambda)$ is continuous for these values of $\lambda$ as well.

**Remark.** There is an intermediate kneading sequence $\nu$ between $\tau_j \tau_j$ and $\hat{\tau}_{j+1} = \tau_j \hat{\tau}_j$ for which the last entry is $C$. That is, if $K(f_\lambda) = \nu$, the critical point is periodic with period $2^{j+1}$. If we define $\gamma_{j+1}(\lambda) = f_\lambda(c)$ when $K(f_\lambda) = \nu$, then $\gamma_{j+1}$ is continuous here as well (Exercise 5).

Finally, we note that $\gamma_{j+1}(\lambda) \to \gamma_j(\lambda)$ as $\lambda$ approaches a value for which $(f_\lambda^{2^j})'(\gamma_j) = -1$. That is, $\gamma_j$ undergoes a period-doubling bifurcation at this value of $\lambda$ (Exercise 6). This means that the bifurcation diagram for $f_\lambda$ must be at least as complicated as that of Fig. 19.1.

Thus we see that a transition family has virtually no choice regarding the transition to chaotic dynamics. Such a family must follow the period-doubling route, at least until infinitely many periodic points have been born.

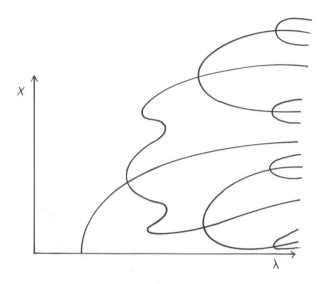

**Fig. 19.1.** A possible bifurcation diagram for a transition family.

Thus the qualitative picture of renormalization that we described in §1.17 is at least qualitatively correct.

**Remark.** Our assumptions do not eliminate the birth and subsequent death of one of the periodic points. This would happen if the kneading sequence of $f_\lambda$ first increased, then decreased, and finally increased again. The resulting bifurcation diagram is depicted in Fig. 19.2. Recent research has shown that this pathology does not occur in the quadratic family $F_\mu(x) = \mu x(1 - x)$.

   This is just the beginning of a long and detailed story. There are many other periodic points in a transition family besides the $\gamma_j$'s that we have discussed. However, the mechanism by which these other periodic points arise is similar to that described above. One may describe completely the "genealogy" of any periodic point in a transition family—where it is "born," which sequences are its (period-doubling) "ancestors," and which are its "descendants". We relegate these facts to the exercises (Exercises 9-13) but recommend them to the reader as a nice method to tie together many of the ideas in this entire Chapter.

   To summarize, a pair of periodic orbits is born in a saddle node bifurcation. This bifurcation may of course be degenerate, but at least two periodic orbits are produced. One of these orbits eventually becomes repelling with derivative $> 1$; the other orbit becomes attracting and eventually attains negative derivative. Thereafter, the period-doubling story unfolds. Infinitely

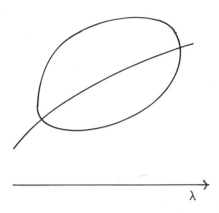

**Fig. 19.2.** The birth and death of a fixed point as $\lambda$ increases in a transition family.

many orbits successively bifurcate away as in the case of the $\gamma_j$, and all persist until the transition family reaches the stage where the map is topologically conjugate to the shift.

**Exercises**

**1.** Prove Theorem 19.1 in case $K(f)$ is repeating. Also show that, if the kneading sequence of a unimodal map $f$ is repeating, then there exists a periodic point for $f$ whose itinerary is the kneading sequence.

**2.** Prove Corollary 19.2 in case two of the three periodic orbits have period $2n$.

**3.** Suppose $z_s$ is periodic with period $n$ and itinerary s. Suppose $(f^n)'(z_s) < -1$ and $K(f) = s$. Prove that there exists a pair of periodic points of period $2n$ for $f$ which has itinerary s.

In the next three exercises, suppose that $f_\lambda$ is a transition family of maps.

**4.** Suppose $K(f_\lambda) = \tau_j$ and $(f^{2^j})'(\gamma_j) < -1$. By the results of this section, the periodic point $\gamma_{j+1}(\lambda)$ exists. Prove that

$$(f^{2^{j+1}})'(\gamma_{j+1}) \neq 1.$$

**5.** Prove that $\gamma_{j+1}(\lambda)$ is continuous at the $\lambda$-value for which $\gamma_{j+1}(\lambda) = f_\lambda(c)$.

**6.** Prove that $\gamma_{j+1}(\lambda)$ approaches $\gamma_j(\lambda)$ as $\lambda$ approaches a value for which $(f^{2^j})'(\gamma_j) = -1$.

The following exercises describe the "genealogy" of any periodic point in a transition family by describing how the periodic point is "born" and which sequences are "related" to it.

**7.**   Let $s = (s_1 \ldots s_n)$ be any regular repeating sequence of period $n$. Prove that there exists a continuous curve $\gamma_s(\lambda)$ defined for $\lambda_s \leq \lambda \leq \lambda_1$ such that $f_\lambda^n(\gamma_s(\lambda)) = \gamma_s(\lambda)$ and $S(\gamma_s(\lambda_1)) = s$.

**8.**   Prove that $\gamma_s(\lambda)$ may be extended to a point $\lambda_s$ at which

$$(f_{\lambda_s}^n)'(\gamma_s(\lambda_s)) = 1.$$

The parameter value $\lambda_s$ is called the "birthplace" of s. We now assume that $M(s) = s$ and that $\sum_{i=1}^{n} s_i$ is odd.

**9.**   Prove that if s is of the form $s = u\hat{u}$, then

   a. $\mathbf{u} > \hat{\mathbf{u}}$
   
   b. $\gamma_s(\lambda_0)$ is born in a period-doubling bifurcation at $\lambda = \lambda_s$ along the family $\gamma_u(\lambda)$.
   
   c. Let $\mathbf{t} = \hat{\mathbf{u}}\mathbf{u}$. Prove that $\lim_{\lambda \to \lambda_s} \gamma_t(\lambda) = \gamma_s(\lambda_s)$ so that t and s are related. The sequence u is called an "ancestor" of s and t. Similarly, s and t are "descendants" of u.

**10.**   Using the notation of the previous exercise,

   a. Prove that, if s is not of the form $\mathbf{u}\hat{\mathbf{u}}$ for some sequence u, then $\gamma_s(\lambda)$ is born in a saddle node bifurcation at $\lambda = \lambda_s$.
   
   b. Let $\mathbf{t} = \hat{\mathbf{s}}$. Prove that

$$\lim_{\lambda \to \lambda_s} \gamma_t(\lambda) = \gamma_s(\lambda_s)$$

   Thus ŝ is related to s via this bifurcation.

**11.**   List the six maximal period 5 orbits. Which of these orbits are related?

**12.**   Prove that there are 9 orbits of prime period 6. Which of these orbits are born in saddle nodes and which are born in period doublings?

**13.**   Prove that there are 30 orbits of prime period 8. Identify those that are related to the repeating sequence (1) via period doublings. Which others arise out of period doublings?

The following exercises apply to the family of tent maps of the form

$$T_\mu(x) = \begin{cases} \mu x & 0 \leq x \leq 1/2 \\ \mu(1-x) & 1/2 < x \leq 1. \end{cases}$$

**14.** Prove that $T_\mu$ has a unique fixed point and no other periodic points if $0 < \mu < 1$. Prove that $T_\mu$ has periodic points of period $2^j$ for each $j$ if $\mu > 1$.

**15.** Consider $T_{\sqrt{2}}$. Show that the "critical point" $x = 1/2$ is eventually fixed for this map. Prove that there is a subinterval of $[0, 1]$ on which $T_{\sqrt{2}}$ is topologically conjugate to $T_2$.

**16.** Prove that $T_{\sqrt{2}}$ has periodic points of period $2k$ for any $k > 0$, but no periodic points of odd period $> 1$.

**17.** Prove that, if $\mu > \sqrt{2}$, then $T_\mu$ has a periodic point of period 3 and hence of all periods.

**18.** Prove that there is an interval on which $T_{\sqrt{2}}$ is chaotic.

## FOR FURTHER READING:

There are a number of advanced texts in Dynamical Systems which extend or complement the material presented here. These texts also provide good references to the current research literature. Among them are:

Collet, P. and Eckmann, J.-P. *Iterated Maps of the Interval as Dynamical Systems.* Birkhäuser, Boston, 1980.

This text delves more deeply into the kneading theory than we do. It also emphasizes the renormalization theory, and gives an overview of the work of Feigenbaum in this area. It also summarizes the recent important work of Guckenheimer, Misiurewicz, Jonker, Rand and others which aims at classifying one-dimensional maps.

Guckenheimer, J. and Holmes, P. *Nonlinear Oscillations, Dynamical Systems, and Bifurcations of Vector Fields.* Springer-Verlag, New York, 1983.

As the title suggests, the scope of this text is much wider than Chapter One. It is aimed primarily at graduate students, and stresses continuous dynamical systems (vector fields) and applications. There is considerable overlap in our treatment of bifurcations, circle maps, and the kneading theory, but in general this text is most suitable for an advanced course in Dynamical Systems.

Nitecki, Z. *Differentiable Dynamical Systems.* M.I.T. Press, Cambridge, Mass., 1971.

Although the principal emphasis in this text is on higher dimensional systems, there is an introductory section on circle maps which extends the results in §1.14 − 15.

We have emphasized discrete dynamical systems in this chapter. However, many of the dynamical systems which occur in applications are continuous systems, such as vector fields or ordinary differential equations. The following texts provide introductory treatments of these systems, which should be more accessible once the fundamental concepts of discrete dynamics are mastered. Both are suitable for advanced undergraduate and beginning graduate courses in Dynamical Systems.

Hirsch, M.W. and Smale, S. *Differential Equations, Dynamical Systems, and Linear Algebra.* Academic Press, New York, 1974.

Arnol'd, V.I. *Ordinary Differential Equations.* M.I.T. Press, Cambridge, Mass., 1973.

One major topic that we have omitted in this chapter is the Ergodic Theory of one-dimensional maps. A quick introduction to a variety of topics in this field is contained in the following text.

Sinai, J.G. *Introduction to Ergodic Theory.* Princeton University Press, Princeton, New Jersey, 1976.

An interesting pictorial approach to Dynamical Systems is contained in the series of monographs:

Abraham, R. and Shaw, C. *Dynamics: The Geometry of Behavior. Part One: Periodic Behavior. Part Two: Chaotic Behavior.* Aerial Press, Santa Cruz, Calif., 1982.

# Chapter Two

# Higher Dimensional Dynamics

In this chapter, we investigate dynamical phenomena which are higher dimensional in nature. In accordance with our desire to keep the presentation as simple as possible, we will treat only two- and three-dimensional dynamical systems. Higher dimensional systems are, of course, important but there are relatively few dynamical phenomena that are currently understood which occur in dimensions four or more and which are not already present in dimensions two and three. Most of what we say goes over to higher dimensions without any difficulty.

One of the main differences between higher dimensional systems and those treated in the previous chapter is the possibility of both expansion and contraction in the same invariant set. We will illustrate this with three important examples whose dynamics will dominate most of this chapter. They are the horseshoe map, the hyperbolic toral automorphisms and the attractors. Each of these maps illustrate a different higher dimensional phenomenon, but they all share several fundamental properties. The horseshoe map is the higher dimensional analogue of the quadratic map (for large $\mu$-values) which occupied so much of our attention in Chapter One. Here we see that the addition of both contracting and expanding directions in the same map presents no additional difficulties; a minor modification to the symbolic dynamics of Chapter One enables us to analyze these types of maps completely.

The hyperbolic toral automorphisms are quite different. They are chaotic everywhere. Also the expansion and contraction occurs at every point in the space, not just on a Cantor set. To handle this situation we introduce a modified type of symbolic dynamics generated by a Markov partition.

Finally, attractors present another facet of higher dimensional dynamics. Basically, an attractor is a set toward which most other points tend under iteration. As we shall see, the dynamics on the attractor itself may be quite chaotic, so that iteration of virtually any point eventually leads to seemingly random behavior. Using the branched manifold construction of Williams, we will nevertheless be able to describe this situation satisfactorily.

A number of themes which arose in Chapter One will return in slightly different format in this chapter. For example, a new type of bifurcation occurs in higher dimensions, namely the Hopf bifurcation. This bifurcation occurs when the derivative at a fixed point is a rotation, i.e., the multipliers are complex of modulus one.

As we mentioned above, the main difference between one- and higher dimensional systems is the possibility of both expansion and contraction at the same point in the latter case. We illustrate what this means for linear maps in §2.2, when we introduce the notion of stable and unstable sets. Each of the three main examples also feature stable and unstable sets through each point. So this sets the stage for the general concept of a stable and unstable manifold, which we discuss in §2.6. Finally, in §2.9, we consolidate many of these ideas in a section that consists almost entirely of Exercises. All of these Exercises deal with a single family of maps of the plane, the Hénon maps. As we will see, this family is the natural generalization of the quadratic family that played such a prominent role in the previous chapter.

This chapter assumes a slightly more advanced mathematical background on the part of the student than the previous chapter. The main new requirement is Linear Algebra. We assume throughout that the reader is familiar with such concepts as linear transformations, matrices, eigenvalues, and eigenvectors. As we work primarily in dimensions two and three, an elementary course in Linear Algebra which emphasizes these dimensions should be sufficient. We also use a number of concepts from multi-dimensional calculus. Many of these topics are reviewed in the next section.

## §2.1 PRELIMINARIES

For the study of higher dimensional dynamical systems, the most important new ingredients are techniques from linear algebra. Here we review some of the standard techniques and introduce a few more advanced topics. Later we review some multi-dimensional calculus.

We denote Euclidean $n$-dimensional space by $\mathbf{R}^n$. Elements of $\mathbf{R}^n$ are vectors which we will write either in column form

$$\mathbf{x} = \begin{pmatrix} x_1 \\ x_2 \\ \vdots \\ x_n \end{pmatrix}$$

or as a row vector $\mathbf{x} = (x_1, x_2, \ldots, x_n)$ where $x_i \in \mathbf{R}$. As our interest is mainly in low dimensional systems, $n$ will usually be 2 or 3.

**Definition 1.1.** A map $L: \mathbf{R}^n \to \mathbf{R}^n$ is *linear* if $L(\alpha\mathbf{v} + \beta\mathbf{v}) = \alpha L(\mathbf{v}) + \beta L(\mathbf{w})$, where $\alpha, \beta \in \mathbf{R}$ and $\mathbf{v}, \mathbf{w} \in \mathbf{R}^n$.

Linear maps provide the basic local models for higher dimensional systems. Recall that an $n \times n$ matrix $A$ is a square of real or complex numbers of the form

$$A = (a_{ij}) = \begin{pmatrix} a_{11} & \cdots & a_{1n} \\ \vdots & & \vdots \\ a_{n1} & \cdots & a_{nn} \end{pmatrix}.$$

Here, $a_{ij}$ denotes the entry in the $i^{th}$ row and $j^{th}$ column of $A$. We need a few concepts from matrix algebra. If $\mathbf{x} \in \mathbf{R}^n$ and $A = (a_{ij})$ is $n \times n$, then the product $A\mathbf{x} \in \mathbf{R}^n$ is the vector given by

$$A\mathbf{x} = \begin{pmatrix} a_{11}x_1 + \ldots + a_{1n}x_n \\ a_{21}x_1 + \ldots + a_{2n}x_n \\ \vdots \\ a_{n1}x_1 + \ldots + a_{nn}x_n \end{pmatrix}.$$

If $B = (b_{ij})$ is another $n \times n$ matrix, then the product $A \cdot B$ is a new $n \times n$ matrix $(c_{ij})$ given by

$$c_{ij} = \sum_{k=1}^{n} a_{ik} b_{kj}.$$

Note that $B \cdot A = (d_{ij})$ where

$$d_{ij} = \sum_{k=1}^{n} b_{ik} a_{kj}$$

so that $A \cdot B \neq B \cdot A$ in general.

**Example 1.2** Let

$$A = \begin{pmatrix} 1 & 2 \\ 1 & 1 \end{pmatrix} \qquad B = \begin{pmatrix} 1 & 2 \\ 3 & 4 \end{pmatrix}.$$

Then

$$A \cdot B = \begin{pmatrix} 7 & 10 \\ 4 & 6 \end{pmatrix} \qquad B \cdot A = \begin{pmatrix} 3 & 4 \\ 7 & 10 \end{pmatrix}.$$

There is a close relationship between linear maps and the algebra of matrices. Indeed, let $\mathbf{e}_i, \ldots, \mathbf{e}_n$ denote the standard basis of $\mathbf{R}^n$, i.e., $\mathbf{e}_j$ is the vector whose entries are all zero except the $j^{th}$, which is one. $L$ is completely determined by what it does to the $\mathbf{e}_j$, by linearity. That is, if $L(\mathbf{e}_j) = \mathbf{v}_j$ for $j = 1, \ldots, n$, then we know $L(\mathbf{v})$ for all vectors $\mathbf{v} \in \mathbf{R}^n$.

Indeed, we may write $\mathbf{v} = \sum_{j=1}^{n} \alpha_j \mathbf{e}_j$ where $\alpha_j \in \mathbf{R}$. That is

$$\mathbf{v} = \begin{pmatrix} \alpha_1 \\ \vdots \\ \alpha_n \end{pmatrix}.$$

Hence $L(\mathbf{v}) = \Sigma \alpha_j (L(\mathbf{e}_j)) = \Sigma \alpha_j \mathbf{v}_j$ is completely determined.

Let $A$ be the $n \times n$ matrix whose columns are $\mathbf{v}_1, \ldots, \mathbf{v}_n$. We write

$$A = [\mathbf{v}_1, \ldots, \mathbf{v}_n].$$

Then it follows immediately that

$$L(\mathbf{x}) = A\mathbf{x}$$

for all $\mathbf{x}$ which shows that linear maps and matrices are intimately related. The matrix $A$ is called the *matrix representation of $L$* (in the standard basis).

**Example 1.3** Suppose $L(\mathbf{e}_1) = 2\mathbf{e}_1$ and $L(\mathbf{e}_2) = 3\mathbf{e}_2$. Then $L(\mathbf{x})$ is given by

$$L(\mathbf{x}) = \begin{pmatrix} 2 & 0 \\ 0 & 3 \end{pmatrix} \mathbf{x}.$$

Similarly, if $L(\mathbf{e}_1) = \mathbf{e}_1 + \mathbf{e}_2$ and $L(\mathbf{e}_2) = 2\mathbf{e}_1 + 3\mathbf{e}_2$, then

$$L(\mathbf{x}) = \begin{pmatrix} 1 & 2 \\ 1 & 3 \end{pmatrix} \mathbf{x}.$$

Composition of maps is most important in dynamical systems. For linear maps, composition is intimately related to matrix multiplication, as shown by the following proposition.

**Proposition 1.4.** *Let $L$ and $P$ be linear maps with matrix representations $A$ and $B$ respectively. Then $P \circ L(\mathbf{v}) = (B \cdot A)\mathbf{v}$ for all $\mathbf{v} \in \mathbf{R}^n$.*

A linear map $L$ is invertible if it is one-to-one and onto. In this case, $L$ has a unique inverse which we denote by $L^{-1}$. It is easy to check that $L^{-1}$ is also a linear map.

We denote the $n \times n$ identity matrix by $I_n$:

$$I_n = [\mathbf{e}_1, \ldots, \mathbf{e}_n] = \begin{pmatrix} 1 & 0 & \cdots & 0 \\ 0 & 1 & \cdots & 0 \\ \vdots & & \ddots & \\ 0 & 0 & \cdots & 1 \end{pmatrix}.$$

The inverse of an $n \times n$ matrix $A$ is the (unique) matrix $B$ which satisfies

$$A \cdot B = B \cdot A = I_n.$$

It is well known that $B$ exists iff det $A \neq 0$, where det$(A)$ is the determinant of $A$. We denote the inverse of a matrix $A$ by $A^{-1}$. Clearly, $L(\mathbf{x}) = A\mathbf{x}$ is invertible iff $A$ has an inverse, and $L^{-1}(\mathbf{x}) = A^{-1}\mathbf{x}$.

**Definition 1.5.** Let $L_1$ and $L_2$ be linear maps of $\mathbf{R}^n$. $L_1$ and $L_2$ are *linearly conjugate* if there is an invertible linear map $P$ such that

$$L_1 = P^{-1} \circ L_2 \circ P.$$

In terms of matrices, if $L_1(\mathbf{x}) = A_1\mathbf{x}$ and $L_2(\mathbf{x}) = A_2\mathbf{x}$ and $P(x) = G\mathbf{x}$, then we must have

$$A_1\mathbf{x} = (G^{-1} \cdot A_2 \cdot G)\mathbf{x}.$$

The matrices $A_1$ and $A_2$ are called *similar* when $A_1 = G^{-1}A_2G$.

**Definition 1.6.** Let $A$ be an $n \times n$ matrix. An eigenvalue of $A$ is a root of the characteristic polynomial of $A$ given by $p(\lambda) = \det(A - \lambda I)$. An eigenvector of $A$ associated to the eigenvalue $\lambda$ is a non-zero vector $\mathbf{v}$ for which

$$A\mathbf{v} = \lambda\mathbf{v}.$$

The multiplicity of $\lambda$ as an eigenvalue is the multiplicity of $\lambda$ as a root of the characteristic polynomial.

**Example 1.7.** If

$$A = \begin{pmatrix} 1 & -4 \\ -1 & 1 \end{pmatrix}$$

then $p(\lambda) = \lambda^2 - 2\lambda - 3$, so $A$ has eigenvalues $3, -1$. The eigenvector associated to $\lambda = 3$ is found by solving the system of equations

$$A\begin{pmatrix} x \\ y \end{pmatrix} = 3\begin{pmatrix} x \\ y \end{pmatrix}$$

or, equivalently,

$$x - 4y = 3x$$
$$-x + y = 3y.$$

This yields an eigenvector of the form

$$\mu\begin{pmatrix} 2 \\ -1 \end{pmatrix}$$

for any $\mu \neq 0$. Similarly, the vector

$$\mu\begin{pmatrix} 2 \\ 1 \end{pmatrix}$$

is an eigenvector of $A$ corresponding to $\lambda = -1$, for any $\mu \neq 0$.

**Remark.** If $A$ is a diagonal or upper triangular matrix, then the eigenvalues of $A$ are displayed along the diagonal.

**Example 1.8.** If

$$A = \begin{pmatrix} 0 & 1 \\ -1 & 0 \end{pmatrix}$$

then $A$ has eigenvalues $\pm i$. The eigenvector corresponding to the eigenvalue $i$ is given by the solution of the equations

$$y = ix$$
$$-x = iy.$$

That is, any vector of the form

$$\mu \begin{pmatrix} 1 \\ i \end{pmatrix}$$

is an eigenvector corresponding to the eigenvalue $i$ when $\mu \neq 0$.

The above examples show that a real matrix may have eigenvalues and eigenvectors that are complex. Such eigenvalues always occur in complex conjugate pairs, however. That is, if $\alpha + i\beta$ is an eigenvalue of a real matrix, then so is $\alpha - i\beta$. This is true since the characteristic polynomial has real coefficients.

**Proposition 1.9.** *If $L_1(\mathbf{x}) = A_1 \mathbf{x}$ and $L_2(\mathbf{x}) = A_2 \mathbf{x}$ are linearly conjugate, then $A_1$ and $A_2$ have the same eigenvalues.*

*Proof.* Recall that if $A$ and $B$ are $n \times n$ matrices, then

$$\det(AB) = \det A \cdot \det B.$$

Thus we have

$$\det(A_1 - \lambda I) = \det (G^{-1} A_2 G - \lambda I)$$
$$= \det (G^{-1} A_2 G - \lambda G^{-1} G)$$
$$= \det (G^{-1}[A_2 - \lambda I]G)$$
$$= \det (A_2 - \lambda I)$$

so that the characteristic polynomials of $A_1$ and $A_2$ are the same.

q.e.d.

To study the dynamics of a linear map, it is usually most effective to put the matrix representation of the map in simple form. For low dimensional systems, this simple form is given by the following theorem.

**Theorem 1.10.** *Let* $L: \mathbf{R}^3 \to \mathbf{R}^3$ *be linear with* $L(\mathbf{x}) = A\mathbf{x}$. *There exists a real* $3 \times 3$ *matrix* $G$ *such that* $G^{-1}AG$ *assumes one of the four forms*

$$1. \begin{pmatrix} \alpha & -\beta & 0 \\ \beta & \alpha & 0 \\ 0 & 0 & \lambda \end{pmatrix} \qquad 2. \begin{pmatrix} \lambda & 0 & 0 \\ 0 & \mu & 0 \\ 0 & 0 & \eta \end{pmatrix}$$

$$3. \begin{pmatrix} \lambda & 1 & 0 \\ 0 & \lambda & 0 \\ 0 & 0 & \mu \end{pmatrix} \qquad 4. \begin{pmatrix} \lambda & 1 & 0 \\ 0 & \lambda & 1 \\ 0 & 0 & \lambda \end{pmatrix}$$

*where all entries are real and* $\beta \neq 0$.

*Proof.* First assume that $A$ has a complex eigenvalue $\alpha + i\beta$. Since $A$ is a real matrix, $\alpha - i\beta$ is also an eigenvalue. The remaining eigenvalue $\lambda$ of $A$ is necessarily a real number. Let $\mathbf{w}$ be the eigenvector associated to $\alpha + i\beta$. Write $\mathbf{w} = \mathbf{v}_1 + i\mathbf{v}_2$ where $\mathbf{v}_i$ are real vectors. Since $A\mathbf{w} = (\alpha + i\beta)\mathbf{w}$, it follows that

$$A\mathbf{v}_1 = \alpha\mathbf{v}_1 - \beta\mathbf{v}_2$$
$$A\mathbf{v}_2 = \beta\mathbf{v}_1 + \alpha\mathbf{v}_2.$$

If $\mathbf{v}_3$ is the eigenvector associated to $\lambda$, then also

$$A\mathbf{v}_3 = \lambda\mathbf{v}_3.$$

Now let $G$ be the matrix which satisfies $G\mathbf{e}_j = \mathbf{v}_j$. That is, the columns of $G$ are just the vectors $\mathbf{v}_1, \mathbf{v}_2$, and $\mathbf{v}_3$. Then one may check easily that $G^{-1}AG$ assumes the desired form 1.

For the remaining cases, we first note that all of the eigenvalues are real. Let us assume that all eigenvalues are equal to $\lambda$. We first assume that there is a non-zero vector $\mathbf{v}$ which satisfies

$$(A - \lambda I)^3\mathbf{v} = \mathbf{0}$$

$$(A - \lambda I)^2\mathbf{v} \neq \mathbf{0}.$$

Let

$$\mathbf{w}_3 = (A - \lambda I)^2\mathbf{v}$$
$$\mathbf{w}_2 = (A - \lambda I)\mathbf{v}$$
$$\mathbf{w}_1 = \mathbf{v}.$$

Since $(A - \lambda I)\mathbf{w}_3 = \mathbf{0}$, we have $A\mathbf{w}_3 = \lambda\mathbf{w}_3$. Similarly, $A\mathbf{w}_2 = \lambda\mathbf{w}_2 + \mathbf{w}_3$ and $A\mathbf{w}_1 = \lambda\mathbf{w}_1 + \mathbf{w}_2$. Consequently, if $G = [\mathbf{w}_1, \mathbf{w}_2, \mathbf{w}_3]$, then $G^{-1}AG$ assumes form 3.

If $(A - \lambda I)^2 \mathbf{v} = 0$ for all $\mathbf{v}$, but there exists $\mathbf{v}$ with $(A - \lambda I)\mathbf{v} \neq 0$, then we argue as follows. Let $\mathbf{w}_2 = (A - \lambda I)\mathbf{v}$ and $\mathbf{w}_1 = \mathbf{v}$. Then $A\mathbf{w}_2 = \lambda \mathbf{w}_2$ and $A\mathbf{w}_1 = \lambda \mathbf{w}_1 + \mathbf{w}_2$. We claim that there exists a third vector $\mathbf{w}_3$ which satisfies $(A - \lambda I)\mathbf{w}_3 = 0$ but $\mathbf{w}_3$ and $\mathbf{w}_2$ are not collinear. If no such vector exists, then the null space of the linear map $A - \lambda I$ is one-dimensional. Hence the range space must be two-dimensional. But the range space of $A - \lambda I$ must in turn be in the null space of $A - \lambda I$, since we know that $(A - \lambda I)^2 \mathbf{v} = 0$ for all $\mathbf{v}$. That is, we have a contradiction.

Thus, as before, we let $G$ be the matrix which has columns $\mathbf{w}_1, \mathbf{w}_2$ and $\mathbf{w}_3$. One checks easily that $G^{-1}AG$ assumes form 2.

In case not all of the eigenvalues of $A$ are identical, the methods above produce forms 2 or 3 even more easily. We leave the details to the reader.

q.e.d.

**Corollary 1.11.** *Let $A$ be a $2 \times 2$ matrix. Then there exists a real matrix $G$ such that $G^{-1}AG$ assumes one of the three forms with $\beta \neq 0$*

$$1. \begin{pmatrix} \alpha & -\beta \\ \beta & \alpha \end{pmatrix} \qquad 2. \begin{pmatrix} \lambda & 1 \\ 0 & \lambda \end{pmatrix}$$

$$3. \begin{pmatrix} \lambda & 0 \\ 0 & \mu \end{pmatrix}.$$

We call the simple matrices in Theorem 1.6 and its Corollary the *standard form* for the linear map. Thus all linear maps in dimensions two and three are linearly conjugate to a map whose matrix representation is in standard form. We remark that these forms are usually called normal forms. However, this term has another connotation which we will encounter later. Thus we will adopt this non-standard terminology and hope that this causes no confusion later.

**Remark.** Note that the eigenvalues of a matrix in standard form are displayed on the diagonal, except in the case of a $2 \times 2$ block of the form

$$\begin{pmatrix} \alpha & -\beta \\ \beta & \alpha \end{pmatrix},$$

which has eigenvalues $\alpha \pm i\beta$.

For later use, we note that the standard forms with 1's above the diagonal can be modified by a linear conjugacy.

**Proposition 1.12.** *The linear map*

$$L(\mathbf{x}) = \begin{pmatrix} \lambda & 1 & 0 \\ 0 & \lambda & 1 \\ 0 & 0 & \lambda \end{pmatrix} \mathbf{x}$$

*is linearly conjugate to*

$$L_\epsilon(\mathbf{x}) = \begin{pmatrix} \lambda & \epsilon & 0 \\ 0 & \lambda & \epsilon \\ 0 & 0 & \lambda \end{pmatrix} \mathbf{x}$$

*for any* $\epsilon \neq 0$.

*Proof.* Let

$$S_\epsilon(\mathbf{x}) = \begin{pmatrix} \epsilon^2 & 0 & 0 \\ 0 & \epsilon & 0 \\ 0 & 0 & 1 \end{pmatrix}.$$

Then one computes easily that $S \circ L \circ S^{-1}$ assumes the desired form.

q.e.d.

Similarly, the linear maps with matrix representations

$$\begin{pmatrix} \lambda & 1 \\ 0 & \lambda \end{pmatrix} \quad \text{and} \quad \begin{pmatrix} \lambda & 1 & 0 \\ 0 & \lambda & 0 \\ 0 & 0 & \mu \end{pmatrix}$$

are linearly conjugate to those with representations

$$\begin{pmatrix} \lambda & \epsilon \\ 0 & \lambda \end{pmatrix} \quad \text{and} \quad \begin{pmatrix} \lambda & \epsilon & 0 \\ 0 & \lambda & 0 \\ 0 & 0 & \mu \end{pmatrix}$$

for $\epsilon \neq 0$.

The situation for general linear maps in higher dimensions is more or less similar to that described above. Since we will primarily work in two and three dimensions in the sequel, we will only state the basic facts without proof. These facts will not be used in the sequel, but are basic for any extension of our results to higher dimensions.

**Definition 1.13.** A matrix $A$ is a $\lambda$-Jordan block if it is of the form

$$A = \begin{pmatrix} \lambda & 1 & 0 & 0 & \cdots & 0 \\ 0 & \lambda & 1 & 0 & \cdots & 0 \\ 0 & 0 & \lambda & 1 & \cdots & 0 \\ 0 & 0 & 0 & \lambda & \cdots & 0 \\ & & & & \ddots & 1 \\ 0 & & \cdots & & & \lambda \end{pmatrix}.$$

That is, the matrix $A$ has $\lambda$'s on the diagonal, 1's above the diagonal, and 0's elsewhere.

**Definition 1.14.** A matrix $A$ is in Jordan form if

$$A = \begin{pmatrix} A_1 & & & \\ & A_2 & & \\ & & \ddots & \\ & & & A_k \end{pmatrix}$$

where each $A_i$ is a $\lambda_i$-Jordan block and all other entries are 0.

**Remarks.**

**1.** There exists a complex matrix $G$ such that, for any matrix $A$, $G^{-1}AG$ is in Jordan form.

**2.** The $\lambda$'s in the Jordan blocks may be complex.

**3.** The eigenvalues of a matrix in Jordan form are displayed along the diagonal.

When the $\lambda$'s in a $\lambda$-Jordan block are complex, then the Jordan form differs from the forms introduced in Theorem 1.10 and its Corollary. However, we can again make a linear conjugacy to put it in real Jordan form where

$$J = \begin{pmatrix} A & I & & & \\ & A & I & & \\ & & A & I & \\ & & & \ddots & \\ & & & & A \end{pmatrix}$$

where $A$ is a $2 \times 2$ matrix of the form

$$\begin{pmatrix} \alpha & -\beta \\ \beta & \alpha \end{pmatrix}$$

and $I$ is the identity $2 \times 2$ matrix. One simply breaks the complex vectors into their real and imaginary parts to achieve this form. We omit the tedious details.

Let us now review several facts from advanced calculus. Let $F: \mathbf{R}^2 \to \mathbf{R}^2$ be a map. We will most often write such a map in the form

$$x_1 = f_1(x, y)$$
$$y_1 = f_2(x, y)$$

where the vector $(x_1, y_1)$ is the image of $(x, y)$ under $F$. In vector notation, we may also write

$$\begin{pmatrix} x_1 \\ y_1 \end{pmatrix} = F\begin{pmatrix} x \\ y \end{pmatrix} = \begin{pmatrix} f_1(x, y) \\ f_2(x, y) \end{pmatrix}.$$

Calculus in higher dimensions necessitates the use of linear algebra. Recall that the Jacobian matrix of the map $F$ at the point $\mathbf{x}$ is given by

$$DF(\mathbf{x}) = \begin{pmatrix} \dfrac{\partial f_1}{\partial x}(\mathbf{x}) & \dfrac{\partial f_1}{\partial y}(\mathbf{x}) \\ \dfrac{\partial f_2}{\partial x}(\mathbf{x}) & \dfrac{\partial f_2}{\partial y}(\mathbf{x}) \end{pmatrix}.$$

This matrix will play a crucial role in the sequel, much the same as $f'(x)$ did in one dimension.

**Example 1.15.** Let $H: \mathbf{R}^2 \to \mathbf{R}^2$ be given by

$$x_1 = a - by - x^2$$
$$y_1 = x$$

where $a$ and $b$ are parameters. One computes readily that

$$DH\begin{pmatrix} x \\ y \end{pmatrix} = \begin{pmatrix} -2x & -b \\ 1 & 0 \end{pmatrix}.$$

Note that $\det(DH) = b$, no matter at which point we evaluate the Jacobian matrix. That is, the Jacobian determinant of $H$ is constant. The map $H$ is called the Hénon map. We will return to $H$ in the final section of this chapter.

A map $F: \mathbf{R}^2 \to \mathbf{R}^2$ is $C^1$ if all of its first partial derivatives exist and are continuous. $F$ is $C^\infty$ if its mixed $k^{th}$ partial derivatives exist and are continuous for all $k$. We will mainly consider $C^\infty$ maps in this chapter, although we will occasionally use piecewise linear maps as examples.

**Definition 1.16.** $F: \mathbf{R}^2 \to \mathbf{R}^2$ is a diffeomorphism if $F$ is one-to-one, onto, and $C^\infty$, and its inverse is also $C^\infty$.

**Example 1.17.** The Hénon map of Example 1.15 is a diffeomorphism of $\mathbf{R}^2$ as long as $b \neq 0$. The inverse map is given by

$$x_1 = y$$
$$y_1 = (a - x - y^2)/b.$$

Unlike in Chapter One, where many of our maps were non-invertible, we will concentrate mainly on diffeomorphisms in this chapter. Many of the phenomena that were encountered in the first chapter also occur in higher dimensions. We may raise the dimension of the space by considering the Cartesian product of two spaces.

**Definition 1.18.** Let $X$ and $Y$ be arbitrary sets. The Cartesian product $X \times Y$ is the set of ordered pairs of elements of $X$ and $Y$, i.e.,

$$X \times Y = \{(x, y) \mid x \in X, y \in Y\}.$$

For example, $\mathbf{R}^2 = \mathbf{R} \times \mathbf{R}$ and $\mathbf{R}^3 = \mathbf{R}^2 \times \mathbf{R}$. Also, we may identify a cylinder with $\mathbf{R} \times S^1$ and a torus (or surface of a doughnut) with $S^1 \times S^1$. The solid torus is $S^1 \times B^2$ where

$$B^2 = \{\mathbf{x} \in \mathbf{R}^2 \mid |\mathbf{x}| \leq 1\}.$$

Here the absolute value means the Euclidean distance from the origin.

We need higher dimensional generalizations of three important results from advanced calculus.

**Theorem 1.19.** The Implicit Function Theorem. *Suppose* $F \colon \mathbf{R}^3 \to \mathbf{R}^2$ *is given by*

$$x_1 = f_1(x, y, z)$$
$$y_1 = f_2(x, y, z).$$

*Suppose that* $F(\mathbf{0}) = \mathbf{0}$ *and that the matrix of partial derivatives*

$$\begin{pmatrix} \dfrac{\partial f_1}{\partial x}(\mathbf{0}) & \dfrac{\partial f_1}{\partial y}(\mathbf{0}) \\[2ex] \dfrac{\partial f_2}{\partial x}(\mathbf{0}) & \dfrac{\partial f_2}{\partial y}(\mathbf{0}) \end{pmatrix}$$

*is invertible (i.e., has non-zero determinant). Then there exists* $\epsilon > 0$ *and a smooth curve* $\varsigma(z)$ *of the form*

$$x = \varsigma_1(z)$$
$$y = \varsigma_2(z)$$

*defined for* $|z| < \epsilon$ *and such that* $F(\varsigma_1(z), \varsigma_2(z), z) = 0.$

**Remark.** As in Chapter One, we will use the Implicit Function Theorem to guarantee the existence of "nice" solutions to equations, i.e., a nice curve of solutions. This will arise when we deal with bifurcation theory in higher dimensions.

**Theorem 1.20.** The Inverse Function Theorem. *Let $F: \mathbf{R}^2 \to \mathbf{R}$. Suppose $F(0) = 0$ and $DF(0)$ is an invertible matrix. Then there exists a neighborhood $U$ of $0$ and a $C^\infty$ map $G: U \to \mathbf{R}^2$ such that $F \circ G(\mathbf{x}) = \mathbf{x}$ for all $\mathbf{x} \in U$.*

That is, if the Jacobian matrix of $F$ is invertible at $0$, then there exists a local inverse for $F$.

**Theorem 1.21.** The Contraction Mapping Theorem. *Let $F: B^2 \to B^2$ where $B^2$ is the closed unit disk*

$$B^2 = \{\mathbf{x} \in \mathbf{R}^2 \,|\, |\mathbf{x}| \le 1\}.$$

*Suppose $|F(\mathbf{x}_1) - F(\mathbf{x}_2)| < \lambda|\mathbf{x}_1 - \mathbf{x}_2|$ for all vectors $\mathbf{x}_i \in B^2$ and some $\lambda < 1$. Then there exists a unique fixed point $\mathbf{x}_* \in B^2$. Moreover,*

$$\lim_{n \to \infty} F^n(\mathbf{x}) = \mathbf{x}_*$$

*for all $\mathbf{x} \in B^2$.*

**Exercises**

1.  Find all of the eigenvalues and eigenvectors of the following matrices:

    a. $\begin{pmatrix} 1 & 1 \\ 1 & 0 \end{pmatrix}$.

    b. $\begin{pmatrix} 3 & 1 \\ -1 & 1 \end{pmatrix}$.

    c. $\begin{pmatrix} 1 & 1 \\ -1 & 1 \end{pmatrix}$.

    d. $\begin{pmatrix} -1 & -2 \\ 4 & 3 \end{pmatrix}$.

2.  Use the eigenvalues and eigenvectors computed in Exercise 1 to construct the standard forms for each of these matrices.

## §2.2 THE DYNAMICS OF LINEAR MAPS: TWO AND THREE DIMENSIONS

Using the classification theorems of the previous section, it is relatively straightforward to describe the dynamical behavior of all linear maps in dimensions two and three.

**Example 2.1.** Let

$$L_1(\mathbf{x}) = \begin{pmatrix} 2 & 0 \\ 0 & \frac{1}{2} \end{pmatrix} \mathbf{x}.$$

That is, if

$$\mathbf{x} = \begin{pmatrix} x \\ y \end{pmatrix},$$

then

$$L_1 \begin{pmatrix} x \\ y \end{pmatrix} = \begin{pmatrix} 2x \\ \frac{1}{2}y \end{pmatrix}.$$

Note that the matrix representation of $L_1$ has eigenvalues 2 and 1/2. Also, $L_1$ preserves both the $x$- and $y$-axes. Points on the $x$-axis move away from the origin under iteration of $L_1$, while points on the $y$-axis converge toward $\mathbf{0}$ under iteration. Points not on either axis tend toward $\infty$ under both forward and backward iteration.

**Example 2.2** Let

$$L_2(\mathbf{x}) = \begin{pmatrix} 2 & 0 \\ 0 & -\frac{1}{2} \end{pmatrix} \mathbf{x}.$$

$L_2$ also contracts the $y$-axis, but this time, points hop from one side of the origin to the other under iteration of $L_2$. The eigenvalues here are 2 and $-\frac{1}{2}$.

**Example 2.3.** Let

$$L_3(\mathbf{x}) = \begin{pmatrix} \frac{1}{2} & 0 \\ 0 & \frac{1}{3} \end{pmatrix} \mathbf{x}.$$

In this case, the eigenvalues are $1/2$ and $1/3$. All points tend toward the origin under iteration of $L_3$. Points move more quickly toward $\mathbf{0}$ in the $y$-direction since the rate of contraction in that direction is stronger.

**Example 2.4.** Let

$$L_4(\mathbf{x}) = \begin{pmatrix} 0 & \frac{1}{2} \\ -\frac{1}{2} & 0 \end{pmatrix} \mathbf{x}.$$

Again, all points move toward the origin under iteration of $L_4$. This follows since each vector in $\mathbf{R}^2$ is contracted by a factor of $\frac{1}{2}$ by each application of $L_4$. This time, however, there is no invariant line; the polar angle of each point is decreased by $\pi/2$ each time $L_4$ is iterated. Hence points tend to spiral into the origin. The eigenvalues are $\pm i/2$.

**Example 2.5.** Let

$$L_5(\mathbf{x}) = \begin{pmatrix} 0 & 1/2 & 0 \\ -1/2 & 0 & 0 \\ 0 & 0 & 2 \end{pmatrix}.$$

In this example, the $x, y$-plane is invariant and $L_5$ behaves on this plane exactly as in the previous example. Points off the $x, y$-plane tend to $\infty$ under iteration of $L_5$. Note that the $z$-axis is invariant. and that $L_5$ expands vectors on the $z$-axis by a factor of 2. Here the eigenvalues are $\pm i/2$ and 2.

The phase portraits of the linear maps in Examples 2.1-2.5 are sketched in Fig. 2.1. While the linear maps described in these examples do not by any means exhaust all possibilities, they nevertheless highlight some of the similarities and differences between one- and higher dimensions. The major difference is that the origin may have both expanding and contracting directions simultaneously. As in the one-dimensional case, it is the eigenvalues which govern whether or not the map has contracting or expanding directions. Eigenvalues larger than one (in absolute value) lead to expansion, whereas eigenvalues smaller than one lead to contraction. This motivates the definition of hyperbolicity.

**Definition 2.5.** An invertible linear map is hyperbolic if none of its eigenvalues have absolute value one.

Note that all of the above examples are hyperbolic. We also note that, whenever a linear map has an eigenvalue of absolute value less than one, then there is a corresponding direction in which points are attracted toward the origin, which is always fixed. When there is a pair of complex conjugate

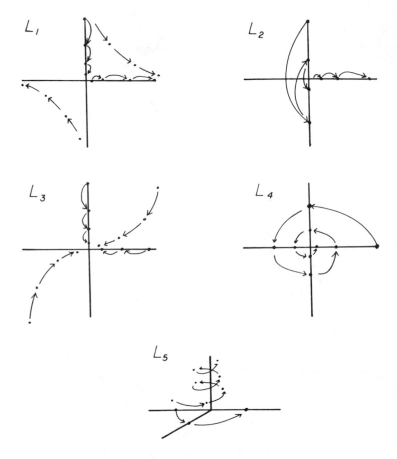

**Fig. 2.1.** The dynamics of the linear maps $L_1, \ldots, L_5$.

eigenvalues of absolute value less than one, then there is a two-dimensional set of points attracted to **0**. This is a general fact.

**Proposition 2.6.** *Suppose* $L: \mathbf{R}^3 \to \mathbf{R}^3$ *has all eigenvalues less than one in absolute value. Then* $L^n(\mathbf{x}) \to 0$ *as* $n \to \infty$ *for all* $\mathbf{x} \in \mathbf{R}^3$.

*Proof.* Recall that the matrix representation $A$ of $L$ assumes one of four standard forms as shown in Theorem 1.6 and Proposition 1.8:

$$1. \begin{pmatrix} \alpha & -\beta & 0 \\ \beta & \alpha & 0 \\ 0 & 0 & \lambda \end{pmatrix} \quad 2. \begin{pmatrix} \lambda & 0 & 0 \\ 0 & \mu & 0 \\ 0 & 0 & \eta \end{pmatrix}$$

$$3. \begin{pmatrix} \lambda & \epsilon & 0 \\ 0 & \lambda & 0 \\ 0 & 0 & \mu \end{pmatrix} \quad 4. \begin{pmatrix} \lambda & \epsilon & 0 \\ 0 & \lambda & \epsilon \\ 0 & 0 & \lambda \end{pmatrix}$$

where all entries are real and $\epsilon \neq 0$.

Consider the real-valued function $V(x, y, z) = x^2 + y^2 + z^2$. We claim that there exists $\nu < 1$ such that, if $\epsilon > 0$ is small enough, we have

$$V \circ L(\mathbf{x}) \leq \nu V(\mathbf{x})$$

with equality if and only if $\mathbf{x} = 0$. This is a simple computation which we will make in case 4 only.

$$\begin{aligned} V \circ L(\mathbf{x}) &= \lambda^2(x^2 + y^2 + z^2) + \epsilon^2(y^2 + z^2) + 2\lambda\epsilon(xy + yz) \\ &\leq (\lambda^2 + \epsilon^2)(x^2 + y^2 + z^2) + 2|\lambda\epsilon|(|xy| + |yz| + |xz|) \\ &\leq (\lambda^2 + \epsilon^2 + 4|\lambda\epsilon|)(V(\mathbf{x})) \end{aligned}$$

since $|xy| \leq x^2 + y^2$. Consequently, we may choose $\epsilon$ small enough so that the inequality holds with $\nu = \lambda^2 + 4|\lambda\epsilon| + \epsilon^2$.

It follows that if $\mathbf{x} \neq 0$, then

$$V \circ L^n(\mathbf{x}) \leq \nu^n V(\mathbf{x}).$$

Hence $V \circ L^n(\mathbf{x}) \to 0$ as $n \to \infty$. But $V(\mathbf{x}) = 0$ iff $\mathbf{x} = 0$. Thus $L^n(\mathbf{x}) \to 0$ as required. The other cases are handled similarly.

q.e.d.

The function $V$ constructed in the proof of Proposition 2.6 is called a Liapounov function. We formalize this notion with a definition.

**Definition 2.7.** Let $F: \mathbf{R}^n \to \mathbf{R}^n$ be a diffeomorphism. $V: \mathbf{R}^n \to \mathbf{R}$ is a Liapounov function for $F$ centered at $\mathbf{p}$ if

1. $V(\mathbf{x}) > 0$ for $\mathbf{x} \neq \mathbf{p}$
2. $V(\mathbf{p}) = 0$
3. $V \circ F(\mathbf{x}) \leq V(\mathbf{x})$ with equality iff $\mathbf{x} = \mathbf{p}$.

Note that $F(\mathbf{p}) = \mathbf{p}$ is forced by this definition. If $V$ is a *strict* Liapounov function, i.e., $V \circ F(\mathbf{x}) < V(\mathbf{x})$ if $\mathbf{x} \neq \mathbf{p}$, then it follows easily as in the previous proof that $F^n(\mathbf{x}) \to \mathbf{p}$ as $n \to \infty$ for all $\mathbf{x}$ in a neighborhood of $\mathbf{p}$.

In case all of the eigenvalues of $L$ are larger than one in absolute value, then the arguments above may be altered to yield the following result.

**Corollary 2.8.** *Suppose* $L: \mathbf{R}^3 \to \mathbf{R}^3$ *is linear and all eigenvalues of* $L$ *have absolute value larger than one. Then* $L^n(\mathbf{x}) \to 0$ *as* $n \to -\infty$.

We now turn our attention to the case of mixed eigenvalues: some with absolute value larger than one and some smaller.

**Proposition 2.9.** *Suppose the eigenvalues of* $L$ *are* $\lambda_1, \lambda_2$ *and* $\lambda_3$ *with*

1. $|\lambda_1|, |\lambda_2| < 1$
2. $|\lambda_3| > 1$.

*Then there is a plane* $W^s$ *and a line* $W^u$ *on which*

1. *if* $\mathbf{x} \in W^s$, *then* $L(\mathbf{x}) \in W^s$ *and* $L^n(\mathbf{x}) \to 0$ *as* $n \to \infty$.
2. *if* $\mathbf{x} \in W^u$, *then* $L(\mathbf{x}) \in W^u$ *and* $L^{-n}(\mathbf{x}) \to 0$ *as* $n \to \infty$.
3. *if* $x \notin W^u \cup W^s$ *then* $|L^n(\mathbf{x})| \to \infty$ *as* $n \to \pm\infty$.

*Proof.* The normal form of $L(\mathbf{x}) = \mathbf{A}\mathbf{x}$ may be written in one of the two forms

$$A = \begin{pmatrix} \alpha & \beta & 0 \\ -\beta & \alpha & 0 \\ 0 & 0 & \lambda_3 \end{pmatrix} \quad \text{or} \quad A = \begin{pmatrix} \lambda_1 & * & 0 \\ 0 & \lambda_2 & 0 \\ 0 & 0 & \lambda_3 \end{pmatrix}$$

where $*$ is either positive (only if $\lambda_1 = \lambda_2$) or zero. (In the first case, $\lambda_1 = \alpha + i\beta$). In this form, $W^s$ is the $x, y$-plane and $W^u$ is the $z$-axis. Application of Proposition 2.6 and its Corollary to either of these subspaces yields the result.

q.e.d.

**Remarks.**

**1.**  In the case where two of the eigenvalues are larger than one in absolute value, the above result may clearly be modified to yield a plane of points which tend to zero under iteration of $L^{-1}$. There is a line through $\mathbf{0}$ on which $L$ is a contraction.

**2.**  In higher dimensions, there is an analogous result which may be proved by using the Jordan form of the map.

The invariant subspaces given by Proposition 2.9 will play an important role in the sequel. Hence they deserve a name.

**Definition 2.10.** $W^s$ is called the stable subspace of $L$; $W^u$ is the unstable subspace.

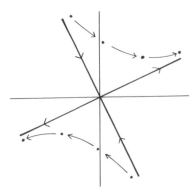

**Fig. 2.2.**

In the simple case where the matrix representations are in standard form, the stable and unstable subspaces are the coordinate planes and/or axes. This is not true in general.

**Example 2.11.** Consider the linear map

$$L(\mathbf{x}) = \begin{pmatrix} 1 & 1 \\ 1 & 0 \end{pmatrix} \mathbf{x}.$$

This linear map has eigenvalues $\frac{1}{2} + \frac{\sqrt{5}}{2} > 1$ and $-1 < \frac{1}{2} - \frac{\sqrt{5}}{2} < 0$. The eigenvectors corresponding to these eigenvalues are

$$y = \left( \frac{\sqrt{5} - 1}{2} \right) x$$

$$y = -\left( \frac{\sqrt{5} + 1}{2} \right) x$$

respectively. The phase portrait is given in Fig. 2.2. We will encounter this particular linear map in a vastly different setting when we discuss the hyperbolic toral automorphisms in §2.4.

In one dimension, the lack of hyperbolicity was often a signal that a bifurcation might take place. This is true in higher dimensions as well, as we shall see in §2.8. Without discussing the possible bifurcations which occur, let us simply note by several linear examples that the lack of hyperbolicity yields quite different phase portraits than those described in the hyperbolic case above. For simplicity, we will deal only with the invertible case.

**Example 2.12.** Let

$$L(\mathbf{x}) = \begin{pmatrix} 1 & 0 \\ 0 & 2 \end{pmatrix} \mathbf{x}.$$

In this case, the entire $x$-axis is fixed while there is expansion away from the $x$-axis.

**Example 2.13.** Let

$$L(\mathbf{x}) = \begin{pmatrix} -1 & 0 \\ 0 & \frac{1}{2} \end{pmatrix} \mathbf{x}.$$

In this case, **0** is the only fixed point, but all other points on the $x$-axis have period 2. All other points are contracted toward the $x$-axis.

**Example 2.14.** Let

$$L(\mathbf{x}) = \begin{pmatrix} 0 & -1 \\ 1 & 0 \end{pmatrix} \mathbf{x}.$$

$A$ has eigenvalues $\pm i$ and $L$ is a rotation through $90^{\circ}$. Hence all points except **0** are periodic with period 4. In general, if $A$ is of the form

$$A = \begin{pmatrix} \alpha & -\beta \\ \beta & \alpha \end{pmatrix}$$

with $\alpha^2 + \beta^2 = 1$, then $L$ is a rotation of the plane through angle $\arctan(\beta/\alpha)$.

**Exercises**

**1.** Describe the dynamics of the linear maps whose matrix representation is

a. $\begin{pmatrix} -2 & 0 \\ 0 & 2 \end{pmatrix}$.

b. $\begin{pmatrix} -\frac{1}{2} & 0 \\ 0 & 2 \end{pmatrix}$.

c. $\begin{pmatrix} \frac{1}{2} & 0 \\ 0 & -2 \end{pmatrix}$.

d. $\begin{pmatrix} -\frac{1}{2} & 0 \\ 0 & -\frac{1}{2} \end{pmatrix}$.

e. $\begin{pmatrix} 2 & 0 \\ 0 & 2 \end{pmatrix}$.

**2.** Describe the dynamics of the linear maps whose matrix representation is given below. Identify precisely the stable and unstable sets.

a. $\begin{pmatrix} 2 & 1 \\ 0 & \frac{1}{2} \end{pmatrix}$

b. $\begin{pmatrix} 1 & 2 \\ 3 & 4 \end{pmatrix}$

c. $\begin{pmatrix} 2 & 1 \\ 1 & 1 \end{pmatrix}$

d. $\begin{pmatrix} \frac{1}{2} & 0 & 0 \\ 1 & 2 & 0 \\ 0 & 0 & 3 \end{pmatrix}$

e. $\begin{pmatrix} 1 & 1 & 0 \\ 1 & 2 & 0 \\ 0 & 0 & 2 \end{pmatrix}$

**3.** Describe the dynamics of each of the following linear maps, indicating which are non-hyperbolic.

a. $\begin{pmatrix} 0 & 1 \\ 1 & 0 \end{pmatrix}$

b. $\begin{pmatrix} 0 & 0 & 1 \\ 0 & 1 & 0 \\ -1 & 0 & 0 \end{pmatrix}$

c. $\begin{pmatrix} 1 & 0 \\ 0 & -1 \end{pmatrix}$

d. $\begin{pmatrix} \frac{1}{\sqrt{2}} & \frac{1}{\sqrt{2}} \\ -\frac{1}{\sqrt{2}} & \frac{1}{\sqrt{2}} \end{pmatrix}$

e. $\begin{pmatrix} 1 & 0 & 0 \\ 0 & 2 & 2 \\ 0 & -2 & 2 \end{pmatrix}$

**4.** Consider the linear map

$$L(\mathbf{x}) = \begin{pmatrix} \frac{1}{2} & 0 \\ 0 & \frac{1}{3} \end{pmatrix} \mathbf{x}.$$

Prove that $L^n \mathbf{x} \to \mathbf{0}$ for all $\mathbf{x} \in \mathbf{R}$? Prove that, if $\mathbf{x}$ does not lie on the $y$-axis, then the orbit of $\mathbf{x}$ tends to $\mathbf{0}$ tangentially to the $x$-axis.

5.   A function $F: \mathbf{R}^n \to \mathbf{R}$ is called an integral for a linear map $L$ if $F \circ L(\mathbf{x}) = F(\mathbf{x})$, i.e., $F$ is constant along orbits of $L$. Show that

$$F \begin{pmatrix} x \\ y \end{pmatrix} = x^2 + y^2$$

is an integral for

$$L(\mathbf{x}) = \begin{pmatrix} 0 & 1 \\ -1 & 0 \end{pmatrix} \mathbf{x}.$$

6.   Construct (non-trivial) integrals for each of the following linear maps.

a. $L(\mathbf{x}) = \begin{pmatrix} 0 & 1 \\ 1 & 0 \end{pmatrix} \mathbf{x}.$

b. $L(\mathbf{x}) = \begin{pmatrix} 2 & 0 \\ 0 & \frac{1}{3} \end{pmatrix} \mathbf{x}.$

## §2.3 THE HORSESHOE MAP

Symbolic dynamics, which played such a crucial role in our understanding of the one-dimensional quadratic map, can also be used to study higher dimensional phenomena. In this section, we will study a now-classical example due to Smale, the horseshoe map. This was the first example of a diffeomorphism which had infinitely many periodic points and yet was structurally stable. We will see that this map has much in common with the quadratic map which motivated so much of the material in Chapter One.

To define the map, we first consider a region $D$ consisting of three components: a central square $S$ with side length 1 and two semicircles $D_1$ and $D_2$ at either end. See Fig. 3.1. $D$ is shaped like a "stadium."

The horseshoe map $F$ takes $D$ inside itself according to the following prescription. First, linearly contract $S$ in the vertical direction by a factor $\delta < 1/2$ and expand it in the horizontal direction by a factor $1/\delta$ so that $S$ is long and thin. Then put $S$ back inside $D$ in a horseshoe-shaped figure as in Fig. 3.2.

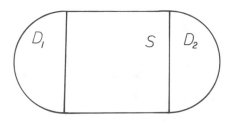

**Fig. 3.1.**

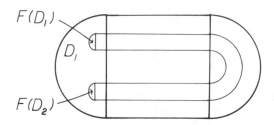

**Fig. 3.2.** The Smale horseshoe map.

The semicircular regions $D_1$ and $D_2$ are contracted and mapped inside $D_1$ as depicted. We remark that $F(D) \subset D$ and that $F$ is one-to-one. However, since $F$ is not onto, $F^{-1}$ is not globally defined. The remainder of this section is devoted to the study of the dynamics of $F$ in $D$.

Note first that the preimage of $S$ consists of two vertical rectangles $V_0$ and $V_1$ which we may assume are mapped linearly onto the two horizontal components $H_0$ and $H_1$ of $F(S) \cap S$. The width of $V_0$ and $V_1$ is $\delta$, as is the height of $H_0$ and $H_1$. See Fig. 3.3.

By linearity of $F$ on $S$, it follows that $F$ preserves horizontal and vertical lines *in* $S$. For later use, we note that if $h$ is a horizontal line segment in $S$ whose image also lies in $S$, then the length of $F(h)$ is expanded by a factor $1/\delta$ times the length of $h$. Similarly. if both $v$ and $F(v)$ are vertical line segments in $S$, then the length of $F(v)$ is shrunk by a factor of $\delta$.

We claim that the dynamics of $F$ are very similar to those of the quadratic map studied in §1.5. Note first that. since $F$ is a contraction on $D_1$. $F$ has a unique fixed point $p$ in $D_1$ and $\lim_{n \to \infty} F^n(q) = p$ for all $q \in D_1$. This follows immediately from the Contraction Mapping Theorem. Since $F(D_2) \subset D_1$, all forward orbits in $D_2$ behave likewise. Similarly, if $q \in S$ but $F^k(q) \notin S$ for some $k > 0$, then we must have that $F^k(q) \in D_1 \cup D_2$ so that $F^n(q) \to p$

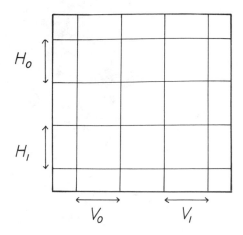

**Fig. 3.3.**

as $n \to \infty$. Consequently, to understand the forward orbits of $F$, it suffices to consider the set of points whose forward orbits lie for all time in $S$. We will do more: we will describe

$$\Lambda = \{q \in S \,|\, F^k(q) \in S \text{ for all } k \in \mathbf{Z}\}.$$

Now, if the forward orbit of $q$ lies in $S$, we must have, first of all, that $q \in V_0$ or $q \in V_1$, for all other points in $S$ are mapped out of $S$ and into $D_1 \cup D_2$. If $F(q) \in S$, then, similarly, we must have $F(q) \in V_0 \cup V_1$, i.e., $q \in F^{-1}(V_0) \cup F^{-1}(V_1)$. Here $F^{-1}(V_0)$ means the inverse image of $V_0$ in $S$. Clearly, there are substrips in both $V_0$ and $V_1$ which map into $V_0$ as depicted in Fig. 3.4.

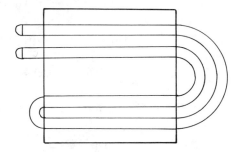

**Fig. 3.4.** The forward image $F^2(S)$.

This is the inductive step: if $V$ is any vertical rectangle connecting the upper and lower boundaries of $S$ with width $w$, then $F^{-1}(V)$ is a pair of smaller vertical rectangles of width $\delta w$, one in each $V_i$. Consequently, $F^{-1}(F^{-1}(V_i)) = F^{-2}(V_i)$ consists of four vertical rectangles, each of width $\delta^2$, $F^{-3}(V_i)$ consists of eight vertical rectangles of width $\delta^3$, etc. Hence the same procedure we used in §1.6 shows that

$$\Lambda_+ = \{q | F^k(q) \in S \text{ for } k = 0, 1, 2, \ldots\}$$

is the product of a Cantor set with a vertical interval. Arguing entirely analogously, it is easy to check that

$$\Lambda_- = \{q | F^{-k}(q) \in S \text{ for } k = 1, 2, 3, \ldots\}$$

consists of a product of a Cantor set with an interval. In this case, the intervals are horizontal. Finally,

$$\Lambda = \Lambda_+ \cap \Lambda_-$$

is the intersection of these two sets.

To introduce symbolic dynamics into the system, we first choose any vertical interval $\ell$ in $\Lambda_+$. Note that $F^k(\ell)$, is a vertical line segment of length $\delta^k$ in either $V_0$ or $V_1$. Hence we may attach an infinite sequence $s_0 s_1 s_2 \ldots$ of 0's or 1's to any point in $\ell$ according to the rule $s_j = \alpha$ iff $F^j(\ell) \subset V_\alpha$. The number $s_0$ tells us in which vertical strip the line $\ell$ is located, $s_1$ tells where its image is located, etc. We can similarly attach a sequence of integers to any horizontal line segment $h$. For convenience, we write this sequence $s_{-3} s_{-2} s_{-1}$, where $s_{-j} = \alpha$ iff

$$F^{-j}(h) \subset V_\alpha \text{ for } j = 1, 2, 3, \ldots.$$

Note again that $F^{-1}(h)$, $F^{-2}(h), \ldots$ are horizontal line segments of decreasing lengths.

Consequently, if $p$ is any point in $\Lambda_+ \cap \Lambda_-$, we may associate a pair of sequences of 0's and 1's to $p$. One sequence gives the itinerary of the forward orbit of $p$; the other describes the backward orbit. Let us amalgamate both of these sequences into one, doubly-infinite sequence of 0's and 1's. That is, we define the itinerary $S(p)$ by the rule

$$S(p) = (\ldots s_{-2} s_{-1} \cdot s_0 s_1 s_2 \ldots)$$

where $s_j = k$ if and only if $F^j(p) \in V_k$.

This then gives the symbolic dynamics on $\Lambda$. Let $\Sigma_2$ denote the set of all doubly-infinite sequences of 0's and 1's:

$$\Sigma_2 = \{(s) = (\ldots s_{-2}s_{-1} \cdot s_0 s_1 s_2 \ldots) | s_j = 0 \text{ or } 1\}.$$

Impose a metric on $\Sigma_2$ by defining $d[(s),(t)] = \displaystyle\sum_{i=-\infty}^{\infty} \frac{|s_i - t_i|}{2^{|i|}}$ exactly as before. Define the shift map $\sigma$ by

$$\sigma(\ldots s_{-2}s_{-1} \cdot s_0 s_1 s_2 \ldots) = (\ldots s_{-2}s_{-1}s_0 \cdot s_1 s_2 \ldots).$$

That is, $\sigma$ simply shifts each sequence in $\Sigma_2$ one unit to the left (equivalently, $\sigma$ shifts the decimal point one unit to the right). Unlike our previous shift map, this map has an inverse. Clearly, shifting one unit to the right gives this inverse. It is easy to check that $\sigma$ is a homeomorphism on $\Sigma_2$ (see Exercise 2).

The shift map is now the model for the restriction of $F$ to $\Lambda$. Indeed, the map $S$ gives a topological conjugacy between $F$ on $\Lambda$ and $\sigma$ on $\Sigma_2$. We leave the details of this proof to the reader (see Exercise 3).

All of the properties which held for the old one-sided shift hold for $\sigma$ as well. For example, there are precisely $2^N$ periodic points of period $N$ for $\sigma$. There is a dense orbit for $\sigma$ as well (see Exercises 4,5). But there are new phenomena present as well.

**Definition 3.1.** Two points $p_1$ and $p_2$ are forward (respectively backward) asymptotic if $F^n(p_1), F^n(p_2) \in D$ for all $n \geq 0$ (resp. $n \leq 0$ ) and

$$\lim_{n \to \infty} |F^n(p_1) - F^n(p_2)| = 0$$

(resp. $n \to -\infty$ ).

Intuitively, two points in $D$ are forward asymptotic if their orbits approach each other as $n \to \infty$. Note that any point which leaves $S$ under forward iteration of $F$ is forward asymptotic to the fixed point $p \in D_1$. Also, if $p_1$ and $p_2$ lie on the same vertical line in $\Lambda_+$, then $p_1$ and $p_2$ are forward asymptotic. If $p_1$ and $p_2$ lie on the same horizontal line, then they are backward asymptotic.

As in the linear theory, the notion of forward and backward asymptotic orbits allows us to define the stable and unstable sets of a point.

**Definition 3.2.** The stable set of $p$ is given by

$$W^s(p) = \{z || F^n(z) - F^n(p)| \to 0 \text{ as } n \to \infty\}.$$

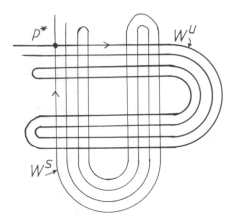

**Fig. 3.5.** The stable and unstable sets associated to $p^*$.

The unstable set of $p$ is given by

$$W^u(p) = \{z | |F^{-n}(p) - F^{-n}(z)| \to 0 \text{ as } n \to \infty\}.$$

Equivalently, a point $z$ lies in $W^s(p)$ if $p$ and $z$ are forward asymptotic. For example, any point in $S$ which leaves $S$ under forward iteration of the horseshoe map lies in the stable set of the fixed point in $D_1$.

The stable and unstable sets of points in $\Lambda$ are more complicated. For example, consider the fixed point $p^*$ which lies in $V_0$ and therefore has the sequence $(\ldots 00.000 \ldots)$ attached. Any point which lies on the vertical segment $\ell_s$ through $p^*$ lies in $W^s(p^*)$. But there are many other points in this stable set. Suppose the point $q$ eventually maps into $\ell_s$. Then there is an integer $n$ such that $|F^n(q) - p^*| < 1$. Hence

$$|F^{n+k}(q) - p^*| < \delta^k$$

and it follows that $q \in W^s(p^*)$. Thus, the union of vertical intervals given by $F^{-k}(\ell_s)$ for $k = 1, 2, 3, \ldots$ all lie in $W^s(p^*)$. The reader may easily check that there are $2^k$ such intervals. See Fig. 3.5.

Since $F(D) \subset D$, the unstable manifold of $p^*$ assumes a somewhat different form. The horizontal line segment $\ell_u$ through $p^*$ in $D$ clearly lies in $W^u(p^*)$. As above, all of the forward images of $\ell_u$ also lie in $D$. The reader may easily check that $F^k(\ell_u)$ is a "snake-like" curve in $D$ which cuts across $S$ exactly $2^k$ times in a horizontal segment. See Fig. 3.5.

These stable and unstable sets are easy to describe on the shift level. Let

$$\mathbf{s}^* = (\ldots s^*_{-2} s^*_{-1} \cdot s^*_0 s^*_1 s^*_2 \ldots) \in \Sigma_2.$$

Clearly, if $\mathbf{t}$ is a sequence whose entries agree with those of $\mathbf{s}^*$ to the right of some entry, then $\mathbf{t} \in W^s(\mathbf{s}^*)$. The converse of this is also true, as is shown in Exercise 6.

A natural question that arises is our use of the term "Cantor set" to describe the set $\Lambda = \Lambda_+ \cap \Lambda_-$ for the horseshoe map and the similar set $\Lambda$ for the quadratic map of Chapter One. Intuitively, it may appear that the $\Lambda$ for the horseshoe has "twice" as many points. However, both $\Lambda$'s are actually homeomorphic! This is best seen on the shift level.

Let $\Sigma_2^1$ denote the set of one-sided sequences of 0's and 1's and $\Sigma_2$ the set of two-sided such sequences. Define a map

$$\Phi \colon \Sigma_2^1 \to \Sigma_2$$

by $\Phi(s_0 s_1 s_2 \ldots) = (\ldots s_5 s_3 s_1 \cdot s_0 s_2 s_4 \ldots)$. It is easy to check that $\Phi$ is a homeomorphism between $\Sigma_2^1$ and $\Sigma_2$ (see Exercise 11).

**Remarks.**

**1.** We have now seen stable and unstable sets in two guises: the stable and unstable subspaces of linear maps and the above collection of horizontal and vertical line segments. This will become a common pattern for higher dimensional systems that are "hyperbolic" in a sense to be made precise later. Each point in a hyperbolic set will come equipped with contracting and expanding directions which will play the role of stable and unstable sets

**2.** Unlike the quadratic map, the horseshoe example was defined geometrically rather than algebraically. This is often the case with higher dimensional maps: it is easier to present and work with examples defined geometrically. It is important to realize that it is possible to write down an explicit algebraic expression which gives a map similar to the horseshoe. This map is the Hénon map which we will discuss later in §2.9.

**Exercises**

**1.** Prove that $d[(\mathbf{s}), (\mathbf{t})] = \sum\limits_{i=-\infty}^{\infty} \dfrac{|s_i - t_i|}{2^{|i|}}$ is a metric on $\Sigma_2$.

**2.** Prove that the shift $\sigma$ is a homeomorphism.

**3.** Prove that $S \colon \Lambda \to \Sigma_2$ gives a topological conjugacy between $\sigma$ and $F$.

**4.** Construct a dense orbit for $\sigma$.

**5.**  Prove that periodic points are dense for $\sigma$.

**6.**  Let $s^* \in \Sigma_2$. Prove that $W^s(s^*)$ consists of precisely those sequences whose entries agree with those of $s^*$ to the right of some entry of $s^*$.

**7.**  Let $(0) = (\ldots 00.000 \ldots) \in \Sigma_2$. A sequence $s \in \Sigma_2$ is called homoclinic to $(0)$ if $s \in W^s(0) \cap W^u(0)$. Describe the entries of a sequence which is *homoclinic* to $(0)$. Prove that sequences which are homoclinic to $(0)$ are dense in $\Sigma_2$.

**8.**  Let $(1) = (\ldots 11.111 \ldots) \in \Sigma_2$. A sequence $s$ is a *heteroclinic* sequence if $s \in W^s(0) \cap W^u(1)$. Describe the entries of such a heteroclinic sequence. Prove that such sequences are dense in $\Sigma_2$.

**9.**  Generalize the definitions of homoclinic and heteroclinic points to arbitrary periodic points for $\sigma$ and reprove Exercises 7 and 8 in this case.

**10.**  Prove that the set of homoclinic points to a given periodic point is countable.

**11.**  Let $\Sigma_2^1$ denote the set of one-sided sequences of 0's and 1's. Define $\Phi \colon \Sigma_2^1 \to \Sigma_2$ by

$$\Phi(s_0 s_1 s_2 \ldots) = (\ldots s_5 s_3 s_1 \cdot s_0 s_2 s_4 \ldots).$$

Prove that $\Phi$ is a homeomorphism.

**12.**  Consider the map $F$ on $D$ defined geometrically as in Fig. 3.6. Assume that $F$ linearly contracts vertical lengths and linearly expands horizontal lengths in $S$ exactly as in the case of the Smale horseshoe. Let

$$\Lambda = \{p \in D \,\big|\, F^n(p) \in S \text{ for all } n \in \mathbf{Z}\}.$$

Use the techniques of §1.13 to show that $F$ on $\Lambda$ is topologically conjugate to a two-sided subshift of finite type generated by a $3 \times 3$ matrix $A$. Identify $A$. Discuss the dynamics of $F$ off $\Lambda$.

**13.**  Rework Exercise 12, this time with the map defined geometrically in Fig. 3.7.

**14.**  Let $R \colon \Sigma_2 \to \Sigma_2$ be defined by

$$R(\ldots s_{-2} s_{-1} . s_0 s_1 s_2 \ldots) = (\ldots s_2 s_1 s_0 . s_{-1} s_{-2} \ldots).$$

Prove that $R \circ R = id$ and that $\sigma \circ R = R \circ \sigma^{-1}$. Conclude that $\sigma = U \circ R$ where $U$ is a map which satisfies $U \circ U = id$. Maps which are their own inverse are called *involutions*. They represent very simple types of dynamical systems. Hence the shift may be decomposed into a composition of two such maps.

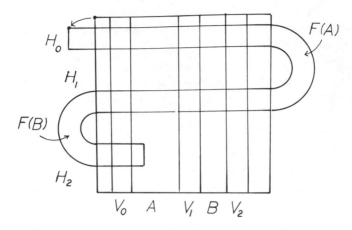

**Fig. 3.6.**

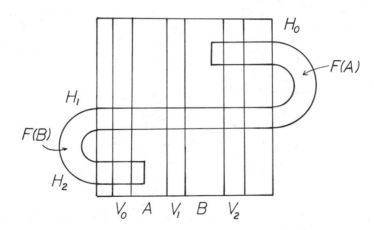

**Fig. 3.7.**

**15.** Let s be a sequence which is fixed by $R$. Suppose that $\sigma^n(s)$ is also fixed by $R$. Prove that s is a periodic point of $\sigma$ of period $2n$.

**16.** Rework the previous exercise, assuming that $\sigma^n(s)$ is fixed by $U$, where $U$ is given as in Exercise 13. What is the period of s?

## §2.4 HYPERBOLIC TORAL AUTOMORPHISMS

In this section, we introduce a completely different class of dynamical system, the Anosov systems or hyperbolic toral automorphisms. These maps are important in that they are chaotic everywhere that they are defined. Nevertheless, their dynamics can be described completely. One difference between these maps and those discussed previously is that these maps are naturally defined on a torus or "doughnut" rather than on Euclidean space. Even though the maps are induced by linear maps on Euclidean space (which have extremely simple dynamics), the maps on the tori have extremely rich dynamical structure.

To describe the torus, let us begin with the plane. We will consider as identical all points whose coordinates differ by integers. That is to say, the point $(\alpha, \beta)$ in the plane is to be regarded as the same as the points $(\alpha + 1, \beta)$, $(\alpha + 5, \beta + 3)$, and, in general, $(\alpha + M, \beta + N)$, where $M$ and $N$ are integers. We let $[\alpha, \beta]$ denote the set of all points equivalent to $(\alpha, \beta)$ under this relation. To be somewhat more formal, the relation $(x, y) \sim (x', y')$ if and only if $x - x'$ and $y - y'$ are integers gives an equivalence relation on points in the plane. The torus is thus the set of all equivalence classes under this relation.

Geometrically, this procedure can be visualized as follows. Consider the unit square in the plane $0 \leq x, y \leq 1$. Under the above identifications, only points on the boundary of the square need be considered. Indeed, the top boundary $y = 1$ should be considered the same as the bottom boundary $y = 0$, and similarly the left and right boundaries $x = 0$ and $x = 1$ should be identified. When this occurs, the square becomes first a cylinder and then a torus, as in Fig. 4.1.

### Remarks

**1.** This procedure is not limited to two dimensions; one may define an $n$-dimensional torus using the same equivalence relation on $\mathbf{R}^n$. This is shown in Exercise 2.

**2.** The torus may also be regarded as the Cartesian product of two circles. See Exercise 3.

Let $T$ denote the torus, and let $\pi$ be the natural projection of $\mathbf{R}^2$ onto

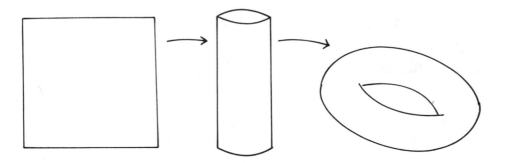

**Fig. 4.1.** Construction of a torus from a square.

$T$, i.e.,

$$\pi(x, y) = [x, y] = \pi(x + M, \ y + N).$$

Certain dynamical systems on a torus can be described most efficiently in the plane and then projected onto the torus. For example, suppose $F: \mathbf{R}^2 \rightarrow \mathbf{R}^2$ has the property that

$$F\binom{x}{y} - F\binom{x+M}{y+n}$$

belongs to the integer lattice for all points in the plane and all integers $M$ and $N$. It follows that

$$\pi \circ F\binom{x}{y} = \pi \circ F\binom{x+M}{y+N}$$

so that $F$ induces a well-defined map $\hat{F}$ on the torus. $\hat{F}$ is defined by the diagram

$$\mathbf{R}^2 \ \overset{F}{\rightarrow} \ \mathbf{R}^2$$

$$\downarrow \pi \qquad \downarrow \pi$$

$$T \ \overset{\hat{F}}{\rightarrow} \ T.$$

As an example, if $L$ is a linear map whose matrix representation is an integer matrix, then $\hat{L}$ is clearly well-defined on $T$. $\hat{L}$ is called a *toral automorphism*. For our purposes, we need a few more hypotheses on $L$.

**Definition 4.1.** Let $L(\mathbf{x}) = A \cdot \mathbf{x}$ where $A$ is a $2 \times 2$ matrix satisfying
   1. All entries of $A$ are integers.
   2. $\det(A) = \pm 1$.

3. $A$ is hyperbolic.

The map induced on $T$ by $A$ is called a hyperbolic toral automorphism and is denoted by $L_A$. $L_A$ is clearly differentiable, since its Jacobian matrix is simply the matrix $A$. Moreover, since $\det(A) = \pm 1$, the inverse of $A$ is also an integer matrix which is hyperbolic. Hence $A^{-1}$ also induces a hyperbolic toral automorphism which is, of course, the inverse of $L_A$. It follows that $L_A$ is a diffeomorphism of $T$.

The following proposition shows that $L_A$ is dynamically quite different from its linear counterpart.

**Proposition 4.2.** Per $(L_A)$ *is dense in* $T$.

*Proof.* Let $p$ be any point in $T$ with rational coordinates. By finding a common denominator, we may assume that $p$ is of the form $[\frac{\alpha}{k}, \frac{\beta}{k}]$, where $\alpha, \beta$ and $k$ are integers. Such points are clearly dense in $T$, for we may take $k$ arbitrarily large. We claim that $p$ is periodic with period less than or equal to $k^2$.

To see this, we note that there are exactly $k^2$ points in $T$ of the form $[\frac{\alpha}{k}, \frac{\beta}{k}]$ with $0 \leq \alpha, \beta < k$. Moreover, the image of any such point under $L_A$ may also be written in this form, since the entries of $A$ are integers. This means that $L_A$ permutes these points. Therefore there exist integers $i$ and $j$ such that $L_A^i(p) = L_A^j(p)$ and $|i - j| \leq k^2$. Applying $L_A^{-i}$ to this equation shows that $p$ is periodic of period less than or equal to $k^2$.

**Example 4.3.** Consider the map $L_A : T \to T$ where

$$A = \begin{pmatrix} 2 & 1 \\ 1 & 1 \end{pmatrix}.$$

Clearly, $[0,0]$ is a fixed point. $[0,0]$ is in fact the only fixed point, as we see by solving the equations for a fixed point:

$$2x + y = x + M$$
$$x + y = x + N$$

for $M, N \in \mathbf{Z}$. We have $L_A[1/2, 1/2] = [1/2, 0]$, $L_A[1/2, 0] = [0, 1/2]$ and $L_A[0, 1/2] = [1/2, 1/2]$, so that $[1/2, 1/2]$ is periodic with period 3. One may readily compute other periodic points for this map.

The density of the periodic points is just the beginning of the story of the chaotic nature of $L_A$. Since $A$ is hyperbolic with determinant $\pm 1$, the eigenvalues must both be real. Moreover, one of the eigenvalues, $\lambda_s$, must satisfy $|\lambda_s| < 1$ and the other, $\lambda_u$, must satisfy $|\lambda_u| > 1$. By the results of section 2.1, the stable and unstable subspaces $W^s$ and $W^u$ must be lines through the origin in $\mathbf{R}^2$. Now let $[x, y] \in T$. Let $\ell_s$ and $\ell_u$ be lines in $\mathbf{R}^2$ which intersect at $(x, y)$ and which are parallel to $W^s$ and $W^u$ respectively. We denote the projections of these straight lines in $T$ by

$$W^s[x, y] = \pi(\ell_s)$$

$$W^u[x, y] = \pi(\ell_u).$$

The notation is meant to suggest that these lines project to the stable and unstable sets associated to $[x, y]$ in $T$, as we see from the following Proposition.

**Proposition 4.4.**
1. $W^s[x, y]$ *is the stable set associated to* $[x, y]$*, i.e., if* $[x', y'] \in W^s[x, y]$ *then* $d(L_A^n[x', y'], L_A^n[x, y]) \to 0$ *as* $n \to \infty$ *where* $d$ *is the distance in* $T$ *induced by the Euclidean distance along the stable set.*
2. *Similarly,* $W^u[x, y]$ *is the unstable set associated to* $[x, y]$*.*

*Proof.* We prove part 1; the proof of part 2 is similar.

Let $L(\mathbf{x}) = A \cdot \mathbf{x}$ be the linear map on $\mathbf{R}^2$. Let $(x, y)$ and $(x', y')$ lie on a line parallel to $W^s$ in $\mathbf{R}^2$. Let $\ell$ denote the line segment connecting $(x, y)$ to $(x', y')$. By linearity, $L^n(\ell)$ is a segment of a line which is also parallel to $W^s$. Moreover, length $(L^n(\ell)) = \lambda_s^n \cdot$ length $(\ell)$. Hence $d|L^n(x, y) - L^n(x', y')| \to 0$ as $n \to \infty$. It follows that $d(L_A^n[x, y], L_A^n[x', y']) \to 0$ as well.

**Proposition 4.5.** $W^s[x, y]$ *and* $W^u[x, y]$ *are dense in* $T$ *for each* $[x, y] \in T$*.*

*Proof.* First consider $W^s$ in $\mathbf{R}^2$. We claim that $W^s$ is a line with irrational slope in $\mathbf{R}^2$. For if this were not the case, $W^s$ would necessarily pass through a point with coordinates $(M, N)$ for $M, N \in \mathbf{Z}$. But then all of the $L$-iterates of $(M, N)$ would have integer coordinates, since $A$ is an integer matrix. But this is impossible, since $L^n(M, N) \to \mathbf{0}$ as $n \to \infty$.

Now consider the successive intersections of $W^s$ with the lines $y = N$ in $\mathbf{R}^2$. Let $x_N$ be the $x$-coordinate of the point on $W^s$ and $y = N$. Note that $x_1$ is the reciprocal of the slope of $W^s$, which is irrational. Also, $x_2 = 2x$, and, in general $x_N = Nx_1$.

On the torus, the point $(x_j, j)$ projects to a point of the form $[\alpha_j, 0]$, where $0 \le \alpha_j < 1$. The line $y = 0$ defines a circle in $T$ and the $\alpha_j$ are the successive images of [0] under an irrational translation of this circle. By Jacobi's Theorem (Theorem 3.13 of Chapter One), these points are dense in the circle. The result then follows easily.

In the general case, the lines $\ell_s$ similarly project to curves that wind densely about the torus, and $W^u$ is handled in an analogous fashion.

q.e.d.

The stable and unstable sets of a point have some special properties. They are preserved by $L_A$ in the sense that if $[x', y'] \in W^s[x, y]$, then $L_A[x', y'] \in W^s(L_A[x, y])$. Moreover, through each point $[x, y] \in T$, there is a unique stable and unstable set. Thus the stable and unstable sets give examples of a *foliation*. In two dimensions, a foliation is simply a collection of curves whose union is the entire space. The curves in this collection cannot cross each other, although they may be closed like a circle.

The stable and unstable set of any point in $T$ is the image of a straight line in $T$ under the projection $\pi$. By Proposition 4.5, each of these curves must wind densely about the torus. Since $W^s$ and $W^u$ have different slopes, it follows that their projections must meet at a dense set of points in $T$ as well. These points of intersection generalize the notion of a homoclinic point introduced in §1.16.

**Definition 4.6.** Let $[x, y] \in T$ be a periodic point for $L_A$. A homoclinic point to $[x, y]$ is a point $p \neq [x, y]$ which lies in $W^s[x, y] \cap W^u[x, y]$.

We remark that, for a hyperbolic toral automorphism, $W^s[x, y]$ and $W^u[x, y]$ always meet at a non-zero angle at a homoclinic point. When this happens. the homoclinic point is called *transverse*. Hence we have:

**Proposition 4.7.** *Transverse homoclinic points are dense in $T$.*

**Remark.** If $[x, y]$ is a periodic point for $L_A$, then any homoclinic point to $[x, y]$ tends to the orbit of $[x, y]$ under both forward and backward iteration of $L_A$. These points cannot be recurrent in the sense that their forward orbits continually return to any prescribed neighborhood. See Exercise 1.

These ideas also allow us to show that $L_A$ is topologically transitive. Let $U$ and $V$ be any two open sets in $T$, meaning, by definition, that the preimages under $\pi$ of $U$ and $V$ are open in $\mathbf{R}^2$. We will produce a point $[p]$ in $U$ and an integer $k$ such that $L_a^k[p] \in V$. We may select points $[r] \in U$ and $[s] \in V$ that are homoclinic to $[0]$.

Now let $\epsilon > 0$. Choose an open interval $I_u$ of length $\delta > 0$ in $W^u[0]$ and containing $[r]$. Similarly, choose $I_s$ in $W^s[0]$ containing $[s]$. $L_A^n$ expands $I_u$

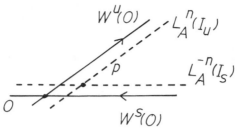

Fig. 4.2

by a factor of $|\lambda_u|^n$ and $L_A^{-n}$ expands $I_s$ by the same factor. Now choose $n$ large enough so that

1. $d(L_A^n[r], 0) < \epsilon/2$
2. $d(L_A^{-n}[s], 0) < \epsilon/2$
3. $|\lambda_u|^n \delta > \epsilon$.

Here the distance $d$ is the Euclidean distance defined in a neighborhood of $[0]$. Since $L_A^n(I_u)$ and $L_A^{-n}(I_s)$ are parallel to $W^u[0]$ and $W^s[0]$ respectively, it follows that $L_A^n(I_u) \cap L_A^{-n}(I_s) \neq \emptyset$. Let $[q]$ be a point in this intersection. See Fig. 4.2. Then $[p] = L_A^{-n}[q] \in U$ and $L_A^n[q] \in V$. Consequently $L_A^{2n}[p] \in V$, giving the required point.

We finally note that $L_A$ has sensitive dependence on initial conditions. Indeed, if $[p] \in T$ and $[q] \in W^u[p]$, then each iteration of $L_A$ lengthens the distance between images of $[p]$ and $[q]$, at least along $W^u[p]$. As a consequence, $L_A$ is chaotic on the entire torus. We single this fact out as a Theorem.

**Theorem 4.8.** *Let $L_A$ be a hyperbolic toral automorphism of $T$. Then*

1. *Periodic points of $L_A$ are dense in $T$.*
2. *$L_A$ is topologically transitive.*
3. *$L_A$ has sensitive dependence on initial conditions.*

Thus, a hyperbolic toral automorphism is chaotic on all of $T$. To study the dynamics of $L_A$ one can again invoke symbolic dynamics. The answer in this case is only partially satisfactory for, unlike the case of the horseshoe, we do not get a conjugacy with a shift map. Rather, we get some ambiguity in the choice of sequences.

To describe the symbolic dynamics, we need to introduce the concept of a *Markov partition*. For definiteness, let us fix the matrix

$$A = \begin{pmatrix} 1 & 1 \\ 1 & 0 \end{pmatrix}$$

and work with the hyperbolic linear automorphism induced by $A$ which we denote simply by $L$.

The eigenvalues of $A$ are $\frac{1}{2}(1 + \sqrt{5})$ and $\frac{1}{2}(1 - \sqrt{5})$. The eigenvector corresponding to the unstable eigenvalue is the line $y = \frac{1}{2}(-1+\sqrt{5})x$, whereas the stable eigenvalue has eigenvector $y = -\frac{1}{2}(1 + \sqrt{5})x$.

As in the case of the horseshoe, we will first construct some rectangles with sides on the stable and unstable set of $[0]$. To be precise, consider the interval from $a$ to $b$ in $W^s[0]$ and the interval from $c$ to $d$ in $W^u[0]$ as depicted in Fig. 4.3. These two intervals define three rectangles in the torus which we denote by $R_1, R_2$, and $R_3$. Two sides of each rectangle lie in the interval from $a$ to $b$. We call these sides the stable boundaries. Note that $L$ maps this interval inside itself. As a consequence, if $[p]$ is any point on a stable boundary of one of the $R_i$, then the entire forward orbit of $[p]$ lies in this interval.

Similarly, the unstable boundaries of the $R_i$ lie in the interval from $c$ to $d$. This interval is contracted by $F^{-1}$, so that the entire backward orbits of points in the unstable boundary lie in this set. See Fig. 4.4.

To study the dynamics on $T$, we note first that whenever $L(R_i)$ meets the interior of $R_j$, the image cuts completely across $R_j$ in the unstable direction. Similarly, whenever $L^{-1}(R_i)$ meets the interior of $R_j$, the image cuts completely across $R_j$ in the stable direction. Rectangles which have this property and whose boundaries lie in the stable and unstable sets are said to form a Markov partition for the map. This partition allows us to define the symbolic dynamics just as in the horseshoe map, since forward images of the $R_i$ always give "unstable" rectangles and backward images give "stable" rectangles.

Note that $L(R_1)$ cuts across the interior of both $R_2$ and $R_3$, $L(R_2)$ cuts across the interior of $R_1$ and $R_3$ and $L(R_3)$ meets only the interior of $R_2$. For the moment, we ignore the fact that $L(R_1)$ meets $R_1$ along the boundary of $R_1$ and $R_2$. Similarly, the intersections $L(R_2) \cap R_2$, $L(R_3) \cap R_1$ and $L(R_3) \cap R_2$ are non-empty but are contained in the boundaries of the rectangles. Thus one would hope to set up an equivalence with the subshift of finite type $\Sigma_B$ where the transition matrix $B$ is given by

$$B = \begin{pmatrix} 0 & 1 & 1 \\ 1 & 0 & 1 \\ 0 & 1 & 0 \end{pmatrix}.$$

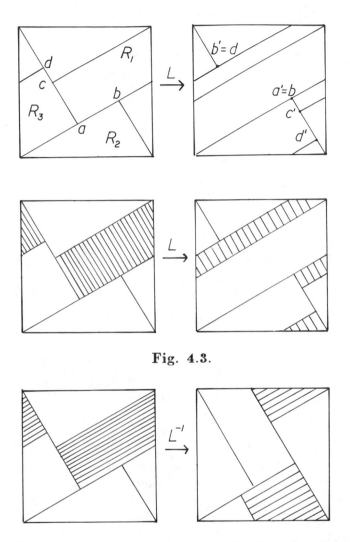

Fig. 4.3.

**Fig. 4.4.** The action of $L$ on the Markov partition.

There are obvious problems with this, however. For one thing, none of the fixed points $(\dots 111 \dots)$, $(\dots 222 \dots)$, and $(\dots 333 \dots)$ are allowed sequences in $\Sigma_B$, yet we know that there is a fixed point in $T$, namely $[0]$. Moreover, there is an ambiguity in our assignment of sequences when the point or one of its images lies on one of the boundaries of a rectangle.

To remedy these problems, we will work with a *quotient* of the subshift. Suppose a point $p$ lies on the stable boundary of $R_2 \cap R_3$. Let $S(p) =$

$(\ldots s_0 s_1 s_2 \ldots)$ be the sequence naturally associated to $p$. Since $p \in R_2 \cap R_3$, we must have either $s_0(p) = 2$ or $s_0(p) = 3$. Now $L(p)$ lies in the intersection $R_1 \cap R_2$, as we see in Fig. 4.5.

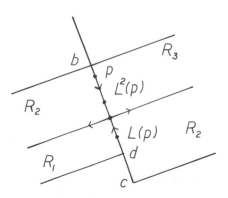

**Fig. 4.5.**

Subsequent images of $p$ hop back and forth between $R_2 \cap R_3$ and $R_1 \cap R_2$. Now let us return to $S(p)$. If $s_0(p) = 2$, then $s_1(p)$ is either 1 or 2. But 2 cannot follow 2, so $s_1 = 1$. Continuing this argument, we must have $S(p) = (\ldots s_{-1}.2121\ldots)$. On the other hand, if $s_0(p) = 3$, then we must have $S(p) = (\ldots s_{-1}.3232\ldots)$. This means that the two possible choices $(\ldots s_{-2}s_{-1}.2121\ldots)$ and $(\ldots s_{-2}s_{-1}.3232\ldots)$ must represent the same point in $T$, i.e., these sequences should be identified.

More generally, sequences of either form

$$(\ldots s_{k-1}s_k 2121 \ldots)$$

$$(\ldots s_{k-1}s_k 3232 \ldots)$$

should also be identified, as they represent points which eventually land on $W^s[0]$.

There are two ambiguities in $W^u[0]$. We leave it to the reader to check, using Fig. 4.6, that the pair of sequences of the form

$$(\ldots 1212 s_k s_{k+1} \ldots)$$

$$(\ldots 2121 s_k s_{k+1} \ldots)$$

should be identified, as should sequences of the form

$$(\ldots 2323 s_k s_{k+1} \ldots)$$

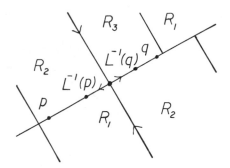

**Fig. 4.6.**

$$(\ldots 3232 s_k s_{k+1} \ldots).$$

Now, let $\tilde{\Sigma}_B$ denote the "quotient" of the subshift of finite type obtained by making all of the above identifications. Note that $\sigma$ is naturally defined on $\tilde{\Sigma}_B$. We leave it to the reader to check that $S$ gives a one-to-one and onto map from $T$ to $\tilde{\Sigma}_B$ which conjugates $L$ with $\sigma$. That is,

$$
\begin{array}{ccc}
T & \overset{L}{\to} & T \\
S \downarrow & & \downarrow S \\
\tilde{\Sigma}_B & \overset{\sigma}{\to} & \tilde{\Sigma}_B.
\end{array}
$$

**Remarks.**

**1.**   All points of the form $(\ldots 1212 \ldots)$, $(\ldots 2121 \ldots)$, $(\ldots 2323 \ldots)$, and $(\ldots 3232 \ldots)$ should be regarded as identical in $\tilde{\Sigma}_B$. Since $\sigma(\ldots 1212 \ldots) = (\ldots 2121 \ldots)$, it follows that these are the sequences which represent the fixed point at $[0]$.

**2.**   The only identifications in the above procedure occur on $W^s[0]$ and $W^u[0]$, where the dynamics of $L$ are relatively straightforward. All of the other periodic points for $L$ occur in the complement of these two sets. Admittedly, $W^s[0]$ and $W^u[0]$ are dense in $T$. Nevertheless, $S$ is well-behaved on the complement and completely describes the dynamics there.

**3.**   We will not discuss the continuity of $S$ as the identifications in $\tilde{\Sigma}_B$ mean that the topology on the sequence space is different from the usual one.

The Markov partition constructed above is simultaneously quite general and quite special. It is special because the elements of the partition are actual rectangles; all we really needed was that the boundaries of the elements of the partition lie in appropriate stable and unstable sets. The use of known stable and unstable sets as above to construct Markov partitions is a completely general operation. All that is necessary is that the map preserve these sets. For example, the vertical rectangles used in the construction of the horseshoe is a Markov partition for the associated Cantor set $\Lambda$. Note that no identifications in the sequence space are necessary in this case since there are no overlapping rectangles.

## Exercises

1.  Let $L_A$ be a hyperbolic toral automorphism. Prove that:
    a. transverse homoclinic points are dense in $T$.
    b. all points in $T$ are nonwandering (in the sense of Exercise 1.7.2).
    c. homoclinic points are not recurrent points (in the sense of Exercise 1.7.3).

2.  One may define an $n$-dimensional torus $T^n$ in exact analogy with our construction of the two-dimensional torus in this section. That is, let $[x_1, \ldots, x_n]$ denote the set of all equivalence classes of points in $\mathbf{R}^n$ under the equivalence relation

$$(x_1, \ldots, x_n) \sim (y_1, \ldots, y_n)$$

if and only if $x_j - y_j$ is an integer for each $j$. The $n$-torus is then simply the set of all such equivalence classes of points in $\mathbf{R}^n$. Similarly, one may define a hyperbolic toral automorphism on $T^n$ by starting with a matrix $A$ which satisfies the conditions in Definition 4.1. Note that the stable and unstable sets need no longer be curves in $T^n$.
    a. Prove that the induced hyperbolic toral automorphism on $T^n$ has dense periodic points.
    b. Prove that if $[p] \in T^n$, then $W^s[p]$ and $W^u[p]$ are dense in $T^n$.
    c. Prove that a hyperbolic toral automorphism is chaotic on $T^n$.

3.  Prove that $T^n$ is homeomorphic to the $n$- fold cross product

$$S^1 \underbrace{\times \ldots \times}_{n\ factors} S^1.$$

4.  Consider the map $A: \mathbf{R}^n \to \mathbf{R}^n$ given by $A(\mathbf{x}) = 2\mathbf{x}$. $A$ induces a map on $T^n$ exactly as in the case of a hyperbolic toral automorphism, but the induced map is no longer a diffeomorphism.
    a. Prove that periodic points are dense for this map.

   b. Prove that eventually fixed points are dense.

   c. Prove that this map is chaotic on $T^n$.

5.  Let
$$A = \begin{pmatrix} 2 & 1 \\ 1 & 1 \end{pmatrix}.$$

Construct a Markov partition for $L_A$.

6.  Let $L_A$ be a hyperbolic toral automorphism on $T$. Let $[p] \in W^s[0] \cap W^u[0]$ be a homoclinic point. Let $\ell_s$ be the segment in $W^s[0]$ connecting $[0]$ to $[p]$ and let $\ell_u$ be a similar segment in $W^u[0]$. Construct a rectangle $R$ containing $\ell_s$ with sides in stable and unstable sets.

   a. Show that there is an integer $n$ such that $L_A^n(R) \supset \ell_u$.

   b. Prove that we may choose $[p]$ so that $L_A^n : R \to L_A^n(R)$ is topologically conjugate to the linear map which produced the horseshoe in §2.3.

## §2.5 ATTRACTORS

   In this section, we introduce a third type of dynamical phenomenon which is higher dimensional in nature, the attractor. Roughly speaking, an attractor is an invariant set to which all nearby orbits converge. Hence attractors are the sets that one "sees" when a dynamical system is iterated on a computer. Thus far, all of the attractors we have encountered have been fixed or periodic points. Here we introduce two new and much more complicated attractors, the solenoid and the Plykin attractor. These are examples of a special type of attractor known as a transitive or hyperbolic attractor. We will see that these attractors are similar in many respects to the horseshoe map and the hyperbolic toral automorphisms. For example, there is a set on which the map is chaotic and, through each point in this set, there passes a stable and an unstable set. Since these are familiar phenomena, we will leave many of the details in the verification to the reader.

   The solenoid is an attractor which is contained in a "solid" torus. This space is defined as follows. Let $S^1$ be the unit circle and let $B^2$ be the unit disk in the plane; that is

$$B^2 = \{(x, y) \in \mathbf{R}^2 | x^2 + y^2 \leq 1\}.$$

The Cartesian product $D = S^1 \times B^2$ is a solid torus in $\mathbf{R}^3$. Its boundary is a torus as described in the previous section. To define the solenoid, we consider the map $F$ which maps $D$ strictly inside itself by the formula:

$$F(\theta, p) = (2\theta, \frac{1}{10}p + \frac{1}{2}e^{2\pi i\theta})$$

where $p \in B^2$ and $e^{2\pi i\theta} = (\cos(2\pi\theta), \sin(2\pi\theta)) \in S^1$.

Geometrically, $F$ may be described as follows. Let $\theta^* \in S^1$. The disk $B(\theta^*)$ which is given by $\theta = \theta^*$ and $p$ arbitrary is mapped by $F$ into another disk given by $B(2\theta^*)$. The image of this disk is a disk of radius $1/10$ with center at the point $\frac{1}{2}(\cos(2\theta^*), \sin(2\theta^*))$ in $B(2\theta^*)$. See Fig. 5.1. The disk located at $\theta = \theta^* + \pi$ is also mapped into the disk given by $\theta = 2\theta^*$, but its image is a small disk of radius $1/10$ diametrically opposite the image of $B(\theta^*)$ in $B(2\theta^*)$.

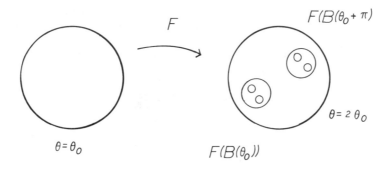

**Fig. 5.1.** Construction of the solenoid.

Globally, $F$ may be interpreted as follows. In the $\theta$ coordinate, $F$ is simply the doubling map of the circle discussed in Example 3.4 of Chapter One. In the $B^2$-direction, $F$ is a strong contraction, with image a disk whose center depends on $\theta$. The image of this disk is one-tenth the size of the original disk. Thus the image of $D$ is another solid torus inside $D$ which wraps twice around $D$. See Fig. 5.2.

The fact that $F$ stretches in one direction and contracts in the others is, by now, a familiar phenomenon, reminiscent of both the horseshoe and the hyperbolic toral automorphisms.

Strictly speaking, $F$ is not a diffeomorphism, since it is not onto. We think of $D$ as a piece of a larger space and the action of $F$ on $D$ as just a portion of the dynamics. Since $F(D) \subset D$, it follows that all forward orbits of points in $D$ lie in $D$. Regions like $D$ have a special name.

**Definition 5.1.** A closed region $N \subset \mathbf{R}^n$ is a trapping region in $F$ if $F(N)$ is contained in the interior of $N$.

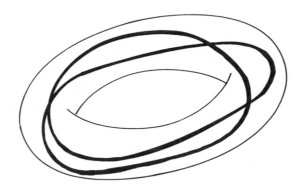

**Fig. 5.2.** The image of the solid torus under $F$ is a
solid torus which wraps twice around itself.

Since $F(N)$ is closed and $F(N) \subset N$, it follows that the sets $F^n(N)$ are
all closed and nested for $n \geq 0$. Therefore

$$\Lambda = \bigcap_{n \geq 0} F^n(N)$$

is a closed, nonempty set. $\Lambda$ is the set of points whose full orbits, both
forward *and* backward, remain in $N$ for all time. $\Lambda$ will be our attractor.

**Proposition 5.2.** $\Lambda$ *is an invariant set.*

*Proof.* We have

$$F(\Lambda) = F(\bigcap_{n \geq 0}^{\infty} F^n(N)) = \bigcap_{n \geq 1}^{\infty} F^n(N) \subset N.$$

But

$$\bigcap_{n \geq 0}^{\infty} F^n(N) = \bigcap_{n \geq 1}^{\infty} F^n(N)$$

since the intersections are nested. Hence $F(\Lambda) = \Lambda$ and $\Lambda$ is invariant.
Invariance under $F^{-1}$ follows as well.

q.e.d.

**Definition 5.3.** A set $\Lambda$ is called an attractor for $F$ if there is a neighborhood
$N$ of $\Lambda$ for which the closure of $N$ is a trapping region and

$$\Lambda = \bigcap_{n \geq 0} F^n(N)$$

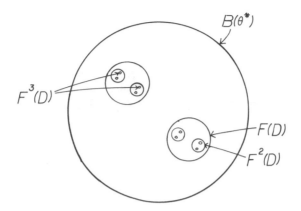

**Fig. 5.3.** The intersection of the $F^n(D)$ yields a Cantor set.

There are other definitions of attractors in common use. Ours is by no means standard, although it is perhaps the simplest. This definition suffers the defect that it does not produce a single, indecomposable attractor. For example, the "stadium" $D$ for the horseshoe map of §2.3 is a trapping region. The attractor is easily seen to consist of two pieces, the fixed point in $D_1$ and the invariant Cantor set together with all of its unstable sets. On the other hand, the region $D_1 \subset D$ is also a trapping region, but this time the attractor is quite different; it is simply the fixed point in $D_1$.

To remedy this, we introduce the following terminology:

**Definition 5.4.** $\Lambda$ is a transitive attractor for $F$ if $F$ is topologically transitive on $\Lambda$.

Our goal is to show that the attractor $\Lambda = \cap_{n\geq 0}F^n(D)$ for the above map is a transitive attractor and that, moreover, the dynamics of $F$ on $\Lambda$ are chaotic.

Let us investigate the nature of the set $\Lambda$. Since $F$ stretches $D$ in the $S^1$-direction and contracts it by a factor of $1/10$ in the $B^2$-direction, it follows that $F(D)$ is a torus of radius $1/10$ which wraps around $D$ twice. Applying $F$ to $F(D)$, we see that $F^2(D)$ is a torus of radius $1/100$ in the $B^2$-direction which wraps around $D$ four times and which is properly contained in $F(D)$. Inductively, $F^n(D)$ is a torus of radius $1/10^n$ which wraps around $D$ exactly $2^n$ times and which is contained in $F^{n-1}(D)$.

In each $B(\theta^*)$, we therefore see that $F^n(D)$ is a nested collection of $2^n$ disks, as in Fig. 5.3. We have seen this process before: the nesting of the $F^n(D)$ yields a Cantor set in each disk $B(\theta^*)$.

If we perform the above construction in a cylindrical piece of $D$ of the form

$$C = \{(\theta, p)|\theta_1 \leq \theta \leq \theta_2\}.$$

we see that $C \cap \Lambda$ is locally the Cartesian product of a Cantor set and an arc in the $S^1$-direction. The arcs are given by the nested intersection of the $2^n$ tubes in $F^n(D) \cap C$. Since each iteration of $F$ contracts the radius of these tubes by $1/10$, it is intuitively clear that these arcs are continuous. Nevertheless, we will prove this later by completely different methods. In fact, it may be shown that these curves are smooth. The set $\Lambda$ is called a *solenoid*.

We now turn to the dynamics of $F$ on and near $\Lambda$. Let $x \in \Lambda$. Suppose $x = (\theta_0, p_0)$ where $\theta_0 \in S^1$ and $p_0 \in B^2$. Let $F^n(x) = (\theta_n, p_n)$. Consider the disk $B(\theta_0)$. Since $F$ maps $B(\theta_0)$ inside $B(2\theta_0)$, it follows that $F^n(B(\theta_0)) \subset B(\theta_n)$. Moreover, each application of $F$ contracts $B(\theta_0)$ by a factor of $1/10$. Therefore, if $y \in B(\theta_0)$, it follows that $F^n(y) \in B(\theta_n)$ and $|F^n(x) - F^n(y)| < 1/10^n$, where the absolute value is the usual one in $\mathbf{R}^2$. Consequently, $B(\theta_0)$ is part of the stable set $W^s(x)$ associated to $x$.

Similarly, the arc constructed above as the nested intersection of tubes about $x$ is part of the unstable set for $x$ which we denote by $W^u(x)$. This follows since $F^{-1}$ contracts distances along the arc by a factor of $1/2$. We thus see that all of the points in $\Lambda$ come equipped with stable and unstable sets, just as in the cases of the horseshoe and the hyperbolic toral automorphisms.

**Proposition 5.5.**

1. *$F$ has sensitive dependence on initial conditions on $\Lambda$.*
2. *$Per(F)$ is dense in $\Lambda$.*
3. *$F$ is topologically transitive on $\Lambda$.*

*Proof.* For sensitive dependence on initial conditions, we simply note that any point on the unstable arc associated to $x \in \Lambda$ separates from $x$ by a factor of 2 in the $\theta$-direction when $F$ is iterated. To prove density of periodic points, let $U$ be any neighborhood of $x = (\theta_0, p_0)$. There exists $\delta > 0$ and $n \in \mathbf{Z}$ such that the tube $C$ in $F^n(D)$ defined by

$$C = \left\{(\theta, z)||\theta - \theta_0| < \delta, |z - p_0| < \frac{1}{10^n}\right\}$$

is completely contained in $U$. We will produce a periodic point in $C$. To accomplish this, recall that $F^n(D)$ wraps around $D$ exactly $2^n$ times. We may choose $m$ so that $2^m \delta > 2^{n+1} \cdot 4\pi$. Hence $F^m(C)$ is a tube lying in

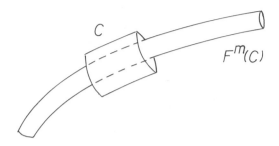

**Fig. 5.4.** The image $F^m(C)$ cuts through $C$.

$F^n(D)$ and wrapping around $D$ at least $2 \cdot 2^n$ times. It follows that $F^m(C)$ cuts completely across $C$ at least once as shown in Fig. 5.4. Hence there exists $\theta^*$ with $|\theta^* - \theta_0| < \delta$ such that $F^m(B(\theta^*) \cap C) \subset B(\theta^*) \cap C$. It follows that $F^m$ has a fixed point in $B(\theta^*) \cap C$.

Similar arguments also prove topological transitivity. For if $x, y, \in \Lambda$ and $U$ and $V$ are neighborhoods of $x$ and $y$, we may then produce tubes as above in $F^n(D)$ about $x$ and $y$ which are completely contained in $U$ and $V$ respectively. Sufficiently many iterations of these tubes produce a $\theta^*$ such that $B(\theta^*) \cap U$ is a disk which is mapped into $V$. It is easy to check that there is a point in $\Lambda$ inside $B(\theta^*) \cap U$.

q.e.d.

As in our previous examples, we may use symbolic dynamics to model the dynamics of $F$ on $\Lambda$. This time we use a different construction first introduced by R.F. Williams. Let $g: S^1 \to S^1$ be the doubling map $g(\theta) = 2\theta$. Our model for $\Lambda$ will be the *inverse limit space*

$$\Sigma = S^1 \overset{g}{\leftarrow} S^1 \overset{g}{\leftarrow} S^1 \dots .$$

More precisely

$$\Sigma = \{\theta = (\theta_0 \theta_1 \theta_2 \dots) | \theta_j \in S^1 \text{ and } g(\theta_{j+1}) = \theta_j\}.$$

Thus $\Sigma$ consists of all infinite sequences of points of $S^1$ subject to the restriction that $\theta_{j+1}$ is one of the two preimages of $\theta_j$ for each $j$. Unlike our previous sequence spaces, elements of $\Sigma$ are not sequences whose entries are integers. Rather, the entries in this case are points in the circle. For example, the sequences

$$(0\,0\,0\dots)$$

$$(0\,\pi\,\frac{\pi}{2}\,\frac{\pi}{4}\,\frac{\pi}{8}\dots)$$

$$\left(\frac{\pi}{3} \ \frac{2\pi}{3} \ \frac{\pi}{3} \ \frac{2\pi}{3} \ \frac{\pi}{3} \cdots\right)$$

all belong to $\Sigma$. Using the doubling map $g$, it is helpful to think of these sequences as backward orbits:

$$0 \xleftarrow{g} 0 \xleftarrow{g} 0 \xleftarrow{g} \cdots$$

$$0 \xleftarrow{g} \pi \xleftarrow{g} \frac{\pi}{2} \xleftarrow{g} \frac{\pi}{4} \xleftarrow{g} \cdots$$

$$\frac{\pi}{3} \xleftarrow{g} \frac{2\pi}{3} \xleftarrow{g} \frac{\pi}{3} \xleftarrow{g} \frac{2\pi}{3} \xleftarrow{g} \cdots$$

We define a metric on $\Sigma$ much as we did on $\Sigma_n$. If $\Theta = (\theta_0 \theta_1 \theta_2 \ldots)$ and $\Psi = (\psi_0 \psi_1 \psi_2 \ldots)$ are points in $\Sigma$, we define the distance between them to be

$$d[\Theta, \Psi] = \sum_{j=0}^{\infty} \frac{\left|e^{2\pi i \theta_j} - e^{2\pi i \psi_j}\right|}{2^j}$$

where $|\alpha - \beta|$ denotes the usual Euclidean distance in the plane. It is easy to check that $d$ is a metric on $\Sigma$. Moreover, two points are "close" if each of their first few entries are close together.

On $\Sigma$, we have a natural map, a version of the shift given by

$$\sigma(\theta_0 \theta_1 \theta_2 \ldots) = (g(\theta_0) \theta_0 \theta_1 \theta_2 \ldots).$$

As in previous sections, $\sigma$ is easily seen to be a homeomorphism. The inverse of $\sigma$ is given by a map that resembles our previous shift (but which is a homeomorphism)

$$\sigma^{-1}(\theta_0 \theta_1 \theta_2 \ldots) = (\theta_1 \theta_2 \theta_3 \ldots).$$

As with our previous models, this map is also easy to understand dynamically. If $\theta$ is a periodic point for $g$, with period $n$, then the repeating sequence $(\theta, g^{n-1}(\theta), g^{n-2}(\theta), \ldots, g(\theta), \theta, \ldots)$ is clearly periodic for $\sigma$ with period $n$ as well. As with our other examples, it is easy to check that $\sigma$ has periodic points which are dense in $\Sigma$ and that $\sigma$ has a dense orbit. See Exercises 3-4.

How are $\sigma$ and $F$ related? Let $\pi: D \to S^1$ be the natural projection, i.e., $\pi(\theta, p) = \theta$. For any point $x \in \Lambda$, the map $S: \Lambda \to \Sigma$ given by

$$S(x) = (\pi(x), \ \pi F^{-1}(x), \ \pi F^{-2}(x), \ldots$$

is well defined. This follows since we can invert $F$ on $\Lambda$ even though $F^{-1}$ is not defined on all of $D$. Clearly, $S \circ F = \sigma \circ S$, since $F$ is the doubling map in the $S^1$- direction.

We leave it as an exercise for the reader to prove that:

**Theorem 5.6.** *S gives a topological conjugacy between F on Λ and σ on Σ.*

Let us use this conjugacy to fill in the gap above where we failed to prove that the unstable sets in Λ were curves. For simplicity, let us prove this only for the fixed point which corresponds to the sequence $\mathbf{0} = (000\ldots)$. One checks easily that this is the point $\theta = 0$ and $p = (\frac{5}{9}, 0) \in \mathbf{R}^2$.

**Proposition 5.7.** *The unstable set of $\mathbf{0}$ consists of precisely those sequences of the form*

$$(x, \frac{x}{2}, \frac{x}{2^2}, \frac{x}{2^3}, \ldots)$$

*for any $x \in \mathbf{R}$.*

*Proof.* Since $\sigma^{-1}(x, \frac{x}{2}, \frac{x}{4}, \ldots) = (\frac{x}{2}, \frac{x}{4}, \frac{x}{8}, \ldots)$ it follows that $\sigma^{-n}(x, \frac{x}{2}, \frac{x}{4}, \ldots) \to \mathbf{0}$ as $n \to \infty$. For the converse, we first recall that if $\theta \in S^1$, then $g^{-1}(\theta)$ is one of $\frac{\theta}{2}$ or $\frac{\theta}{2} + \pi$. Now let $\Theta = (\theta_0 \theta_1 \theta_2, \ldots) \in W^u(\mathbf{0})$. There exists $N$ such that if $n \geq N$, $|\theta_n| < 1$. Hence $\theta_N, \theta_{N+1}, \theta_{N+2}, \ldots$ all lie in the right hand semicircle in $S^1$. It follows that $\theta_{N+1} = \theta_N/2$, for the other preimage $(\theta_N/2) + \pi$ lies in the left semicircle. Continuing, we find $\theta_{N+k} = \frac{\theta_N}{2^k}$ and $\theta_{N-k} = 2^k \theta_N$ so that $\Theta$ assumes the desired form.

q.e.d.

Consequently, the unstable set of $\mathbf{0}$ in $\Sigma$ is parametrized by $\mathbf{R}$. Under the conjugacy given by $S$, the unstable set of the fixed point is the continuous curve which is the image of $W^u(\mathbf{0})$.

The inverse limit construction works well for a class of attractors known as expanding attractors. These attractors are characterized by uniform expansion within the attractor itself. As in the case of the solenoid, such attractors can be suitably modeled by an inverse limit of a lower dimensional expanding map like $\theta \to 2\theta$ on $S^1$. The main difference in the general case is that the model space is more complicated than $S^1$; usually it is a "branched manifold." This concept was introduced by R.F. Williams. We will illustrate it via an example of an attractor due to Plykin. Rather than give a formula for this map, we will define it geometrically, exactly as we did for the horseshoe.

Consider the region $R$ in the plane depicted in Fig. 5.5. $R$ is a region with three open half-disks removed. We equip $R$ with a foliation whose leaves are intervals as shown in Fig. 5.5. Recall that this means that there is a line

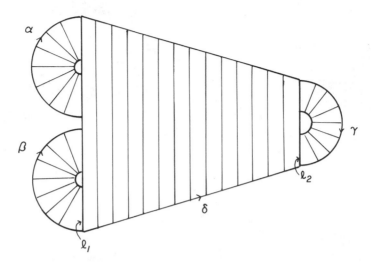

**Fig. 5.5.** The region $R$ for the Plykin attractor.

segment through each point of $R$ (the leaf) and that the leaves are mutually disjoint.

Define a map $P: R \to R$ as shown in Fig. 5.6. We require that $P$ preserve and contract the leaves of the foliation. Note that $P(R)$ is contained in the interior of $R$ so that $R$ is a trapping region. The set $\Lambda = \cap_{n \geq 0} P^n(R)$ is the Plykin attractor.

To understand the dynamics of $P$, we first note that any two points on the same leaf behave identically under iteration of $P$. Since the leaves are contracted, any two such points tend to the attractor in the same asymptotic manner. Thus, to understand the action of $P$ globally, it suffices to understand the action of $P$ on the leaves. We thus collapse each leaf to a point as in Fig. 5.7., and examine the induced map on this space. Observe that the collapsed space $\Gamma$ has "branch" points along the singular leaves $\ell_1$ and $\ell_2$. It is called the branched "manifold" for $P$. We may describe the dynamics on $\Gamma$ by describing how each of the four intervals $\alpha, \beta, \gamma$, and $\delta$ are mapped. From Fig. 5.7 we see that the induced map $g$ on $\Gamma$ preserves the two vertices and maps the other intervals this way:

$$\alpha \to \beta$$

$$\beta \to \beta + \delta + \gamma - \delta - \beta$$

$$\gamma \to \alpha$$

$$\delta \to \delta - \gamma - \delta$$

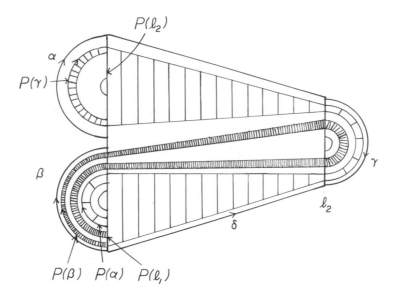

**Fig. 5.6.** The Plykin attractor.

where the signs indicate orientations or directions in which the image crosses the given interval. We may construct such a map so that $g$ expands all distances in the branched manifold $\Gamma$.

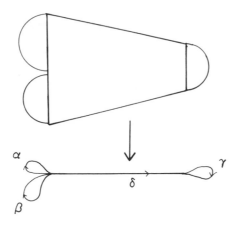

**Fig. 5.7.** The branched manifold for the Plykin attractor.

In the solenoid, a similar construction would have collapsed the $B^2$-directions (the leaves of the foliation of $D$ ) onto a circle (an unbranched manifold) on which the map $g$ is simply $\theta \rightarrow 2\theta$. Since we understand the dynamics of $\theta \rightarrow 2\theta$ completely, we were able to use the inverse limit construction to analyze the solenoid as well. The same process works for the Plykin attractor.

For example, we may prove that $g: \Gamma \rightarrow \Gamma$ has dense periodic points as follows. Let $I$ be any "subinterval" in $\Gamma$. Since $g$ is expanding, it follows that there exists $n$ such that $g^n(I)$ covers one of the four intervals $\alpha, \beta, \gamma, \delta$. Now one may check easily that there is an integer $m$ such that $g^m(\xi) \supset \Gamma$ where $\xi$ is any of the $\alpha, \beta, \gamma$, or $\delta$. Indeed, $g(\alpha) = \beta$, $g(\beta) \supset \gamma$, and $g(\gamma) = \alpha$ so that $g^3(\alpha) \supset \alpha$. Thus we conclude that $g^{m+n}(I) \supset \Gamma$ and so it follows that there is a periodic point in $I$. Using the inverse limit construction, one may then equate the action of $P$ on $\Lambda$ with that of the shift on $\Sigma = \Gamma \xleftarrow{g} \Gamma \xleftarrow{g} \Gamma \ldots$. We leave the details to the reader.

**Remarks.**

**1.**    Much recent research has been devoted to the topic of "strange attractors." These are loosely defined as attractors which are topologically distinct from either a periodic orbit or a "limit cycle" (i.e., an invariant, attracting simple closed curve which arises often in ordinary differential equations). We prefer the term "hyperbolic" attractor for attractors like the solenoid and the Plykin example. Indeed, since we have succeeded in analyzing these maps completely, there is nothing whatsoever "strange" about them.

**2.**    There are, however, some attractors which have thus far defied analysis. One of these is the Hénon attractor as described in Exercise 10. Numerical evidence indicates that this simple quadratic map of the plane possesses a transitive attractor, although this has never been proved rigorously. We urge the reader with access to computer graphics to plot successive iterates of a point under this map. The result is always qualitatively the same (disregarding the first few iterates) and always fascinating! We will return to this map in §2.9, where we will approach it from a different point of view.

**Exercises**

**1.**    Construct a Markov partition for the solenoid.

**2.**    Prove that

$$d[\Theta, \Psi] = \sum_{j=0}^{\infty} \frac{|\theta_j - \psi_j|}{2^j}$$

is a metric on $\Sigma$, where $\Theta$ and $\Psi$ are points in $\Sigma$.

In the following exercises, let $\sigma: \Sigma \rightarrow \Sigma$ be the shift map on the inverse limit space $S^1 \xleftarrow{g} S^1 \xleftarrow{g} S^1 \ldots$ for the solenoid.

3.  Prove that the periodic points of $\sigma$ are dense in $\Sigma$.

4.  Prove that $\sigma$ has a dense orbit.

5.  Prove that $\sigma$ is a homeomorphism.

6.  Prove that $S\colon \Lambda \to \Sigma$ gives a topological conjugacy between $F$ and $\sigma$.

7.  Let $P\colon R \to R$ be the Plykin map as defined in Fig. 5.6 and let $\Lambda \subset R$ be the Plykin attractor with associated branched manifold $\Gamma$.
    a.  Prove that $P$ is chaotic on $\Lambda$.
    b.  Prove that $P$ on $\Lambda$ is topologically conjugate to the shift map on the inverse limit space
    $$\Gamma \xleftarrow{g} \Gamma \xleftarrow{g} \Gamma \xleftarrow{g} \ldots .$$

8.  Let $g\colon \Gamma \to \Gamma$ be the expanding map on the branched manifold as defined in Fig. 5.7. Find a formula for the number of periodic points for $g$ of period $n$.

9.  *The DA map.* In this series of exercises, we show how the hyperbolic toral automorphism of the previous section may be modified to produce a map with a transitive attractor on a torus.
    a.  Consider the linear map $L$ given by
    $$x_1 = \frac{1}{2}x$$
    $$y_1 = 2y.$$

    Explicitly construct a map $\phi$ depending on $x$ alone such that the phase portrait of $\phi \circ L$ is as shown in Fig. 5.8.

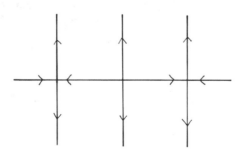

**Fig. 5.8.**

    b.  Use bump functions as described in §1.2 to construct a new map $\psi$ which agrees with $\phi$ on a small neighborhood $U$ of $0$, which is the identity map outside of a neighborhood $V$ which contains $U$, and

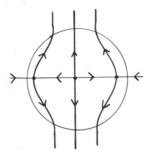

**Fig. 5.9.**

which preserves horizontal lines in $\mathbf{R}^2$. Show that $\psi$ may be chosen so that the phase portrait of $\psi \circ L$ is as depicted in Fig. 5.9.

c. Apply this technique on a neighborhood of $[0]$ in the torus to the hyperbolic linear automorphism generated by the matrix

$$\begin{pmatrix} 1 & 1 \\ 1 & 0 \end{pmatrix}$$

as described in §2.4. The resulting map $F$ is called a $DA$-map. Prove that this map is a diffeomorphism.

d. Show that there is a neighborhood $W$ of $\mathbf{0}$ in $T$ such that $F$ maps $T - W$ strictly inside itself.

e. Prove that $\Lambda = \cap_{u=0}^{\infty} F^n(T - U)$ is a transitive attractor by showing that the original stable foliation for the hyperbolic toral automorphism is still preserved on $T - U$ by $F$.

**10.** *The Hénon Attractor.* Consider the diffeomorphism of the plane given by

$$x_1 = 1 + y - 1.4x^2$$
$$y_1 = 0.3x.$$

This is a specific case of the Hénon map which we will study in §2.9.

a. Consider the quadrilateral $Q$ in the plane whose vertices are given by the four points $(-1.33, 0.42)$, $(1.32, 0.133)$, $(1.245, -0.14)$, and $(-1.06, -0.5)$. Prove that $F(Q) \subset Q$.

b. Using a computer, compute 10,000 iterates of a point in $Q$. Plot the last 9,000 points. Note that the resulting picture appears to be the same no matter which (random) initial point is chosen. This is the "strange attractor" of Hénon whose structure is still not completely understood.

**11.**  *The Lozi Attractor.*  Consider the piecewise linear map of the plane given by

$$L\begin{pmatrix} x \\ y \end{pmatrix} = \begin{pmatrix} 1 + y & -A|x| \\ & \\ Bx & \end{pmatrix}$$

where $A$ and $B$ are parameters. Assume that $A$ and $B$ satisfy $0 < B < 1$, $A > B + 1$ and $2A + B < 4$. Under these conditions,

    a. Prove that $L$ has two fixed points, one of which lies in the first quadrant. We call this point $p$.

    b. Prove that the unstable set $W^u(p)$ contains a straight line which intersects the $x$-axis at a point $q$ and the $y$-axis at $L^{-1}(q)$.

    c. Let $\ell$ denote the straight line segment in $W^u(p)$ connecting $q$ and $L^{-1}(q)$. Sketch $L(\ell)$ and $L^2(q)$.

    d. Construct the triangle $T$ with vertices at $q$, $L(q)$, and $L^2(q)$. Prove that $T$ is a trapping region for $L$.

    e. Use a computer to plot the forward orbits of points in $T$. The result is a picture of the Lozi attractor.

## §2.6 THE STABLE AND UNSTABLE MANIFOLD THEOREM

The examples discussed in the last three sections all share one common feature. Through each point in the "interesting" set where chaotic dynamics is present, there passes both a stable and an unstable set. The crucial property that gives this behavior is *hyperbolicity*, which we investigate in more detail in this and the next section. Recall that a linear map is hyperbolic if it has no eigenvalues on the unit circle. In the hyperbolic case, we distinguished two invariant subspaces through $\mathbf{0}$, the stable and unstable subspaces $W^s$ and $W^u$. Points in $W^s$ converge to $\mathbf{0}$ under forward iteration of the map, whereas points in $W^u$ converge to $\mathbf{0}$ under backward iteration. Our goal in this section is to show that nonlinear dynamical systems behave similarly, at least near hyperbolic fixed and periodic points.

**Definition 6.1.**  A fixed point $p$ for $F: \mathbf{R}^n \to \mathbf{R}^n$ is called hyperbolic if $DF(p)$ has no eigenvalues on the unit circle, where $DF(p)$ is the Jacobian matrix at $p$. If $p$ is periodic of period $n$, then $p$ is hyperbolic if $DF^n(p)$ has no eigenvalues on the unit cricle.

We remark that, for periodic points, the eigenvalues of the Jacobian matrix of $F^n$ are the same at each point on the orbit. Indeed, we have $F^n \circ F^j = F^j \circ F^n$. By the Chain Rule, we therefore have

$$DF^n(F^j(p)) \cdot DF^j(p) = DF^j(F^n(p)) \cdot DF^n(p).$$

If $F^n(p) = p$, this says that

$$(DF^j)^{-1}(p) \cdot DF^n(F^j(p)) \cdot DF^j(p) = DF^n(p).$$

Hence the eigenvalues of $DF^n$ at $p$ and at $F^j(p)$ are the same.

There are three types of hyperbolic periodic points: sinks, sources, and saddles.

**Definition 6.2.** Let $F^n(p) = p$.
   1. $p$ is a sink or attracting periodic point if all of the eigenvalues of $DF^n(p)$ are less than one in absolute value.
   2. $p$ is a source or repelling periodic point if all of the eigenvalues of $DF^n(p)$ are greater than one in absolute value.
   3. $p$ is a saddle point otherwise, i.e., if some of the eigenvalues of $DF^n(p)$ are larger and some are less than one in absolute value.

Case 3 distinguishes higher dimensional systems from the one-dimensional case studied in Chapter One.

For the remainder of this section, we will consider only fixed points in $\mathbf{R}^2$. The extension of the results below to periodic points is straightforward. The extension to higher dimensions is more complicated, but the techniques below do work in $\mathbf{R}^n$. The arguments used in the plane are geometrically much clearer than in higher dimensions, and the technical details are significantly easier.

**Theorem 6.3.** *Suppose $F$ has an attracting fixed point at $p$. Then there is an open set about $p$ in which all points tend to $p$ under forward iteration of $F$.*

**Remark.** The largest such open set in $\mathbf{R}^2$ is called the stable set or the basin of attraction of $p$ and is denoted by $W^s(p)$.

*Proof.* By conjugating with $T(x) = x + p$, we may assume that $p = 0$ and that $DF(0)$ assumes one of the following forms:

$$\begin{pmatrix} \lambda & 0 \\ 0 & \mu \end{pmatrix} \quad \text{with} |\lambda|, |\mu| < 1.$$

$$\begin{pmatrix} \lambda & \epsilon \\ 0 & \lambda \end{pmatrix} \quad \text{with } \epsilon > 0 \text{ but arbitrarily small.}$$

$$\begin{pmatrix} \alpha & -\beta \\ \beta & \alpha \end{pmatrix} \quad \text{with } \alpha^2 + \beta^2 < 1.$$

See Corollary 1.11 and Proposition 1.12. It follows easily that if $\mathbf{v} \neq 0$, then

$$|DF(0)\mathbf{v}| < |\mathbf{v}|.$$

Hence there is a neighborhood $U$ of $0$ in which this inequality holds for each unit vector $\mathbf{e}_1$ and $\mathbf{e}_2$ and thus for all $\mathbf{v} \neq 0$, i.e., $|DF(x)\mathbf{v}| < |\mathbf{v}|$ if $x \in U$.

Now choose $\delta$ so that if $|p| < \delta$, then $p \in U$. We claim that $|F(p)| < |p|$ if $p \neq 0$. Let $\gamma(t) = t \cdot p$. We have $F(\gamma(0)) = 0$, $F(\gamma(1)) = F(p)$, and $\gamma(t) \in U$ for each $t$ which satisfies $0 \leq t \leq 1$. Hence

$$|F(p)| = |\int_0^1 (F \circ \gamma)'(t)dt|$$
$$\leq \int_0^1 |(F \circ \gamma)'(t)|dt$$
$$= \int_0^1 |DF(\gamma(t))\gamma'(t)|dt$$
$$< \int_0^1 |\gamma'(t)|dt$$

since $\gamma'(t) \neq 0$. Hence $|F(p)| < |p|$.

q.e.d.

**Corollary 6.4.** *Suppose $F$ has a repelling fixed point at $p$. Then there is an open set containing $p$ in which all points tend to $p$ under backward iteration of $F$.*

We call the largest such set the unstable set and denote it by $W^u(p)$. We turn now to the case of saddle points. Since one of the eigenvalues is larger than one and one smaller (in absolute value), we expect regions in which $F$

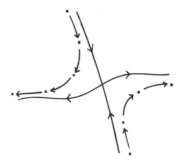

**Fig. 6.1.** The local stable and unstable manifolds.

contracts and expands near **0**. Unlike the situation which occurs for linear maps, this expansion and contraction does not result in a pair of invariant straight lines which pass through the fixed point. However, there are a pair of curves which play this role in the nonlinear case. This is the content of the stable and unstable manifold theorems, whose proof is the main topic of this section.

**Theorem 6.5.** *Suppose $F$ has a saddle point at $p$. There exists $\epsilon > 0$ and a smooth curve, i.e., a $C^1$ curve*

$$\gamma \colon (-\epsilon, \epsilon) \to \mathbf{R}^2$$

*such that*

1. *$\gamma(0) = p$.*
2. *$\gamma'(t) \neq 0$.*
3. *$\gamma'(0)$ is an unstable eigenvector for $DF(p)$.*
4. *$\gamma$ is $F^{-1}$-invariant.*
5. *$F^{-n}(\gamma(t)) \to p$ as $n \to \infty$.*
6. *If $|F^{-n}(q) - F^{-n}(p)| < \epsilon$ for all $n \geq 0$, then $q = \gamma(t)$ for some $t$.*

This complicated statement deserves some explanation. The curve $\gamma$ is called the *local unstable manifold at $p$.* We use the word "manifold" since, in general, $\gamma$ is not a straight line. Intuitively, the local unstable manifold is a curve through the fixed point which is mapped inside itself by $F^{-1}$. All points on the local unstable manifold tend to the fixed point under iteration of $F^{-1}$. See Fig. 6.1.

**Remarks.**

**1.**   The theorem is true for stable sets as well as with the obvious modifications. On the local stable manifold, all points tend to the fixed point under iteration of $F$.

**2.**   In dimensions greater than two, the curve $\gamma$ is replaced by a local "surface" parametrized near $p$ by a smooth map $\phi: U \to \mathbf{R}^n$ where $U$ is an open subset of $\mathbf{R}^k$ and $k$ is the number of eigenvalues less than one in absolute value. Our proof of Theorem 6.5 can be adapted to this higher dimensional setting, but this involves a number of additional technical details. Hence we will content ourselves with the simplest possible case.

**3.** It can be shown that $\gamma$ is $C^\infty$. More generally, if $F$ is $C^r$, then so is $\gamma$.

The local stable and unstable manifolds have global counterparts defined as follows.

**Definition 6.6.** Let $p$ be a hyperbolic fixed point for $F$ and suppose that $\gamma_u$ is the local unstable manifold at $p$. The unstable manifold at $p$, denoted by $W^u(p)$, is given by

$$W^u(p) = \bigcup_{n>0} F^n(\gamma_u).$$

Similarly, if $\gamma_s$ is the local stable manifold at $p$, then the stable manifold is defined by

$$W^s(p) = \bigcup_{n>0} F^{-n}(\gamma_s).$$

Thus the stable and unstable manifolds are invariant curves which emanate from the fixed or periodic point. We have seen these types of curves before; the examples of the previous three sections all featured stable and unstable manifolds through each of the periodic points as well as through certain of the non-periodic points. As these examples show, the stable and unstable manifolds may wind about in very complicated ways. This need not be the case always. Before turning to the proof of Theorem 6.5, let us give several more examples of stable and unstable manifolds. These examples are atypical since we can explicitly compute the invariant manifolds. Usually, this is impossible as there is no formula for these sets: Theorem 6.5 guarantees their existence but gives no prescription for finding them. Nevertheless, these examples are instructive in that they show the global behavior of these sets quite explicitly.

**Example 6.7.** Let $F: \mathbf{R}^2 \to \mathbf{R}^2$ be given by

$$F\begin{pmatrix} x \\ y \end{pmatrix} = \begin{pmatrix} \frac{1}{2}x \\ 2y \ -\frac{15}{8}x^3 \end{pmatrix}.$$

Note that $F(0) = 0$ and that

$$DF(0) = \begin{pmatrix} \frac{1}{2} & 0 \\ 0 & 2 \end{pmatrix}.$$

Consequently, $0$ is a saddle point. Clearly,

$$F\begin{pmatrix} 0 \\ t \end{pmatrix} = \begin{pmatrix} 0 \\ 2t \end{pmatrix}$$

so the $y$-axis serves as the unstable manifold. Also

$$F\begin{pmatrix} t \\ t^3 \end{pmatrix} = \begin{pmatrix} \frac{1}{2}t \\ (\frac{1}{2}t)^3 \end{pmatrix}$$

so the curve $y = x^3$ serves as the stable manifold. Indeed, $F$ is topologically conjugate to the linear map

$$L\begin{pmatrix} x \\ y \end{pmatrix} = \begin{pmatrix} \frac{1}{2} & 0 \\ 0 & 2 \end{pmatrix} \begin{pmatrix} x \\ y \end{pmatrix}$$

via the diffeomorphism

$$h\begin{pmatrix} x \\ y \end{pmatrix} = \begin{pmatrix} x \\ y - x^3 \end{pmatrix}.$$

That is, $F \circ h = h \circ L$. Note that $h$ maps the stable and unstable subspaces for $L$ onto the stable and unstable manifolds for $F$.

**Example 6.8.** Let $T$ be the torus parametrized by $\theta_1, \theta_2$ in the square, $0 \le |\theta_i| \le 2\pi$ with sides identified. Define

$$F\begin{pmatrix} \theta_1 \\ \theta_2 \end{pmatrix} = \begin{pmatrix} \theta_1 + \epsilon \sin \theta_1 \\ \theta_2 + \epsilon \sin \theta_2 \cos \theta_1 \end{pmatrix}.$$

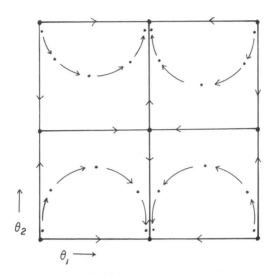

**Fig. 6.2.** The phase portrait of the map
$$F\begin{pmatrix}\theta_1\\\theta_2\end{pmatrix} = \begin{pmatrix}\theta_1+\epsilon\sin\theta_1\\\theta_2+\epsilon\sin\theta_2\cos\theta_1\end{pmatrix}.$$

If $\epsilon$ is sufficiently small, $F$ is a diffeomorphism. There are four fixed points: saddles at $(0,\pi)$ and $(\pi,\pi)$, a sink at $(\pi,0)$, and a source at $(0,0)$. The phase portrait is shown in Fig. 6.2.

Note that the unstable manifold of $(0,\pi)$ matches up exactly with the stable manifold of $(\pi,\pi)$. The stable manifold of $(0,\pi)$ emanates from the repelling fixed point at $(0,0)$, while the unstable manifold of $(\pi,\pi)$ lies in the basin of attraction of the attracting fixed point at $(\pi,0)$.

On the torus, the dynamics may also be pictured as in Fig. 6.3. The stable and unstable manifolds lose their linear character in this presentation.

**Example 6.9.** A simpler diffeomorphism of the torus is given by
$$G\begin{pmatrix}\theta_1\\\theta_2\end{pmatrix} = \begin{pmatrix}\theta_1 - \epsilon\sin\theta_1\\\theta_2 + \epsilon\sin\theta_2\end{pmatrix}.$$

Again there are four fixed points: two saddles at $(0,0)$ and $(\pi,\pi)$, a sink at $(0,\pi)$, and a source at $(\pi,0)$. The phase portrait is shown in Fig. 6.4.

In this case, the unstable manifolds of both saddles tend to the sink, while their stable manifolds come from the source.

The behavior of the stable and unstable manifolds play an important role in the question of the structural stability of a higher dimensional dynamical system. Suppose $F$ and $G$ are diffeomorphisms of the plane which are topologically conjugate via a homeomorphism $h$. If $p$ is a hyperbolic saddle point

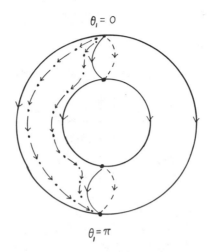

**Fig. 6.3.** The phase portrait of $F$ on the torus.

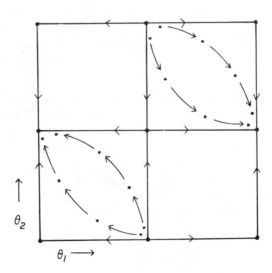

**Fig 6.4.** The phase portrait of
$$G\binom{\theta_1}{\theta_2} = \binom{\theta_1 - \epsilon \sin \theta_1}{\theta_2 + \epsilon \sin \theta_2}.$$

for $F$ which is fixed, then $G$ has a fixed point at $h(p)$. This point need not be hyperbolic, but it must have stable and unstable manifolds. Indeed, it is easy to check that, if $x \in W^s(p)$, then

$$\lim_{n \to \infty} G^n(h(x)) = h(p).$$

Similarly. $h$ also preserves the unstable manifold.

We turn now to the proof of the stable and unstable manifold theorem. The geometric idea behind the proof is quite simple and elegant, although it can be lost among the technical details. Let us illustrate this idea with an example where the result is already known, a linear map. Suppose $F(x) = Ax$ where

$$A = \begin{pmatrix} \lambda & 0 \\ 0 & \mu \end{pmatrix}$$

with $0 < \mu < 1 < \lambda$. From the results of §2.2 we know the unstable set is the $x$-axis, on which vectors are stretched by a factor of $\lambda$.

Let us consider the square $|x|, |y| \le \epsilon$ for some $\epsilon > 0$. Let $\gamma(x) = (x, h(x))$ be a smooth curve in the plane which passes through $\mathbf{0}$ and whose tangent line always has slope between $\pm \frac{1}{2}$, i.e., $|h'(x)| < \frac{1}{2}$. Apply $F$ to such a curve. The result is a new curve which hugs the $x$-axis more closely and which has slope closer to zero. If we restrict this curve to the box $|x|, |y| \le \epsilon$, we see that $\gamma$ has been transformed into another curve which is closer to the $x$-axis. Repeated applications of this procedure yields curves which approach the segment of the unstable set lying in $|x| \le \epsilon$. See Fig. 6.5.

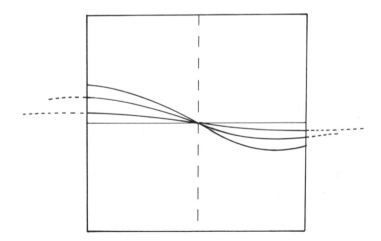

**Fig. 6.5.** Application of the graph transform in the case of a linear map.

Now we turn to the proof of the unstable manifold theorem; the stable manifold theorem will follow by applying this result to $F^{-1}$. We will show that the behavior of a nonlinear map near a hyperbolic fixed point is similar to that of a hyperbolic linear map near $\mathbf{0}$.

Let us make some preliminary simplifications of $F$. First, by conjugating $F$ with the translation $T(x) = x + p$, we may assume that the fixed point is at $\mathbf{0}$. Second, we may assume that

$$DF(\mathbf{0}) = \begin{pmatrix} \lambda & 0 \\ 0 & \mu \end{pmatrix}$$

by conjugating $F$ with a linear map which puts $DF(\mathbf{0})$ into standard form. Third, we will assume that $\lambda > 2$ and $0 < \mu < \frac{1}{2}$. This may be accomplished by considering $F^n$ if necessary. After proving that $F^n$ has an unstable set, we will show how to deduce that this curve is also the unstable set for $F$.

We will denote a point $q$ in $\mathbf{R}^2$ by coordinates $(x_0, y_0)$ and its $F$-image by $(x_1, y_1)$. That is,

$$x_1 = F_1(x_0, y_0)$$
$$y_1 = F_2(x_0, y_0).$$

Similarly, $(x_{-1}, y_{-1}) = F^{-1}(x_0, y_0)$. We denote a tangent vector at $q$ by $(\xi_0, \eta_0)_q$ and its image under the derivative of $F$ by $(\xi_1, \eta_1)_{F(q)}$. That is,

$$DF(p) \begin{pmatrix} \xi_0 \\ \eta_0 \end{pmatrix}_q = \begin{pmatrix} \xi_1 \\ \eta_1 \end{pmatrix}_{F(q)}.$$

In coordinates

$$\xi_1 = \frac{\partial F_1}{\partial x}(q)\xi_0 + \frac{\partial F_1}{\partial y}(q)\eta_0$$
$$\eta_1 = \frac{\partial F_2}{\partial x}(q)\xi_0 + \frac{\partial F_2}{\partial y}(q)\eta_0.$$

We will also need the notion of a sector bundle. Define

$$S^u(q) = \{(\xi_0, \eta_0)_q \mid |\eta_0| \leq \frac{1}{2}|\xi_0|\}$$
$$S^s(q) = \{(\xi_0, \eta_0)_q \mid |\xi_0| \leq \frac{1}{2}|\eta_0|\}.$$

See Fig. 6.6.

Note that $DF(\mathbf{0})$ preserves $S^u(\mathbf{0})$ in the sense that if $\mathbf{v} \in S^u(\mathbf{0})$, then $DF(\mathbf{0})\mathbf{v} \in S^u(\mathbf{0})$. Moreover, if

$$\mathbf{v} = \begin{pmatrix} \xi_0 \\ \eta_0 \end{pmatrix}.$$

we note that $|\xi_1| = \lambda|\xi_0| > 2|\xi_0|$. Similarly, $(DF(\mathbf{0}))^{-1}$ preserves $S^s(\mathbf{0})$ and we have $|\eta_{-1}| = \mu^{-1}|\eta_0| > 2|\eta_0|$.

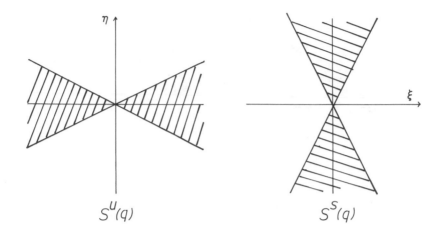

$$S^u(q) \qquad\qquad S^s(q)$$

**Fig. 6.6.** The sector bundles $S^u(q)$ and $S^u(q)$.

Since $F$ is at least $C^1$, the Jacobian matrix $DF(x)$ varies continuously with $x$ and there must therefore be a neighborhood of $\mathbf{0}$ in which the above properties hold. More precisely, there exists $\epsilon > 0$ such that, if $|x|, |y| \le \epsilon$, then

1. $DF(x, y)$ preserves $S^u(x, y)$ and $DF^{-1}(x, y)$ preserves $S^s(x, y)$. i.e., $DF(x, y)\mathbf{v} \in S^u(F(x, y))$ whenever $\mathbf{v} \in S^u(x, y)$.
2. If $(\xi_0, \eta_0) \in S^u(x, y)$, then $|\xi_1| \ge 2|\xi_0|$.
3. If $(\xi_0, \eta_0) \in S^s(x, y)$, then $|\eta_{-1}| \ge 2|\eta_0|$.

The concept of preservation of sector bundles is one that arises whenever hyperbolicity is verified: it is illustrated geometrically in Fig. 6.7.

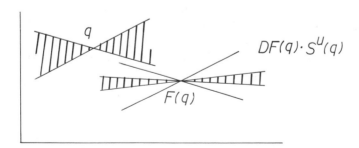

$$q \qquad\qquad DF(q) \cdot S^u(q)$$
$$F(q)$$

**Fig. 6.7.** Preservation of the sector bundles.

We will now concentrate on the square $B$ given by $|x|, |y| \le \epsilon$. We say that the curve $\gamma(x) = (x, h(x))$ is a *horizontal curve* in $B$ if

1. $h$ is defined and continuous for $|x| \leq \epsilon$
2. $h(0) = 0$
3. for any $x_1, x_2$ with $|x_i| \leq \epsilon$, $|h(x_1) - h(x_2)| \leq \frac{1}{2}|x_1 - x_2|$. Note that $\gamma(x)$ is the graph of $h(x)$ which lies in $B$ and is depicted in Fig. 6.8. Reversing the roles of $x$ and $y$ yields a definition of a vertical curve.

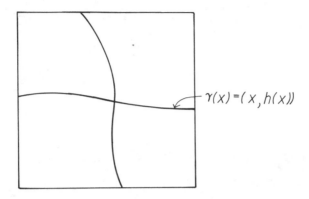

Fig. **6.8.** A horizontal curve in $B$.

**Lemma 6.10.** *If $\gamma(x) = (x, h(x))$ is a horizontal curve, then the image $F(\gamma(x))$ meets $B$ in a horizontal curve.*

*Proof.* We first observe that if $(x_1, y_1) = F(\epsilon, h(\epsilon))$, then $x_1 \geq 2\epsilon$. This follows immediately from the fact that $|\xi_1| > 2|\xi_0|$. Similarly, if $(x_1, y_1) = F(-\epsilon, h(-\epsilon))$, then $x_1 < -2\epsilon$. Clearly, $F(0) = 0$ so that the image curve passes through the origin. Finally, suppose that $(x_0, y_0)$ and $(x_0', y_0')$ lie on $F(x, h(x))$ and $|y_0' - y_0| > \frac{1}{2}|x_0' - x_0|$. Choose $\alpha_1, \alpha_2$ such that $F(\alpha_1, h(\alpha_1)) = (x_0, y_0)$ and $F(\alpha_2, h(\alpha_2)) = (x_0', y_0')$. Consider the straight line segment $\ell$ connecting $(\alpha_1, h(\alpha_1))$ to $(\alpha_2, h(\alpha_2))$. The tangent vector to $\ell$ lies in $S^u$ at each point along $\ell$. Now $F$ maps $\ell$ to a smooth curve connecting $(x_0, y_0)$ to $(x_0', y_0')$. By the Mean Value Theorem, there is a point on this curve where the tangent vector has slope larger than $\frac{1}{2}$. This contradicts the fact that $DF$ preserves the sector bundle $S^u$.

q.e.d.

Thus, for each horizontal curve $\gamma$ in $B$, the action of $F$ defines a new horizontal curve in $B$ which we denote by $\Phi\gamma$. $\Phi$ is called the *graph transform*, since it takes the graph of $h(x)$ into the graph of another function.

Let $H$ denote the set of all horizontal curves in $B$. We may thus regard $\Phi$ as a map $\Phi: H \to H$. A fixed point for $H$ is a horizontal curve which is transformed into itself, i.e., whose $F$-image covers itself. Such a fixed point is therefore our candidate for the unstable set.

Let $\gamma_1(x) = (x, h_1(x))$ and $\gamma_2(x) = (x, h_2(x))$ be horizontal curves in $B$. Define the distance

$$d[\gamma_1, \gamma_2] = \sup_{|x| \leq \epsilon} |h_1(x) - h_2(x)|.$$

**Lemma 6.11.** *If $\gamma_1$ and $\gamma_2$ are horizontal curves, then $d[\Phi\gamma_1, \Phi\gamma_2] < \nu d[\gamma_1, \gamma_2]$ for some $\nu$ with $0 < \nu < 1$.*

*Proof.* We argue geometrically. Suppose that $|\Phi\gamma_2(x) - \Phi\gamma_1(x)| \geq |h_2(z) - h_1(z)|$ for any $z$ with $|z| \leq \epsilon$. We will show that this leads to a contradiction.

Let $\ell = P(x)$ be the vertical line connecting $\Phi\gamma_1(x)$ to $\Phi\gamma_2(x)$. Consider the curve $F^{-1}(\ell)$ which connects the point $(z_1, h_1(z_1)) = \gamma_1(z_1)$ to $\gamma_2(z_2) = (z_2, h_2(z_2))$. Since $DF^{-1}$ preserves the sectors $S^s$ at each point of $\ell$, it follows that the tangent vectors to $F^{-1}(\ell)$ always lie in this sector. As a consequence, $F^{-1}(\ell)$ itself lies in the cone shaped region with vertex at $(z_1, h(z_1))$ and boundary lines of slope $\pm 2$. See Fig. 6.9.

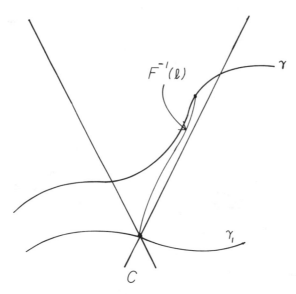

**Fig. 6.9.**

In particular, we have

$$\frac{|h_2(z_2) - h_1(z_1)|}{|z_2 - z_1|} \geq 2.$$

Moreover, since $DF^{-1}$ expands vertical components of these tangent vectors by a factor of at least two, we have

$$|h_2(z_2) - h_1(z_1)| \geq 2|\gamma_2(x) - \gamma_1(x)|.$$

By assumption we have

$$|h_2(z_1) - h_1(z_1)| \leq |\gamma_2(x) - \gamma_1(x)|.$$

Consequently

$$|h_2(z_2) - h_2(z_1)| \geq |h_2(z_2) - h_1(z_1)| - |h_2(z_1) - h_1(z_1)|$$
$$\geq \frac{1}{2}|h_2(z_2) - h_1(z_1)|$$
$$\geq |z_2 - z_1|.$$

By the Mean Value Theorem, there exists a point $z$ with $|h_2'(z)| \geq 1$. This contradicts the fact that $\gamma_2$ is a horizontal curve and proves the lemma.

q.e.d.

It follows that $\Phi$ is a contraction on $H$. Since $H$ is a closed subset of the set of all continuous maps from the interval $|x| \leq \epsilon$ to itself, it is a technical fact that $H$ is a complete metric space and, consequently, $\Phi$ has a unique fixed point in $H$. We refer to any of the standard texts in analysis for a proof of this fact.

Let $\gamma_u$ be the horizontal curve fixed by $\Phi$. This curve clearly passes through the origin, and if $(x_0, y_0)$ is a point on $\gamma_u$ with $x_0 \neq 0$, we have $|x_1| > |x_0|$. Hence points on $\gamma_u$ either leave $B$ under iteration of $F$ or else map into $\gamma_u$ further from $\mathbf{0}$. It follows that $\gamma_u \subset W^u(\mathbf{0})$.

**Lemma 6.12.** *Let $(x_0, y_0) \in B$ and suppose $(x_0, y_0)$ does not lie on $\gamma_u$. Then there is a positive integer $n$ for which $F^{-n}(x_0, y_0)$ does not lie in $B$.*

*Proof.* Let $\ell$ be the vertical line connecting $(x_0, y_0)$ to a unique point in $\gamma_u$. Now apply the cone argument as in the previous lemma: the vertical height of $F^{-1}(\ell)$ must be doubled. Continuing this process yields the result.

q.e.d.

As a consequence of this lemma, $\gamma_u$ is precisely the local unstable set for $F$. We have shown that $\gamma_u$ is a continuous curve in $B$. In fact, $\gamma_u$ is $C^\infty$ if $F$ is. Rather than prove this fact, we will simply sketch a proof that $\gamma_u$ is $C^1$.

To do this, we need more terminology. We define a *horizontal line field* to be a pair of functions $\varsigma(x) = (\gamma(x), M(x))$, where

   1. $\gamma_u$ is a horizontal curve in $B$.

   2. $M$ is a continuous real-valued function with $|M(x)| \leq \frac{1}{2}$ for all $x$ with $|x| \leq \epsilon$.

We view $\varsigma(x)$ geometrically as a horizontal curve together with a collection of straight lines. One straight line passes through each point of $\gamma(x)$, and this line has slope $M(x)$. Since $|M(x)| \leq \frac{1}{2}$, each straight line has a direction vector which lies in $S^u$. See Fig. 6.10.

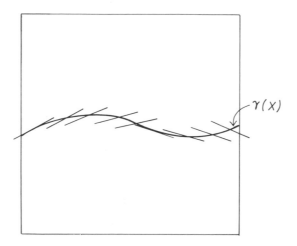

**Fig. 6.10.** A horizontal line field.

We let $H_1$ denote the set of all horizontal line fields in $B$. We define the distance between two horizontal line fields $\varsigma_i = (\gamma_i, M_i)$ for $i = 1, 2$ by

$$d[\varsigma_1, \varsigma_2] = \sup_{|x| \leq \epsilon} (|\gamma_1(x) - \gamma_2(x)|, |M_1(x) - M_2(x)|).$$

A new, fancier graph transform is then given by

$$\Phi_1(\varsigma) = (\Phi\gamma, \hat{M})$$

where $\Phi\gamma$ is the usual graph transform of the horizontal curve $\gamma$, and $\hat{M}$ is the slope of the line field transformed by $DF$. More precisely, if $\gamma(x) = (x, h(x))$ and $\mathbf{v}$ is a vector with slope $M(x)$, then $\hat{M}$ is the slope of the vector $DF(\gamma(x))\mathbf{v}$.

As before, it is clear that $\Phi_1$ maps $H_1$ inside itself. Moreover, $\Phi_1$ is an example of a "fiber" contraction. That is, if $\varsigma_1 = (\gamma, M_1)$ and $\varsigma_2 = (\gamma, M_2)$ are horizontal line fields based on the same horizontal curve $\gamma(x)$, then

$$d[\Phi_1(\varsigma_1), \Phi_2(\varsigma_2)] < d[\varsigma_1, \varsigma_2].$$

When this result is coupled with the contraction in the $\gamma$-direction, it can be shown that there is a unique fixed point for $\Phi_1$ in $H_1$. This fixed point is the horizontal curve $\gamma_u$ found above, together with a preferred direction at each point. One can prove that this direction is in fact tangent to the curve, but we will omit the technical details.

This completes the proof of the Unstable Manifold Theorem in cases where the eigenvalues satisfy appropriate conditions. Recall that we assumed at the outset that the unstable eigenvalue $\lambda$ satisfied $\lambda > 2$ and the stable eigenvector $\mu$ satisfied $0 < \mu < \frac{1}{2}$. If this is not the case, then we have showed that a sufficiently high power of $F$, say $F^{-n}$, has an invariant unstable set given by $\gamma_u$. Clearly, if this set is not also $F^{-1}$-invariant, then at least $F^{-1}(\gamma_u)$ is also invariant under $F^{-n}$. This, however, contradicts the uniqueness of $\gamma_u$.

**Exercises**

**1.**  Consider the diffeomorphism $Q_\lambda$ of the plane given by

$$x_1 = e^x - \lambda$$

$$y_1 = -\frac{\lambda}{2} \arctan y$$

where $\lambda$ is a parameter.

    a. Find all fixed points and periodic points of period 2 for $Q_\lambda$.

    b. Classify each of these points as sinks, sources, or saddles.

    c. If the point is a saddle, identify and sketch the stable and unstable manifolds.

**2.**  Consider the diffeomorphism $F$ of the plane given in polar coordinates by

$$r_1 = \lambda r + \beta r^3$$

$$\theta_1 = \theta + \frac{2\pi}{n} + \epsilon \sin(n\theta)$$

where $\epsilon > 0$ is small, $\lambda > 1$ and $\beta < 0$.

    a. Identify and classify all periodic points of $F$.

    b. Show that the circle $\gamma$ given by $r = \sqrt{(1 - \lambda)/\beta}$ is invariant under $F$.

    c. Identify and sketch the stable and unstable manifolds of the saddle points of $F$.

**3.**  Let $p_1, p_2$ be saddle points for a diffeomorphism $F$. Recall that a point $q$ is a *heteroclinic* point for $F$ if

$$\lim_{n \to \infty} F^n(q) = p_1$$

$$\lim_{n \to \infty} F^{-n}(q) = p_2.$$

That is, heteroclinic points are forward and backward asymptotic to distinct saddle points. If $p_1 = p_2$, $q$ is called *homoclinic* (note how the higher dimensional definition of homoclinic differs from that given in §1.16). See Definition 4.6. Prove that topological conjugacy preserves homoclinic and heteroclinic points.

4.  Identify the heteroclinic points in Example 6.8.

5.  Using a bump function, show that the diffeomorphism in Example 6.8 may be perturbed so that it has a finite number of heteroclinic orbits. Hence intervals of homoclinic or heteroclinic points may be destroyed by a small perturbation. Conclude that this map is not structurally stable.

6.  A homoclinic or heteroclinic point is called *transverse* if the respective stable and unstable manifolds meet at an angle, i.e., their tangent vectors are not collinear at the heteroclinic point. Show that the example in Exercise 5 may be perturbed so that the heteroclinic points are transverse.

*Linear automorphisms of the sphere.* Let $S^2$ denote the two-dimensional sphere is $\mathbf{R}^3$, i.e.,

$$S^2 = \{x \in \mathbf{R}^3 \,|\, |x| = 1\}.$$

Let

$$A = \begin{pmatrix} 1 & 0 & 0 \\ 0 & 2 & 0 \\ 0 & 0 & 3 \end{pmatrix}$$

and define the map

$$F(x) = F_A(x) = \frac{Ax}{|Ax|}.$$

$F_A$ is called a linear automorphism of $S^2$.

7.  Prove that $F$ maps $\mathbf{R}^3 - \{0\}$ onto $S^2$.

8.  Prove that the restriction of $F$ to $S^2$ is a diffeomorphism of the sphere.

9.  Let $e_1 = (1,0,0)$, $e_2 = (0,1,0)$, and $e_3 = (0,0,1)$. Prove that the $\pm e_j$ are the fixed points for $F$.

10.  Compute the Jacobian matrix $DF(\pm e_j)$. Prove that $DF(\pm e_j)$ has an eigenvalue equal to 0 with corresponding eigenvector $e_j$.

11.  Prove that each of the other vectors $e_i, i \neq j$, are also eigenvectors for $DF(\pm e_j)$. Evaluate the corresponding eigenvalues.

12.  Conclude that $\pm e_1$ is a source, $\pm e_2$ is a saddle, and $\pm e_3$ is a sink.

13.  Define $\phi: S^2 \to \mathbf{R}$ by $\phi(x) = |A^{-1}x|^2$. Prove that $\phi(F(x)) \leq \phi(x)$.

**14.** Prove that $\phi(F(x)) = \phi(x)$ if and only if $x = \pm e_j$ for some $j$. The function $\phi$ is called a gradient function since it decreases along all orbits of $F$ except the fixed points. $F$ itself is called *gradient like*.

**15.** Use the gradient function to prove that the phase portrait of $F$ is as depicted in Fig. 6.11.

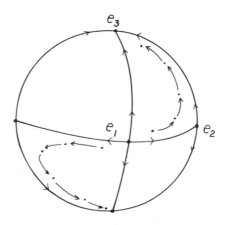

**Fig. 6.11.** The dynamics of $F$.

**16.** Discuss the dynamics of the linear automorphism of the sphere determined by the matrix

$$A = \begin{pmatrix} \cos\theta & \sin\theta & 0 \\ -\sin\theta & \cos\theta & 0 \\ 0 & 0 & 2 \end{pmatrix}.$$

**17.** Let $S^n = \{x \in \mathbf{R}^{n+1} \,\big|\, |x|^2 = 1\}$ be the unit sphere in $\mathbf{R}^{n+1}$. Let $A$ be the diagonal matrix with entries $1, 2, \ldots, n+1$. Describe the dynamics of the linear automorphism of the sphere induced by $A$. Describe the stable and unstable manifolds of each fixed point for this map.

## §2.7 GLOBAL RESULTS AND HYPERBOLIC SETS

The goal of this section is to amalgamate many of the previous notions from dynamical systems theory to present an overview of the global theory of dynamics in higher dimensions. Recall that the structurally stable diffeomorphisms in one dimension were quite simple dynamically. There were only finitely many periodic points, all of which were hyperbolic. All other points simply tended from one of these periodic orbits to another under iteration of the map.

In higher dimensions, the situation is somewhat more complicated. As we have seen in our three fundamental examples, it is entirely possible to have infinitely many periodic points as well as other, more complicated types of recurrence such as dense orbits or recurrent points. All of these points lie in the set on which the map has "interesting" dynamics. This is the chain-recurrent set.

**Definition 7.1.** Let $F$ be a diffeomorphism. A point $x$ is chain recurrent for $F$, if, for any $\epsilon > 0$, there are points $x = x_0, x_1, x_2, \ldots, x_k = x$ and positive integers $n_1, \ldots, n_k$ such that

$$|F^{n_i}(x_{i-1}) - x_i| < \epsilon$$

for each $i$.

The sequence of points $x_0, \ldots, x_n$ is called an $\epsilon$-chain or a pseudo-orbit. Intuitively, an $\epsilon$-chain is almost an orbit in the sense that we allow small jumps or errors at iterations $n_1, n_1 + n_2$, etc. A point $x$ is chain recurrent if we can find $\epsilon$-chains with arbitrarily small jumps. Note that the $\epsilon$-chains always begin and end at $x$.

**Example 7.2.** Any periodic point is chain recurrent. All points in the Cantor set associated to the Smale horseshoe are chain recurrent, as are all points in the attractors discussed in §2.5. See Exercises 1-3.

Note that chain recurrence is a weaker notion than that of *recurrence*. Recall that $x$ is a recurrent point for $F$ if, for any $\epsilon > 0$, there exists $n > 0$ such that $|F^n(x) - x| < \epsilon$. That is, a recurrent point is chain recurrent with an $\epsilon$-chain consisting of just $x$ itself.

There are points which are chain recurrent but not recurrent. For example, if $p$ is a hyperbolic periodic point and $q \in W^u(p) \cap W^s(p)$, then $q$ is chain recurrent but not recurrent. Indeed, both the forward and backward orbit of $q$ tends to $p$, so that the orbit of $q$ never piles up on itself. Points such as $q$ are called *homoclinic points*. If $q$ tends to distinct saddle points, then $q$ is called a *heteroclinic* point.

Let $\Lambda = \Lambda(F)$ denote the set of points which are chain recurrent; $\Lambda$ is called the chain recurrent set. The proof of the following proposition is straightforward.

**Proposition 7.3.**

    *1. $\Lambda$ is a closed subset of $\mathbf{R}^2$*

    *2. If F and G are topologically conjugate via a homeomorphism h, then h maps $\Lambda(F)$ to $\Lambda(G)$.*

Often, the chain recurrent set of a map breaks up into various pieces which may be analyzed by techniques similar to those in §2.2-2.5. There may be transitive attractors or subshifts of finite type embedded as pieces of the dynamics. This leads to the concept of hyperbolic set which we will discuss below. But before that we note that the chain recurrent set may be quite simple. It may in fact be finite. Indeed, Morse-Smale diffeomorphisms of the circle feature a finite number of periodic points as the chain recurrent set. All other points tend from one periodic orbit to another and are not in the chain recurrent set. One major result from Chapter One was the fact that Morse-Smale maps were structurally stable. In higher dimensions, we need to impose additional conditions in the definition of Morse-Smale map to achieve this result.

**Definition 7.4.** Let $p_1$ and $p_2$ be saddle points for $F$. $W^s(p_1)$ and $W^u(p_2)$ are *transverse* if either

    1. $W^s(p_1) \cap W^u(p_2) = \phi$, or

    2. $q \in W^s(p_1) \cap W^u(p_2)$, in which case the tangents to $W^s(p_1)$ and $W^u(p_2)$ at $q$ are not collinear.

Transverse intersections of the stable and unstable manifolds are a necessary condition for structural stability.

**Example 7.5.** Consider the diffeomorphism $F$

$$x_1 = \frac{1}{2}(x + x^3)$$

$$y_1 = y\left(\frac{2}{1 + 2x^2}\right).$$

This map has phase portrait as depicted in Fig. 7.1. Note that there are three saddle points, at $p_+ = (1,0)$, $p_- = (-1,0)$ and $\mathbf{0}$. One branch of

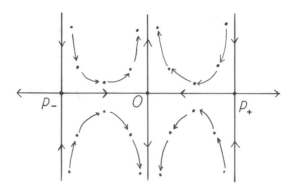

**Fig. 7.1.** One branch of $W^u(p_\pm)$ matches up
with a branch of $W^s(0)$.

$W^u(p_\pm)$ matches up exactly with $W^s(0)$. We may break these connections as follows.

Let $\phi$ be a bump function defined on the closed interval $[0,1]$. That is, $\phi(x) = 0$ if $x \notin (0,1)$ and $\phi(x) > 0$ if $x \in (0,1)$. Consider the diffeomorphism $g$ defined by

$$x_1 = x$$
$$y_1 = y + \phi(|x|).$$

Note that $g$ simply moves points in the strips $0 < |x| < 1$ in a vertical direction. That is, we think of $g$ as giving a gentle push in the direction of the positive $y$-axis. Now consider the perturbed map $\hat{F} = g \circ F$. Let $S$ be the strip $0 < x < 1$ and $y \geq 0$. For the unpeturbed map, the lower boundary of $S$ was preserved. This is not true for $\hat{F}$ however; all points on the lower boundary of $S$ are mapped strictly inside $S$. Moreover, if $0 \leq x < 1/\sqrt{2}$, then $y_1 > y$. Consequently, points in this region tend to $\infty$ under iteration of $\hat{F}$. It follows that the stable manifold of $\mathbf{0}$ does not enter the strip $S$. Arguing similarly, it can easily be shown that the unstable manifold for $\hat{F}$ at $p_+$ lies entirely in $S$ and so the connection has been broken. This is depicted in Fig. 7.2. A similar phenomenon occurs in the strip $-1 < x < 0$ as well.

Thus one necessary condition for structural stability is the transversality of all stable and unstable manifolds of saddle points. One important class of maps which have this property are the higher dimensional Morse-Smale maps. These are defined as follows.

**Definition 7.6.** A diffeomorphism is called Morse-Smale if
  1. The chain recurrent set is a finite set of periodic points, all of which are hyperbolic.

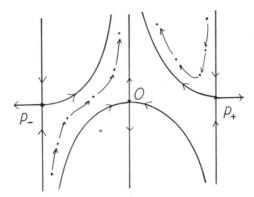

**Fig. 7.2.** The perturbed stable and unstable manifolds.

2. All stable and unstable manifolds of saddle points are transverse.

It is an important result first proved by Palis that Morse-Smale diffeomorphisms on compact surfaces (like the sphere or the torus) are $C^1$-structurally stable.

As we emphasized in Chapter One, hyperbolicity is another necessary ingredient for structural stability. It is an easy exercise using the Implicit Function Theorem to show that hyperbolic fixed and periodic points must persist under small perturbation (see Exercise 7.) But, since a higher dimensional diffeomorphism may involve more than simply periodic recurrence, we need to extend the notion of hyperbolicity to all of the chain recurrent set. That is, we need to introduce the notion of a hyperbolic set.

Let us begin with an example.

**Example 7.7.** Consider the diffeomorphism $F$ of the plane given in polar coordinates by

$$r_1 = 2r - r^3$$
$$\theta_1 = \theta + 2\pi\omega.$$

Clearly, the origin is a repelling fixed point for $F$ and the circle $r = 1$ is invariant. Graphical analysis of the function $r_1 = 2r - r^3$ shows that all non-zero points tend to this circle under iteration. On the circle, the map reduces to rotation by angle $2\pi\omega$. It follows that the chain recurrent set consists of **0** and the unit circle.

This situation is easily destroyed. We first make a preliminary perturbation that makes $\omega$ rational, say $p/q$. Then we perturb again by composing with the map

$$r_1 = r$$
$$\theta_1 = \theta + \epsilon \sin(q\theta).$$

The resulting map is

$$r_1 = 2r - r^3$$
$$\theta_1 = \theta + 2\pi\left(\frac{p}{q}\right) + \epsilon \sin(q\theta).$$

The circle $r = 1$ is still invariant, but this time the map on the circle is a Morse-Smale map as discussed in §1.15. Indeed, there are a pair of periodic orbits of period $q$, one attracting and one repelling. The chain recurrent set is easily seen to consist of these two orbits plus the fixed point at $0$.

To define a hyperbolic set, we first recall the situation near a hyperbolic fixed point. Basically, there were two ingredients to hyperbolicity, a *rate* of expansion or contraction, given by the eigenvalues of the Jacobian matrix at the point, and a *direction* of expansion or contraction, given by the associated eigenvector. We saw that this direction was mirrored in $\mathbf{R}^2$ by the stable and unstable manifolds, on which the map behaved just like its linear counterpart given by the Jacobian matrix. In the horseshoe, the hyperbolic toral automorphism, and the attractors, we saw that a much larger set – not just the periodic points – admitted similar stable and unstable behavior. These are examples of hyperbolic sets which we now define.

For simplicity, we will confine our attention to the plane.

**Definition 7.8.** Let $F: \mathbf{R}^2 \to \mathbf{R}^2$ be a diffeomorphism. A set $\Lambda$ is called a hyperbolic set for $F$ if

1. For each point $p \in \Lambda$, there are a pair of lines $E^s(p)$ and $E^u(p)$ in the tangent plane at $p$ which are preserved by $DF(p)$.
2. $E^s(p)$ and $E^u(p)$ vary continuously with $p$.
3. There is a constant $\lambda > 1$ such that $|DF(p)(\mathbf{v})| \geq \lambda|\mathbf{v}|$ for all $\mathbf{v} \in E^u(p)$ and $|DF^{-1}(p)\mathbf{v}| \geq \lambda|\mathbf{v}|$ for all $\mathbf{v} \in E^s(p)$.

$E^s(p)$ is called a stable line and $E^u(p)$ is an unstable line.

Both the horseshoe and the hyperbolic toral automorphism admit hyperbolic sets: the invariant Cantor set in the case of the horseshoe and the entire torus in the case of the hyperbolic toral automorphism. The solenoid shows that the concept of a hyperbolic set can be extended to higher dimensions: the disks given by $\theta = \theta^*$ play the role of stable planes in this case.

As we have seen in all of these examples as well as in the quadratic map in Chapter One, hyperbolic sets provide the setting on which chaotic dynamics may occur.

As in the case of a fixed point, the dynamics near a point in a hyperbolic set is straightforward. The following Theorem gives the existence of local stable and unstable manifolds for hyperbolic sets and can be verified using the tools of the previous section.

**Theorem 7.9.** *Let $F: \mathbf{R}^2 \to \mathbf{R}^2$ be a diffeomorphism. Let $\Lambda$ be a closed, invariant, hyperbolic set contained in a bounded region of $\mathbf{R}^2$. There exists $\epsilon > 0$ such that, for any $p \in \Lambda$, there is a smooth curve $\gamma_p: [-\epsilon, \epsilon] \to \mathbf{R}^2$ satisfying*

1. $\gamma_p(0) = p$
2. $\gamma_p'(t) \neq 0$
3. $\gamma_p'(0)$ *lies along the unstable line* $E^u(p)$
4. $F^{-1}(\gamma_p) \subset \gamma_{F^{-1}(p)}$
5. $|F^{-n}(\gamma(t)) - F^{-n}(p)| \to 0$ *as* $n \to \infty$.

*Moreover, the curves $\gamma_p$ depend continuously on $p$.*

Thus, the $\gamma_p$ behave exactly the same as the local unstable manifolds of fixed points. The proof of the above Theorem is similar but much more complicated than that of the Unstable Manifold Theorem for a point, so we will omit the details. A similar statement obviously holds for the stable manifolds.

For our examples, verifying hyperbolicity was rather easy since we assumed linearity in at least one of the directions. That is, we wrote down explicitly the stable and unstable lines. Most nonlinear dynamical systems are not given in such a useful fashion. In these cases, verifying hyperbolicity is more complicated. The following ideas give a criterion for hyperbolicity that is often useful.

For a vector $\mathbf{v} \neq \mathbf{0}$ in $\mathbf{R}^2$ or $\mathbf{R}^3$, we define the $\alpha$-cone about $\mathbf{v}$ to be the set of all vectors which make an angle $\leq \alpha$ with $\mathbf{v}$ or $-\mathbf{v}$. For example, the $\pi/4$-cone about $\mathbf{e}_1$ in $\mathbf{R}^2$ consists of all vectors of the form

$$w = \begin{pmatrix} \xi \\ \eta \end{pmatrix}$$

where $|\xi| \geq |\eta|$. The $\pi/4$-cone about $\mathbf{e}_1$ in $\mathbf{R}^3$ consists of all vectors of the form

$$W = \begin{pmatrix} \xi \\ \eta \\ \nu \end{pmatrix}$$

with $|\xi| \geq \sqrt{\eta^2 + \nu^2}$. The vector $\mathbf{v}$ is called the *core* of the $\alpha$-cone.

**Definition 7.10.** Let $U$ be an open set. A cone-field $C_p$ on $U$ is the assignment of an $\alpha$-cone to each point $p \in U$ such that
 1. $\alpha = \alpha(p)$ varies continuously with $p$.
 2. the core vector $\mathbf{v}_p$ varies continuously with $p$.

We visualize a cone field as a collection of tangent vectors at each point $p \in U$. The sector bundles of the previous section are special cases of cone fields with core either $\mathbf{e}_1$ or $\mathbf{e}_2$.

**Example 7.11.** Recall the solid torus $D = S^1 \times B^2$ which carried the solenoid. Any point in $D$ has coordinates $(\theta, x, y)$. Any tangent vector to $D$ may be written in the form

$$\alpha \mathbf{e} + \mathbf{v}$$

where $\mathbf{e}$ is a unit vector tangent to $S^1$ and

$$\mathbf{v} = \begin{pmatrix} \xi \\ \eta \end{pmatrix}$$

is a vector in $\mathbf{R}^2$. A natural cone field on $D$ is defined by

$$\sqrt{\xi^2 + \eta^2} \leq |\alpha|.$$

In the proof of the stable and unstable manifold theorem, we identified two cone fields which were preserved and expanded by either $DF$ or $DF^{-1}$. This condition guaranteed the existence of a unique line through each point which was invariant under $DF$ or $DF^{-1}$. The same is true in general.

**Theorem 7.12.** *Let* $F: \mathbf{R}^2 \to \mathbf{R}^2$. *Let* $\Lambda$ *be a closed, $F$-invariant subset contained in a bounded region of* $\mathbf{R}^2$. *Let* $U$ *be a neighborhood of* $\Lambda$ *and suppose there exists mutually disjoint $\alpha$-cone fields* $C^s$ *and* $C^u$ *for which we have*
 1. $\alpha \leq \pi/4$.
 2. $DF(C^u(p)) \subset C^u(F(p)); \ DF^{-1}(C^s(p)) \subset C^s(F^{-1}(p))$.
 3. *If* $\mathbf{v} \in C^u(p)$, *then* $|DF(p)\mathbf{v}| \geq 2|\mathbf{v}|$.
 4. *If* $\mathbf{w} \in C^s(p)$, *then* $|DF^{-1}(p)\mathbf{w}| \geq 2|\mathbf{w}|$.
*Then* $\Lambda$ *is a hyperbolic set and* $E^u \subset C^u$ *and* $E^s \subset C^s$.

**Remark.** This cone condition is relatively easy to verify in a given nonlinear dynamical system, since it depends on only one iteration of the map, not on all iterations. We will illustrate this in §2.9 when we discuss the Hénon map.

A more general class of diffeomorphisms consists of those whose chain recurrent sets are hyperbolic. It can be shown that the chain recurrent sets of these maps decompose into a union of invariant subsets on which the map is chaotic (at least on bounded surfaces or subsets of the plane). If the stable and unstable manifolds of these sets all meet transversely, we have a natural generalization of the class of Morse-Smale maps. These maps, known as *Axiom A* dynamical systems, can be shown to be structurally stable yet possess chaotic "pieces" like our three basic examples. This class has been the subject of much contemporary research.

**Exercises**

**1.** Prove that all points in the Cantor set associated to the Smale horseshoe map are chain recurrent.

**2.** Prove that every point in the torus is chain recurrent under a hyperbolic toral automorphism.

**3.** Prove that all points in the attractors discussed in §2.5 are chain recurrent.

**4.** Prove that the chain recurrent set is closed and preserved by topological conjugacy.

**5.** Prove that the linear automorphism of $S^2$ induced by the matrix

$$\begin{pmatrix} 1 & 0 & 0 \\ 0 & 2 & 0 \\ 0 & 0 & 3 \end{pmatrix}$$

is Morse-Smale (see Exercises 6.7–6.16).

**6.** Identify the chain recurrent set for the linear automorphism of $S^2$ induced by the matrix

$$\begin{pmatrix} \cos\theta & \sin\theta & 0 \\ -\sin\theta & \cos\theta & 0 \\ 0 & 0 & 2 \end{pmatrix}.$$

**7.** Let $F$ be a diffeomorphism and suppose that $\mathbf{0}$ is a hyperbolic fixed point for $F$. Prove that there is a neighborhood $U$ of $\mathbf{0}$ and an $\epsilon > 0$ such that, if $G$ is $C^1$-$\epsilon$ close to $F$, then $G$ has a unique hyperbolic fixed point in $U$.

## §2.8 THE HOPF BIFURCATION

As in the case of one-dimensional maps, the lack of hyperbolicity is usually a signal for the occurrence of bifurcations. In one-dimensional systems, these occur when the eigenvalue at a periodic point is either $+1$ (the saddle node bifurcation) or $-1$ (the period-doubling bifurcation). For higher dimensional systems, these types of bifurcations also occur, but there are other possible bifurcations of periodic points as well. The most typical of these is the Hopf bifurcation, which we will describe in this section. Before that, however, we give an example of saddle node and period-doubling bifurcations in the plane.

**Example 8.1.** Let $Q_\lambda$ be the map of the plane given by

$$x_1 = e^x - \lambda$$

$$y_1 = -\frac{\lambda}{2} \arctan y.$$

$Q_\lambda$ is really a combination of two one-dimensional maps: one in the $x$-direction and one in the $y$-direction. Using the results of § 1.12, it is easy to see that the map $x \to e^x - \lambda$ undergoes a saddle node bifurcation when $\lambda = 1$ while

$$y \to -\frac{\lambda}{2} \arctan y$$

undergoes a period doubling at $\lambda = 2$. Putting the phase portraits of these two maps together allows us to describe the phase portrait for $Q_\lambda$. When $0 < \lambda < 1$, $Q_\lambda$ moves all points to the right, so that there are no periodic points whatsoever. For $1 < \lambda < 2$, $Q_\lambda$ admits two fixed points, an attracting fixed point and a saddle. Both fixed points undergo period-doubling bifurcations at $\lambda = 2$, so that there are two orbits of period 2 for $\lambda > 2$. As $\lambda$ passes through 2, the attracting fixed point becomes a saddle, while the saddle becomes a repellor. See Fig. 8.1.

This situation is typical. When $\lambda = 1$, a saddle node bifurcation occurs at the fixed point 0. At this $\lambda$-value, $DQ_\lambda(0)$ has an eigenvalue 1 and another eigenvalue less than one in absolute value. When $\lambda = 2$, there are two fixed points for $Q_\lambda$, both of which have one eigenvalue $-1$ and another eigenvalue not equal to one in absolute value.

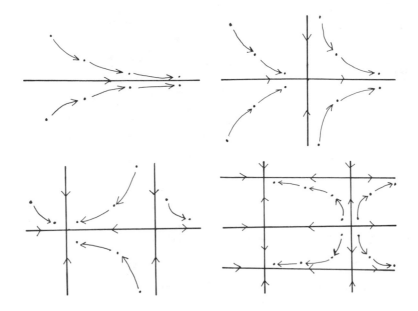

**Fig. 8.1.** The phase portraits of $Q_\lambda^2$

In higher dimensional systems, there is an additional manner in which a fixed or periodic point may fail to be hyperbolic. When the eigenvalues of the Jacobian matrix are complex but of absolute value one, the fixed point is non-hyperbolic. As long as these eigenvalues are not $\pm1$, a different type of bifurcation generally occurs.

Analysis of the dynamics of linear maps shows that a bifurcation must occur when an eigenvalue crosses the unit circle. Consider the family of maps

$$L_\lambda \begin{pmatrix} x \\ y \end{pmatrix} = \lambda \begin{pmatrix} \cos \alpha & -\sin \alpha \\ \sin \alpha & \cos \alpha \end{pmatrix} \begin{pmatrix} x \\ y \end{pmatrix}$$

where $\lambda > 0$ is the parameter. If $\alpha \neq 0$ and $\lambda < 1$, then 0 is an attracting fixed point and all points spiral toward the origin under iteration of $L_\lambda$. If $\lambda > 1$, then 0 is a repellor. Thus a change has occurred at $\lambda = 1$. At the bifurcation value, each circle centered at the origin is invariant under $L_1$. Moreover, the dynamics on these circles are different depending upon whether $\cos \alpha + i \sin \alpha$ induces a rational or irrational rotation of the circle.

While a bifurcation does occur in the linear family $L_\lambda$ when the eigenvalues cross the unit circle, the dynamics of this family are quite special at the bifurcation value. In nonlinear dynamics, the bifurcation occurs in a somewhat different manner. Let us first consider several examples.

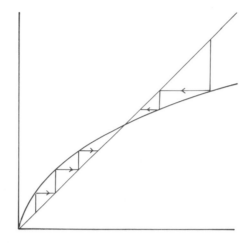

**Fig. 8.2.** The graph of $r \to \lambda r + \beta r^3$.

**Example 8.2.** Let $F_\lambda$ denote the family of planar maps given by

$$F_\lambda \begin{pmatrix} x \\ y \end{pmatrix} = g_\lambda(x, y) \begin{pmatrix} \cos \alpha & -\sin \alpha \\ \sin \alpha & \cos \alpha \end{pmatrix} \begin{pmatrix} x \\ y \end{pmatrix}$$

where $g_\lambda(x, y) = \lambda + \beta(x^2 + y^2)$. Here $\beta$ is a nonzero constant which we may treat as a second parameter. When $\beta = 0$, $F_\lambda$ becomes the linear family $L_\lambda$.

Note that 0 is a fixed point for each $\lambda$, and that the Jacobian matrix of $F_\lambda$ satisfies

$$DF_\lambda(0) = \lambda \begin{pmatrix} \cos \alpha & -\sin \alpha \\ \sin \alpha & \cos \alpha \end{pmatrix}.$$

Hence, at $\lambda = 1$, the eigenvalues cross the unit circle as before. It follows that 0 is an attracting fixed point for $\lambda < 1$ and a repelling fixed point for $\lambda > 1$. To study the bifurcation which occurs at $\lambda = 1$, it is most efficient to change to polar coordinates. In polar coordinates, the map assumes the form

$$r_1 = \lambda r + \beta r^3$$
$$\theta_1 = \theta + \alpha.$$

This map has an invariant circle given by $r = \sqrt{(1 - \lambda)/\beta}$ provided we have $(1 - \lambda)/\beta > 0$. Thus there are two cases. Let us first assume that $\beta < 0$. Then, if $\lambda > 1$, the invariant circle is defined and all points in a neighborhood of the circle are attracted to it. This is easily seen by graphical analysis of the function $r \to \lambda r + pr^3$. See Fig. 8.2. The phase portrait for this map is depicted in Fig. 8.3.

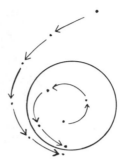

**Fig 8.3.** The phase portrait of the map $r \to \lambda r + \beta r^3$.

Therefore we see that, at the bifurcation value, an invariant circle is born as the attracting fixed point becomes repelling. This is a Hopf bifurcation.

When $\beta > 0$, the situation is somewhat different. For all $\lambda < 1$, 0 is attracting and the invariant circle is repelling. As $\lambda$ passes through 1, the invariant circle and the origin coalesce, and the fixed point at 0 becomes repelling. We leave the details to the reader (see Exercise 1).

A more general example of a Hopf bifurcation is provided by the following family of maps

$$F_\lambda \begin{pmatrix} x \\ y \end{pmatrix} = g_\lambda(x, y) \begin{pmatrix} \cos(\alpha + \gamma r^2) & -\sin(\alpha + \gamma r^2) \\ \sin(\alpha + \gamma r^2) & \cos(\alpha + \gamma r^2) \end{pmatrix} \begin{pmatrix} x \\ y \end{pmatrix}$$

where $\gamma$ is a non-zero constant. In polar coordinates, this family of maps goes over to

$$r_1 = \lambda r + \beta r^3$$
$$\theta_1 = \theta + \alpha + \gamma r^2.$$

One may check that this family undergoes a similar bifurcation as above (see Exercise 2). We will meet this family again, after we discuss normal forms. Note that, in each of the above cases, the dynamics on the invariant circle is simply rotation through a fixed angle. The map on the invariant circle need not be so simple, however, as the following example shows.

**Example 8.3.** Consider the map given in polar coordinates by

$$r_1 = \lambda r + \beta r^3$$
$$\theta_1 = \theta + \nu + \epsilon \sin(k\theta)$$

for an integer $k$ and $\epsilon$ small. As above, the circle $r = \sqrt{(1-\lambda)/\beta}$ is invariant. On the circle, the map is given by $\theta \to \theta + \nu + \epsilon \sin(k\theta)$. This is the standard family which was discussed at length in §1.14. If $\nu = \frac{2\pi}{k}$, one may check easily that there are exactly two periodic orbits of period $k$ for $\epsilon$ small.

The bifurcations in the previous two examples were easy to compute since the map in each case assumed an especially simple form. We cannot expect an arbitrary map to be so simple. We can, however, via a series of transformations, put the map into a form which is much easier to work with in general. This procedure is called transforming a map to normal form. Since we will use similar procedures at several points in the sequel, we will belabor the point somewhat in this section in order to save time later.

We begin with a family $F_\mu$ of nonlinear maps of the form

$$x_1 = \alpha x - \beta y + O(2)$$

$$y_1 = \beta x + \alpha y + O(2)$$

where $\mu = \alpha + i\beta$ is the parameter. The expression $O(2)$ indicates terms of degree two or more, i.e., terms of the form

$$\alpha_1 x^2 + \alpha_2 xy + \alpha_3 y^2 + \beta_1 x^3 + \beta_2 x^2 y + \text{etc.}$$

We think of all of the coefficients $\alpha_j$, $\beta_j$ etc. as depending upon $\mu$. Note that, using Taylor series, any nonlinear map which fixes the origin may be put in such a form with polynomial terms of degree $\leq n$ plus a small remainder.

Note that $DF_\mu(0)$ has eigenvalues $\alpha \pm i\beta = \mu, \bar\mu$. Since we will work primarily with maps for which $\mu$ is complex, it will help to consider this map in complex coordinates rather than in the $x$, $y$-variables.

Define

$$z = x + iy$$

$$\bar z = x - iy.$$

We will work with these complex variables instead. Note that any information given in $z$, $\bar z$ coordinates may be immediately transferred to the $x$, $y$-variables since

$$x = \frac{1}{2}(z + \bar z)$$

$$y = \frac{1}{2i}(z - \bar z)$$

is the inverse transformation.

In these new variables, the map assumes the simple form

$$
\begin{aligned}
z_1 &= x_1 + iy_1 + \cdots \\
&= \alpha x - \beta y + i(\beta x + \alpha y) + \cdots \\
&= (\alpha + i\beta)(x + iy) + \cdots \\
&= \mu z + \cdots
\end{aligned}
$$

and, similarly,

$$
(\bar{z})_1 = \overline{\mu z} + \cdots
$$

where the dots now indicate higher order terms in $z$ and $\bar{z}$. The advantage here is that we now may work with the single equation for $z_1$ rather than the pair of equations for $x_1$ and $y_1$. The equation for $\bar{z}_1$ is obtained from that of $z_1$ by complex conjugation. Then, by the above remark, $x_1$ and $y_1$ are obtained immediately. We remark that the coefficients in the above expression may be complex.

To work successfully with maps in this form, it is almost essential to eliminate some of the higher order terms. This can be achieved by a judicious choice of a conjugacy near the origin. The result is a normal form for the map. In our case, the result is:

**Theorem 8.4.** *Suppose $F_\mu(z) = \mu z + 0(5)$ where $\mu$ is not a $k^{th}$ root of unity for $k = 1, \ldots, 5$. Then there is a neighborhood $U$ of $0$ and a diffeomorphism $L$ on $U$ such that the map $L^{-1} \circ F_\mu \circ L$ assumes the form*

$$
z_1 = \mu z + \beta(\mu) z^2 \bar{z} + 0(5).
$$

Here, the notation $0(5)$ means terms of degree five or more.

**Remark.** Note how simple the nonlinear terms of the normal form are. There are no quadratic and fourth order terms, and only one cubic term. Most of the nonlinearity is confined to the relatively small fifth order terms; the dominant nonlinear term is the remaining third order term.

This Theorem will be proved in a sequence of propositions, each one leading to a simpler form for the map. The proof of each of the propositions is a straightforward calculation; the trick is to guess the desired outcome at the start.

**Proposition 8.5.** *Let $F_\mu$ be a map of the form*

$$
z_1 = \mu z + \alpha_1 z^2 + \alpha_2 z \bar{z} + \alpha_3 \bar{z}^2 + 0(3)
$$

*where $\mu \neq 0$. Then there exists a neighborhood $U_1$ of $0$ and a diffeomorphism $L_1 : U_1 \rightarrow \mathbf{R}^2$ such that $L_1^{-1} \circ F_\mu \circ L_1$ assumes the form $G_\mu$ given by*

$$z_1 = \mu z + 0(3)$$

*provided $\mu$ is not a $k^{th}$ root of unity where $k = 1$ or $3$.*

*Proof.* Let $L_1$ be given by

$$L_1(z) = z + a_1 z^2 + a_2 z\bar{z} + a_3 \bar{z}^2$$

where

$$a_1 = \frac{\alpha_1}{\mu(1-\mu)}$$

$$a_2 = \frac{\alpha_2}{\mu(1-\bar{\mu})}$$

$$a_3 = \frac{\alpha_3}{\mu - \bar{\mu}^2}.$$

Since $DL_1(0) = I$, it follows from the Inverse Function Theorem that $L_1$ is a diffeomorphism in a neighborhood $U_1$ of $0$. Note that the expressions for $a_1$ and $a_2$ presuppose that $\mu \neq 1$, while the expression for $a_3$ necessitates that $\mu$ is not a cube root of $1$. One then computes immediately that

$$F_\mu \circ L = L \circ G_\mu$$

by simply comparing first and second order terms.

q.e.d.

The reader may wonder how the form of $L_1$ was chosen. One finds $L_1$ by simply assuming (or hoping) that $F_\mu$ may be transformed into $G_\mu$, and then solving the conjugacy equation for the terms comprising $L_1$. The algebra is straightforward.

**Proposition 8.6.** *Let $G_\mu$ be a map of the form*

$$z_1 = \mu z + \beta_1 z^3 + \beta_2 z^2 \bar{z} + \beta_3 z\bar{z}^2 + \beta_4 \bar{z}^3 + 0(4)$$

*where $\mu \neq 0$. Then there exists a neighborhood $U_2$ of $0$ and a diffeomorphism $L_2 : U_2 \rightarrow \mathbf{R}^2$ such that $L_2^{-1} \circ G_\mu \circ L_2$ assumes the form $H_\mu$ given by*

$$z_1 = \mu z + \beta_2 z^2 \bar{z} + 0(4)$$

*provided $\mu$ is not a $k^{th}$ root of unity for $k = 2$ or $4$.*

*Proof.* Take $L_2$ in the form

$$L_2(z) = z + b_1 z^3 + b_3 z \bar{z}^2 + b_4 \bar{z}^3$$

where

$$b_1 = \frac{\beta_1}{\mu(1 - \mu^2)}$$

$$b_3 = \frac{\beta_3}{\mu(1 - \bar{\mu}^2)}$$

$$b_4 = \frac{\beta_4}{\mu - \bar{\mu}^3}.$$

Then one may check easily that $G_\mu \circ L_2 = L_2 \circ H_\mu$.

q.e.d.

**Remark.** The term $\beta_2 z^2 \bar{z}$ cannot be removed by adding a term of the form $b_2 z^2 \bar{z}$ to $L_2$ since this leads to an equation of the form

$$b_2 = \frac{\beta_2}{\mu(1 - \mu\bar{\mu})}.$$

The term $1 - \mu\bar{\mu}$ vanishes for any $\mu$ on the unit circle, which is precisely the situation we wish to study. We also note that, if $\mu^4 = 1$, the expression

$$\mu - \bar{\mu}^3 = \frac{\mu^4 - 1}{\mu^3}$$

also vanishes. Hence, in this case, the term $\beta_4 \bar{z}^3$ cannot be removed as well.

The next Proposition is proved exactly as the previous two. We thus leave this proof to the reader.

**Proposition 8.7.** *Let $H_\mu$ be a map of the form*

$$z_1 = \mu z + \beta_2 z^2 \bar{z} \quad + 0(4)$$

*where $\mu \neq 0$. Then there exists a neighborhood $U_3$ of $0$ and a diffeomorphism $L_3 : U_3 \to \mathbf{R}^2$ such that $L_3^{-1} \circ H_\mu \circ L_3$ assumes the form*

$$z_1 = \mu z + \beta_2 z^2 \bar{z} + 0(5)$$

*provided $\mu$ is not a $k^{th}$ root of unity for $k = 5$.*

**Remarks.**

**1.** As in Proposition 8.6, if $\mu^5 = 1$, one may check that the term $\beta_5 \bar{z}^4$ cannot be removed by the above transformation. Hence the normal form in this case becomes

$$z_1 = \mu z + \beta |z|^2 z + \gamma \bar{z}^4 + O(5)$$

where $\beta$ and $\gamma$ are constants and $|z|^2 = z\bar{z}$.

**2.** These three propositions form the basic procedure which allows us to put a map in normal form. Suppose $F: \mathbf{R}^2 \to \mathbf{R}^2$ satisfies $F(0) = 0$ and $DF(0)$ has an eigenvalue $\mu$ where $\mu^k = 1$, $k > 3$. Then a sequence of transformations allows us to put $F$ in the form

$$z_1 = \mu z + \beta_1 |z|^2 z + \beta_2 |z|^4 z + \ldots + \beta_\ell |z|^{2\ell} z + \gamma \bar{z}^{k-1} + O(k)$$

where $\beta_1, \ldots, \beta_\ell, \gamma$ are constants and $\ell$ is the fractional part of $\frac{k-2}{2}$. This generalizes the normal form found above when $k = 5$. See Exercises 4, 5.

The three previous propositions combined yield the proof of Theorem 8.4. Note that, in polar coordinates, the normal form in Theorem 8.4 becomes

$$r_1 = |\mu| r + \beta(\mu) r^3 + 0(5)$$
$$\theta_1 = \theta + \alpha + \gamma(\mu) r^2 + 0(5)$$

where $\mu = |\mu| e^{i\alpha}$ and $\beta$ and $\gamma$ are constants. Here, $0(5)$ indicates terms containing fifth or higher powers of $r$. We remark that, up to fifth order terms, this map is essentially the same as those considered in Example 8.2 above.

We now return to our main goal in this section, the statement of the Hopf Bifurcation Theorem.

**Theorem 8.8.** *Suppose $F_\lambda$ is a family of maps depending on a parameter $\lambda$ and satisfying*
    *i. $F_\lambda(0) = 0$ for all $\lambda$.*
    *ii. $DF_\lambda(0)$ has eigenvalues $\mu(\lambda), \bar{\mu}(\lambda)$ with $|\mu(0)| = 1$ and $\mu(0) \neq k^{th}$ root of unity for $k = 1, \ldots, 5$.*
    *iii. $\frac{d}{d\lambda} |\mu(\lambda)| > 0$ when $\lambda = 0$.*
    *iv. In normal form given by Theorem 8.1, the term $\beta(\mu(0)) < 0$.*
*Then there is an $\epsilon > 0$ and a closed curve $\varsigma_\lambda$ in the form $r = r_\lambda(\theta)$ defined for $0 < \lambda < \epsilon$ and invariant under $F_\lambda$. Moreover, $\varsigma_\lambda$ is attracting in a neighborhood of 0 and $\varsigma_\lambda \to 0$ as $\lambda \to 0$.*

**Remarks.**

**1.**    The assumption that $\frac{d}{d\lambda}|\mu(\lambda)| > 0$ when $\lambda = 0$ means that the eigenvalues cross from the inside to the outside of the unit circle as $\lambda$ increases.

**2.**    If we reverse the inequalities in iii and iv above, the Theorem remains valid. However, after the bifurcation, the invariant circle is repelling while the origin is attracting. See Example 8.3.

We will not prove this theorem since the details are somewhat technical. The basic idea, however, is similar in spirit to our proof of the Stable Manifold Theorem. Hence we will sketch the steps in barest outline.

Let us denote by $N_\lambda$ the map obtained by dropping the higher order terms in the normal form for $F_\lambda$, i.e., $N_\lambda$ is given by

$$r_1 = (1 + \lambda)r + \beta(\lambda)r^3$$
$$\theta_1 = \theta + \alpha(\lambda) + \gamma(\lambda)r^2.$$

As in Example 8.3, $N_\lambda$ admits an attracting invariant circle $C_\lambda$ given by $r = \sqrt{-\lambda/\beta(\lambda)}$ provided $\lambda > 0$, $\beta(\lambda) < 0$. Consider any other simple closed curve $r = r(\theta)$ in a neighborhood of $C_\lambda$. As we showed in Example 8.3, $N_\lambda$ attracts this curve toward $C_\lambda$. Moreover, if we assume that $|r'(\theta)| \leq 1$, then one may check that the image curve has slope strictly less than 1 in absolute value. That is, $N_\lambda$ preserves a sector bundle near $C_\lambda$ as depicted in Fig. 8.4.

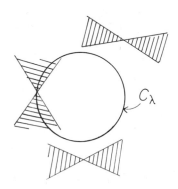

**Fig. 8.4.** The sector bundles near $C_\lambda$.

This allows us to set up a graph transform near $C_\lambda$ similar to that used in the proof of the Stable Manifold Theorem. For $N_\lambda$, this graph transform has a unique fixed graph, which is, of course, the invariant circle $C_\lambda$. One then shows that the same procedure also works when the higher order terms are added to $N_\lambda$. The result is a unique attracting invariant closed curve near $C_\lambda$ for $F_\lambda$ for $\lambda$ sufficiently small and positive.

**Exercises**

**1.** Analyze the bifurcation structure of the map

$$r_1 = \lambda r + \beta r^3$$
$$\theta_1 = \theta + \alpha$$

when $0 < \lambda < 1$ and $\beta > 0$.

**2.** Analyze the bifurcation structure of the map

$$r_1 = \lambda r + \beta r^3$$
$$\theta_1 = \theta + \alpha + \gamma r^2$$

for $\beta$, $\gamma \neq 0$ and $\lambda > 0$.

**3.** Prove Proposition 8.7.

**4.** Suppose $F(0) = 0$ and $DF(0)$ has an eigenvalue $\mu$ which satisfies $\mu^k = 1$. Prove that, if $k = 4$, $F$ may be transformed to

$$z_1 = \mu z + \beta|z|^2 z + \gamma \bar{z}^3 + O(4).$$

If $k = 5$, show that $F$ may be transformed to

$$z_1 = \mu z + \beta|z|^2 z + \gamma \bar{z}^4 + O(5).$$

**5.** Suppose $F(0) = 0$ and $DF(0)$ has an eigenvalue $\mu$ with $\mu^k = 1$, $k > 5$. Show that $F$ may be transformed to normal form:

$$z_1 = \mu z + \beta_1|z|^2 z + \ldots + \beta_\ell|z|^{2\ell} z + \gamma \bar{z}^{k-1} + O(k).$$

where $\ell \leq \frac{k-2}{2}$.

## §2.9 THE HENON MAP

This section serves as an illustration of many of the techniques and ideas introduced in Chapter Two. It consists of a lengthy series of exercises, all of

which deal with the so-called Hénon map. This is a two-parameter family
of maps of the plane which possesses many of the structures and phenomena discussed in this chapter. There are hyperbolic sets, homoclinic points,
bifurcations, horseshoes, "strange attractors" – almost everything we have
discussed *and* a lot more. As an added bonus, the one-dimensional quadratic
family which played such a prominent role in Chapter One is embedded in
the dynamics as well. Thus we view this section as a recapitulation of all
that has come before. The Hénon map is also an important topic in current research, as there are many parameter values for which the map is still
not well understood. Thus we also view this section as an invitation to further research in dynamical systems. We mention some important unsolved
problems at the end of the section.

Let $H = H_{a,b}$ be the map of the plane given by

$$x_1 = a - by - x^2$$
$$y_1 = x.$$

$H$ depends on two real parameters and is called the Hénon map. Note that
there is only one nonlinear term, so that $H$ is indeed one of the simplest
nonlinear maps in higher dimensions.

**1.**   Compute $DH$ and show that $\det(DH) = b$. Prove that, if $b \neq 0$, $H$ is
invertible by exhibiting $H^{-1}$.

**2.**   If $b = 0$, $H$ maps the entire plane onto a parabola $P$ given by $x = a - y^2$.
Prove that the restrictuion of $H$ to $P$ is topologically conjugate to an old
friend, namely, $g(y) = a - y^2$ (*Hint:* Project $P$ onto the $y$-axis and compute
the induced map).

**3.**   Prove that if $0 < |b| \leq 1$, there exists $A, B$ with $|B| > 1$ such that $H_{a,b}$
is topologically conjugate to $H_{A,B}^{-1}$.

As a consequence of these three exercises, it suffices to consider only the
cases $0 < |b| < 1$, since $b = 0$ gives us the quadratic map studied in Chapter
One (see Exercise 1, §1.7). The case $|b| = 1$ will be dealt with later. Hence,
for the moment, we assume that $0 < |b| < 1$.

**4.**   Compute the fixed points for $H_{a,b}$. Prove that for each $b$ there exists
$a_0 = a_0(b)$ such that
   a. if $a < a_0(b)$, $H$ has no fixed points.
   b. if $a = a_0(b)$, $H$ has a unique fixed point.
   c. if $a > a_0(b)$, $H$ has two fixed points of the form $p_\pm(a,b) = p_\pm = (x_\pm, x_\pm)$ where $x_+ > x_-$.

**5.** Prove that $p_-$ is a saddle point for $a > a_0$.

**6.** Prove that there exists $a_1 = a_1(b)$ such that if $a_0 < a < a_1$, the fixed point $p_+$ is attracting.

The previous three exercises show that the saddle node bifurcation occurs in the Hénon map at $a = a_0(b)$.

**7.** By plotting $x_+$ and $x_-$ versus $a = a(b)$, give a picture of the bifurcation diagram for the fixed points of $H$.

**8.** Prove that

    a. if $a = a_1(b)$, $p_+$ has an eigenvalue $-1$.

    b. if $a > a_1(b)$, $p_+$ is a saddle-point. Discuss the behavior of the eigenvalues of $DH(p_+)$ as $a$ varies in both cases $b > 0$ and $b < 0$.

**9.** Compute the periodic points of period two for $H$. Show that there is a unique periodic orbit of period 2 if $a > a_1(b)$ and no such orbit if $a \leq a_1$. Prove that this periodic orbit approaches $p_+$ as $a \to a_1$. This, of course, is the period-doubling bifurcation.

**10.** Prove via the following steps, that if $a < a_0$, and $b > 0$, then $|H^n(p)| \to \infty$ as $n \to \pm\infty$ for all $p \in \mathbf{R}^2$. Conclude that $H$ has no periodic points whatsoever when $a < a_0$.

    a. Let $M_1 = \{(x,y) | x \leq -|y|\}$. Show that if $(x_0, y_0) \in M_1$, then $(x_1, y_1) \in M$ and $x_1 < x_0$.

    b. Let $M_2 = \{(x,y) | x \geq -|y| \text{ and } y \leq 0\}$. Prove that $H(M_1 \cup M_2) \subset$ interior $(M_1)$.

    c. Let $M_3 = \{(x,y) | x \geq -|y| \text{ and } y \geq 0\}$. Prove that $H^{-1}(M_2 \cup M_3) \subset$ interior $(M_3)$.

    d. If $(x_0, y_0) \in M_3$, prove that $|y_{-1}| > |y_0|$ where we have written $H^{-1}(x_0, y_0) = (x_{-1}, y_{-1})$.

**11.** Construct a similar proof for the case $a < a_0$ and $b < 0$.

These two exercises thus prove that there are "no" dynamics for the Hénon map before the saddle node bifurcation. This is similar to the situation encountered for the one-dimensional quadratic map.

**12.** Let $R$ be the larger root of $\rho^2 - (|b|+1)\rho - a = 0$. Let $S$ be the square centered at the origin with vertices $(\pm R, \pm R)$. By using the above partition, prove that all periodic points of $H$ lie in $S$ if $a > a_0$, by showing that if $(x_0, y_0) \notin S$ then either $x_n \to -\infty$ or $|y_{-n}| \to \infty$ as $n \to \infty$.

**13.** Prove that the image of $S$ under $H$ is as depicted in Fig. 9.1. Prove that there exists $a_2 = a_2(b)$ such that the image of $S$ cuts completely across $S$. Compute $a_2$ explicitly.

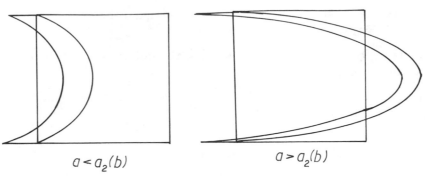

$$a < a_2(b) \qquad\qquad a > a_2(b)$$

**Fig. 9.1.**

Note the great similarity between these figures and those for the horse-shoe map of §2.3. In fact, the set of points whose orbits lie for all time in $S$ is homeomorphic to $\Sigma_2$, as was the case for the horseshoe map. The next few exercises outline a proof of this fact.

**14.**   Consider the cones

$$C^u(\lambda) = \{(\xi, \eta) \,|\, |\xi| \ge \lambda|\eta|\}$$
$$C^s(\lambda) = \{(\xi, \eta) \,|\, |\eta| \ge \lambda|\xi|\}$$

where $\lambda \ge 1$. Prove that, if $|x| \ge \lambda(1 + |b|)/2$, then $C^u(\lambda)$ is invariant under $DH(x, y)$. Similarly, if $|y| \ge \lambda(1 + |b|)/2$, show that $C^s(\lambda)$ is invariant under $DH^{-1}(x, y)$.

**15.**   Suppose $(\xi_0, \eta_0) \in C^u(\lambda)$ and $|x| \ge \lambda(1 + |b|)/2$. Prove that $|\xi_1| > \lambda|\xi_0|$ and $|\eta_{-1}| \ge \lambda|\eta_0|$.

**16.**   Prove that there exists $a_3 = a_3(b)$ such that if $a > a_3$, then $\lambda$ in the above exercises may be chosen larger than 2.

Let $D$ denote the square $S$ with the two strips $|x|, |y| \le \lambda(1 + |b|)/2$ removed. Let $\Lambda = \{(x, y) \in D \,|\, H^n(x, y) \in D \text{ for all } n \in \mathbf{Z}\}$. Provided $a > a_3(b)$, the above three exercises prove that $\Lambda$ is a hyperbolic set.

We may now introduce symbolic dynamics in exactly the same manner as we did for the horseshoe. The requirement that $|x| > \lambda(1 + |b|)/2$ divides $S$ into two vertical strips, which we denote by $V_1$ and $V_2$. Similarly, $|y| > \lambda(1 + |b|)/2$ divides $S$ into two horizontal strips, which we denote by $H_1$ and $H_2$. We say that a subset $V$ of $S$ is a vertical strip if

    a.  $V \subset V_1 \cup V_2$

    b.  $V = \{(x, y) \,|\, v_1(y) \le x \le v_2(y)\}$ where $v_1, v_2$ are vertical curves in $S$.

Horizontal strips are defined analogously. Let $s = (\ldots s_{-2}s_{-1} \cdot s_0 s_1 s_2 \ldots) \in \Sigma_2$. Define

$$V_{s_0 s_1 \ldots s_k} = \{(x,y) \in D \mid H^i(x,y) \in V_{s_i} \text{ for } 0 \leq i \leq k\}$$
$$H_{s_{-1} \ldots s_{-k}} = \{(x,y) \in D \mid H^i(x,y) \in H_{s_i} \text{ for } -k \leq i \leq -1\}.$$

**17.**   Prove that, if $a > a_3(b)$, $V_{s_0 \ldots s_k}$ is a vertical strip and $H_{s_{-1} \ldots s_{-k}}$ is a horizontal strip.

**18.**   Prove that

$$\bigcap_{k=0}^{\infty} V_{s_0 \ldots s_k} = V_s$$

is a vertical curve and

$$\bigcap_{k=-1}^{\infty} H_{s_{-1} \ldots s_{-k}} = H_s$$

is a horizontal curve.

**19.**   Prove that the map $h: \Sigma_2 \to \Lambda$ given by $h(\ldots s_{-2}s_{-1} \cdot s_0 s_1 s_2 \ldots) = V_s \cap H_s$ is a homeomorphism.

**20.**   Prove that $h$ gives a topological conjugacy between $H$ on $\Lambda$ and $\sigma$ on $\Sigma_2$.

Thus the Hénon map is the analogue of the quadratic map in dimension two. For a fixed value of $b$, as the parameter $a$ increases, the dynamics of $H_{a,b}$ become increasingly complex. For $a$ small there are no periodic points for $H_{a,b}$, whereas for $a$ large, there are infinitely many, and, indeed, $H_{a,b}$ admits an invariant set on which $H_{a,b}$ is topologically conjugate to the shift automorphism.

**Remarks.**

**1.**   The transition from simple to complex dynamics in the Hénon map is, to this day, not well understood, although this is a subject of considerable importance in mathematical research.

**2.**   Sarkovskii's Theorem definitely does *not* hold in $\mathbf{R}^2$. In fact, when $b = 1$, we will show below that period two is the *last* period to appear as $a$ increases.

**21.**   *The area preserving cases.* When $b = 1$, the Hénon map has the special property of preserving areas in the plane. That is, if $S$ is a rectangle in $\mathbf{R}^2$ and $H(S)$ is its image, prove that

$$\iint_S dx\, dy = \iint_{H(S)} dx\, dy.$$

This result is also true when $b = -1$, where $H$ preserves area (but reverses orientation since $\det(DH) = -1$). We will concentrate on the case $b = 1$ in exercises 22-34.

**22.**    Compute the eigenvalues at the fixed points $p_\pm$ in case $b = 1$. Show that the eigenvalues of $p_\pm$ are complex and of absolute value 1 if $a_0(1) < a < a_1(1)$. Evaluate $a_0$ and $a_1$ explicitly. The fixed point $p_\pm$ is thus not hyperbolic for $a_0 < a < a_1$.

**23.**    Prove that every periodic point for $H$ has eigenvalues $\lambda, \lambda^{-1}$ which satisfy (when $b = 1$) either

  a. $\lambda \in \mathbf{R}$
  b. $\lambda \in \mathbf{C}$ and $|\lambda| = 1$, $\lambda^{-1} = \overline{\lambda}$.

In case $a$, the periodic point is hyperbolic as usual. In case $b$, the periodic point is called *elliptic*.

  Let $R$ be the linear map given by

$$x_1 = y$$
$$y_1 = x.$$

Note that $R$ fixes the line $\Delta$ given by $y = x$ and satisfies $R \circ R = id$. A map with this property is called an *involution*.

**24.**    Prove that, when $b = 1$, $H \circ R = R \circ H^{-1}$. Conclude that $H = U \circ R$, where $U$ is also an involution. Identify Fix$(U)$. Maps with this special property are called *R-reversible*. This symmetry is present in many dynamical systems which arise in classical mechanics. Its advantage is that it often simplifies the problem of finding certain periodic or homoclinic points.

**25.**    Prove that, if $q \in \Delta$ and $H^k(q) \in \Delta$, then $q$ is periodic of period $2k$.

**26.**    Prove that, if $q \in $ Fix$(U)$ and $H^k(q) \in $ Fix$(U)$, then $q$ is periodic of period $2k$.

**27.**    Prove that, if $q \in \Delta$ and $H^k(q) \in $ Fix$(U)$, then $q$ is periodic of period $2k - 1$.

  Periodic points which arise as in the previous three exercises are called symmetric periodic points.

**28.**    *Symbolic dynamics revisited.* Using the symbolic dynamics introduced above, prove that the action of $R$ induces the map

$$\hat{R}(\ldots s_{-2} s_{-1} \cdot s_0 s_1 s_2) = (\ldots s_2 s_1 s_0 \cdot s_{-1} s_{-2} \ldots)$$

on sequences.

**29.**    Prove that the shift map $\sigma$ may be written as a composition of two involutions, $\sigma = \hat{U} \circ \hat{R}$.

**30.**    Identify all of the symmetric periodic sequences under $\sigma$.

**31.**    *Bifurcation theory for reversible maps.* Recall that, for $a_0 < a < a_1$, the eigenvalues at the fixed point $p_+$ lie on the unit circle. Prove that these eigenvalues traverse the unit circle exactly once as $a$ increases from $a_0$ to $a_1$. Conclude that there is a unique parameter value for which this eigenvalue is a given $n^{th}$ root of unity (where $n \geq 3$).

**32.**    Let $\varsigma$ be an $n^{th}$ root of unity with $n \geq 3$ and suppose $a^*$ is the parameter value for which the eigenvalues of $p_+$ are $\varsigma, \bar{\varsigma}$. Prove that there is a bifurcation at $a^*$ in which at least one symmetric periodic orbit of period $n$ separates away from $p_+$ as $a$ passes through $a^*$ (Hint: watch the behavior of $H^n(\Delta)$ as $a$ passes through $a^*$).

Thus we see that periodic points for the Hénon map arise whenever the eigenvalues pass through an $n^{th}$ root of unity. Actually, one may show that a pair of symmetric periodic orbits arise at each such bifurcation point.

**Remarks.**

**1.**    This shows that there are bifurcations at a dense set of parameter values between $a_0$ and $a_1$.

**2.**    Note that periodic points of all periods $n \geq 3$ must arise as $a$ increases from $a_0$ to $a_1$. At $a_1$ we have the first appearance of a period two point; that is, period two is the last to appear. This is completely different from the Sarkovskii ordering!

**33.**    *Homoclinic Bifurcations.* Let $q \in \Delta$ and suppose that $q \in W^s(p)$. Prove that $q \in W^u(p)$ as well, so that $q$ is a homoclinic point. Such homoclinic points are called symmetric homoclinic points.

**34.**    Prove that, if $a > a_0$, then $H$ admits a symmetric homoclinic point to $p_-$. That is, as soon as the first hyperbolic fixed point for $H$ is born in a saddle node bifurcation, it develops a homoclinic point. Prove that the phase portrait of $H$ is as depicted in Fig. 9.2.

The exercises above show that the character of area-preserving maps is quite different from their dissipative ($|b| < 1$) counterparts. Area-preserving maps form an important class of maps which often arise in applications in mechanics. The structure of such a map near an elliptic fixed point is usually quite intricate and is not yet completely understood. This is the setting for the celebrated Moser Twist Theorem, which asserts that subject to certain differentiability and eigenvalue conditions, there are infinitely many invariant

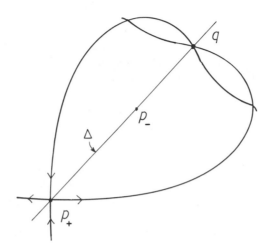

**Fig. 9.2.**

circles around an elliptic fixed point. On these circles, the map is simply irrational rotation. Between these circles, however, the map may be quite chaotic. The study of these maps near an elliptic point could fill another volume!

**Remarks.** The following are two open problems.

**1.**    When $a = 1.4$ and $b = -0.3$, numerical studies indicate that the Hénon map possesses a "strange attractor." See Exercise 5.10. What is happening here?

**2.**    Construct the full bifurcation diagram for $H_{a,b}$ (in the $a, b$-plane with $(x, y) \in \mathbf{R}^2$ ).

## FOR FURTHER READING:

There are a number of texts which can be used to augment and extend the material in Chapter Two. In this text, we have avoided the use of manifolds as the setting for dynamics, but most of the results in this chapter go over easily to this more general case. Also, much of this material may be extended to vector fields or flows without much difficulty. Basic texts which incorporate both of these extensions include:

Palis, J. and de Melo, W. *Geometric Theory of Dynamical Systems*. Springer-Verlag, New York, 1982.

Arnol'd, V.I. *Geometrical Methods in the Theory of Ordinary Differential Equations.* Springer-Verlag, New York, 1977.

We have also avoided applications of the theory of dynamical systems in this text. In recent years, it has become apparent that chaotic dynamics occur in a great number of important physical systems. One text that analyzes a number of these applications is:

Guckenheimer, J. and Holmes, P. *Nonlinear Oscillations, Dynamical Systems, and Bifurcations of Vector Fields.* Springer-Verlag, New York, 1983.

Applications to systems of differential equations which arise in Classical Mechanics have provided a fruitful source of problems in Dynamical Systems. There are a number of advanced texts in this field:

Abraham, R. and Marsden, J. *Foundations of Mechanics.* Second Edition. Benjamin/Cummings, Reading, Mass., 1978.

Arnol'd, V.I. *Mathematical Methods of Classical Mechanics.* Springer-Verlag, New York, 1974.

Moser, J. *Stable and Random Motions in Dynamical Systems.* Princeton University Press, Princeton, New Jersey, 1973.

This last text features an example of how the Smale Horseshoe map arises in the Restricted Three Body Problem of Celestial Mechanics. For a proof of the Stable and Unstable Manifold Theorem for Hyperbolic Sets, the reader may consult:

Nitecki, Z. *Differentiable Dynamical Systems.* M.I.T. Press, Cambridge, Mass., 1971.

The following book contains three survey papers on more advanced topics: bifurcation theory by Guckenheimer, hyperbolic sets by Newhouse, and integrable systems by Moser.

*Dynamical Systems.* CIME Lectures, Bressanone, Italy. Birkhäuser, Boston, 1980.

Finally, the series of four books in the Visual Mathematics Series edited by R. Abraham give an interesting and colorful geometric approach to dynamics:

Abraham, R. and Shaw, C. *Dynamics: The Geometry of Behavior. Part One: Periodic Behavior. Part Two: Chaotic Behavior. Part Three: Global Behavior. Part Four: Bifurcation Behavior.* Aerial Press, Santa Cruz, Calif., 1982.

# Chapter Three

# Complex Analytic Dynamics

The rather specialized subject of the dynamics of complex analytic functions has undergone a remarkable resurgence of interest in recent years. This field, which flourished in the 1920's under the guidance of mathematicians such as Fatou and Julia, lay for the most part dormant until the late seventies. Then, due mainly to the alluring computer graphics of Mandelbrot and the equally enticing mathematical work of Douady, Hubbard, and Sullivan, attention was once again drawn to the rich dynamical behavior of elementary maps of the complex plane.

We will not attempt to survey the most recent work in this chapter. Rather, we will attempt to show how the added assumption of analyticity introduces a new wrinkle into a dynamical system. A complex analytic map always decomposes the plane into two disjoint subsets, the stable set, on which the dynamics are relatively tame, and the Julia set, on which the map is chaotic. We will attempt to describe this chaotic behavior in detail and to sample the types of stable behavior that can occur.

To simplify the exposition, we will concentrate mainly on polynomial maps of the complex plane. Most of the results are true for more general classes of analytic maps, such as rational maps or entire functions. Later in this chapter, we will briefly describe some additional phenomena that occur for these types of maps.

By rights, this chapter should be placed between the chapters on one-dimensional maps and higher dimensional maps. We will see that complex analytic maps share many of the features of one-dimensional systems despite the fact that they live in a higher dimensional setting. We choose to insert this section later, however, since the study of analytic maps necessitates additional mathematical techniques, namely complex analysis. We will summarize the relevant results from this field in the next section.

## §3.1. PRELIMINARIES FROM COMPLEX ANALYSIS

Although we will concentrate mainly on the dynamics of polynomial maps in this chapter, several of the most important techniques that we will use apply to more general complex analytic functions. We will summarize some of the most important results in this section. For proofs and more details, we refer the reader to any of the excellent introductions to the subject of complex analysis such as the books of Conway or Ahlfors.

We denote the complex plane by $\mathbf{C}$. A complex number is written in the form $z = x + iy$, where $i = \sqrt{-1}$. The real part of $z$, denoted by $\mathrm{Re}\,(z)$, is $x$, while the imaginary part, $\mathrm{Im}\,(z)$, is $y$. We denote the modulus of $z$ by $|z|$, i.e.,

$$|z| = \sqrt{x^2 + y^2}$$

**Definition 1.1.** $F : \mathbf{C} \to \mathbf{C}$ is analytic at $z_0$ if

$$F'(z_0) = \lim_{z \to z_0} \frac{F(z) - F(z_0)}{z - z_0}$$

exists.

**Definition 1.2.** Let $U \subset \mathbf{C}$ be an open connected set. $F : U \to \mathbf{C}$ is analytic in $U$ if it is analytic at each $z_0 \in U$.

**Proposition 1.3.** *Let $F(z)$ be analytic at $z_0$. Then there is an $r > 0$ such that*

$$F(z) = \sum_{n=0}^{\infty} a_n (z - z_0)^n$$

*for $|z - z_0| < r$.*

That is, every analytic function may be represented, at least locally, by a power series. Of course, polynomials are a very special case, since the sum above is finite and the convergence question is not present.

Many of the functions encountered in elementary mathematics are analytic. Besides polynomials, all rational functions of the form $P(z)/Q(z)$ where $P$ and $Q$ are polynomials are analytic on their domains of definition. Below we will show how to extend the definition of such a function to points where $Q(z) = 0$. Another class of analytic maps is the class of entire transcendental functions, i.e., those non-polynomial power series which converge in the entire complex plane. Examples of entire functions include the complex exponential, sine, and cosine functions. For later use, we recall the definitions of these complex maps here.

**Definition 1.4.** Let $z = x + iy$

$$\exp(z) = e^x e^{iy} = e^x (\cos y + i \sin y) = \sum_{n=0}^{\infty} \frac{z^n}{n!}$$

$$\sin(z) = \frac{1}{2i}(e^{iz} - e^{-iz}) = \sum_{n=0}^{\infty} (-1)^n \frac{z^{2n+1}}{(2n+1)!}$$

$$\cos(z) = \frac{1}{2}(e^{iz} + e^{-iz}) = \sum_{n=0}^{\infty} (-1)^n \frac{z^{2n}}{(2n)!}.$$

We will have occasion to use open sets which are connected and have no "holes." Such regions are called *simply connected*. Rather than get into a technical description of simply connected sets, we will adopt the following special definition of simple connectivity. In truth, via the extremely important Riemann Mapping Theorem. our definition is completely general (at least for regions in the plane). We emphasize that this is not the standard definition of simple connectivity, but it is sufficient for our purposes.

**Definition 1.5.** An open subset $U$ of $\mathbf{C}$ is simply connected if either $U = \mathbf{C}$ or else there is a one-to-one, onto, analytic map $F : D \to U$ where $D$ is the open unit disk given by $\{z \in \mathbf{C} | \ |z| < 1\}$.

**Example 1.6.** Any half-plane of the form $\text{Re}(z) > a$ or $\text{Im}(z) > a$ is simply connected. Indeed, the map

$$F(z) = \frac{z + 1}{1 - z}$$

maps $D$ onto the right half plane $\text{Re}(z) > 0$.

The following proposition summarizes some of the unique properties of analytic maps.

**Proposition 1.7.** *Let $U \subset \mathbf{C}$ be open and suppose $F : U \to \mathbf{C}$ is analytic. Then*

  1. *$F^{(n)} : U \to \mathbf{C}$ is analytic, where $F^{(n)}$ is the $n$th derivative of $F$.*
  2. *$F(U)$ is open in $\mathbf{C}$, provided $F$ is non-constant.*
  3. *(The Maximum Principle) If $\overline{U}$ is closed and bounded, then $|F(z)|$ assumes its maximum value on the boundary of $\overline{U}$.*

From the Inverse Function Theorem, we have the usual result about the existence of a local (analytic) inverse.

**Proposition 1.8.** *Suppose $F$ is analytic and $F'(z_0) \neq 0$. Then there is an $\epsilon > 0$ and a neighborhood $U$ of $z_0$ such that $F$ maps $U$ onto $D = \{z \in \mathbf{C} | \, |z - F(z_0)| < \epsilon\}$ in a one-to-one fashion. Moreover, the inverse map $F^{-1} : D \to U$ is analytic.*

The assumption of analyticity also allows us to say quite a bit about the behavior of an analytic function near a critical point, i.e., a point $z_0$ where $F'(z_0) = 0$.

**Proposition 1.9.** *Suppose $F$ is analytic and $F^{(j)}(z_0) = 0$ for $1 \leq j < n$, but $F^{(n)}(z_0) \neq 0$. Then there exists $\epsilon > 0$ and a neighborhood $U$ of $z_0$ such that, if $0 < |w - F(z_0)| < \epsilon$, then there are exactly $n$ solutions to the equation $F(z) = w$ in $U$.*

The content of this proposition is best visualized geometrically. We denote the disk of radius $\epsilon$ about $F(z_0)$ by $B_\epsilon(F(z_0))$. Let $\gamma$ be a ray connecting $F(z_0)$ to the boundary of $B_\epsilon$. Then the various preimages of $\gamma$ subdivide $U$ into $n$ sectors. Then $F$ maps the interior of each sector homeomorphically onto $B_\epsilon - \gamma$. See Fig. 1.1. For example, consider $F(z) = z^n$ for $n \geq 2$. There is a unique solution to the equation $F(z) = 0$, while there are exactly

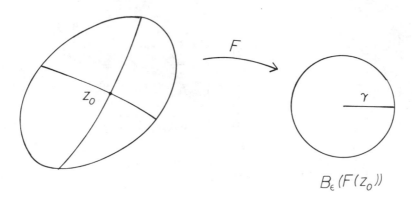

**Fig. 1.1**

$n$ solutions to $F(z) = w$ for all $w \neq 0$. These solutions are, of course, the $n^{th}$ roots of $w$.

The following result is known as the lemma of Schwarz, but its importance in the sequel warrants calling it a Theorem.

**Theorem 1.10.** *Suppose $F$ is analytic in the disk $|z| < 1$ and satisfies*
    *1. $|F(z)| \leq 1$*
    *2. $F(0) = 0$.*
*Then $|F(z)| \leq z$ and $|F'(0)| \leq 1$. Equality holds only if $F(z) = e^{i\theta}z$.*

For a proof, we refer to Ahlfors' text. The Schwarz lemma shows that analytic maps on simply connected regions are very special. The following is a more useful version for our purposes of this result.

**Corollary 1.11.** *Suppose $U$ is a simply connected open subset of $\mathbf{C}$ not equal to $\mathbf{C}$ itself and suppose $F : \overline{U} \to U$ is analytic. If $F$ has a fixed point $z_0$ in $U$, then either*
    *1. $|F'(z_0)| < 1$ and $F^n(z) \to z_0$ for all $z \in U$, or*
    *2. $F'(z_0) = e^{i\theta}$ and $F$ is analytically conjugate to rotation of the unit disk by $e^{i\theta}$.*

**Remark.** The case $U = \mathbf{C}$ must be excluded since the map $F(z) = az$ with $|a| > 1$ is an obvious counterexample. It is one of the remarkable features of complex analytic maps that this cannot happen on any other simply connected open subset of $\mathbf{C}$.

One of the main reasons that we will concentrate on polynomials is the availability of the following theorem known as the Fundamental Theorem of Algebra.

**Theorem 1.12.** *Let $P(z)$ be a polynomial of degree n. If $n > 0$, then $P(z)$ may be written in the form $P(z) = a(z - \alpha_1) \cdot \ldots \cdot (z - \alpha_n)$ where the $\alpha_i$ are not necessarily distinct.*

Complex analytic maps often treat $\infty$ in a fairly straightforward manner. For example, consider $Q(z) = z^2$. If $|z| < 1$, then $Q^n(z) \to 0$ under iteration, whereas if $|z| > 1$, then $Q^n(z) \to \infty$. Hence, for this map, both 0 and $\infty$ may be regarded as attracting "fixed points." Of course, we must define $Q(\infty) = \infty$ in order for this to make sense. But this turns out to be perfectly legitimate. Consider the map $H(z) = 1/z$. $H$ takes 0 to $\infty$ and vice-versa. Note that $H$ is one-to-one and analytic, and that $H \circ Q \circ H^{-1} = Q$, i.e., $H$ conjugates $Q$ to itself! Note, however, that the local behavior of $Q$ near $\infty$ is carried to that of $Q$ near 0. That is, the behavior near $\infty$ for this map is the same as that near 0! That means that there is nothing special whatsoever about $\infty$ and we might as well allow $\infty$ as just another point in the complex plane, or rather, the extended complex plane. All of this may be made precise by introducing the *Riemann sphere*. This is the sphere obtained as follows. We will identify each point in $C \cup \{\infty\}$ with a unique point on a sphere. This is accomplished as follows.

Take the sphere and set its south pole at the origin in $C$. From the north pole, draw a straight line to the point $z$ in $C$. This straight line pierces the sphere in exactly one point which we denote by $S(z)$. Note that $S$ then gives a homeomorphism from $C$ onto the sphere minus the north pole. To complete the picture, we set $S(\infty) = $ north pole. When viewed this way, the "space" $C \cup \{\infty\}$ is called the Riemann sphere and denoted by $\overline{C}$. Intuitively, the Riemann sphere is constructed by wrapping the plane onto the sphere minus the north pole and then gluing it all together by adding in the point "at infinity." See Fig. 1.2.

How do we describe analytic maps on the Riemann sphere? We do precisely what we did for the map $Q(z) = z^2$; we simply conjugate with a map which "moves" $\infty$ elsewhere. More precisely, let $F : \overline{C} \to \overline{C}$ and suppose $F(\infty) = z_0$. For simplicity, we assume that $z_0 \neq 0$. Then $F$ is said to be analytic at $\infty$ if $H \circ F \circ H^{-1}$ is analytic at 0, where $H(z) = 1/z$ as above. (If $F(\infty) = 0$, we replace $H(z)$ by the map $z \to 1/z - a$ where $a \neq 0, \infty$ ).

**Example 1.13.** Let $P(z) = a_n z^n + \ldots + a_0$ be a polynomial. Then $P(\infty) =$

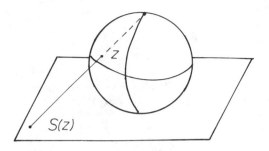

**Fig. 1.2.** The Riemann Sphere.

$\infty$ and $P$ is analytic at $\infty$. Indeed, we have

$$H \circ P \circ H^{-1}(z) = \frac{1}{P(\frac{1}{z})}$$

$$= \frac{z^n}{a_n + a_{n-1}z + \ldots + a_0 z^n}.$$

This rational function vanishes when $z = 0$ as does its derivative. Consequently, $P'(\infty) = 0$ as well.

**Example 1.14.** Let

$$L(z) = \frac{az + b}{cz + d}$$

where the coefficients $a, b, c, d$ are complex and satisfy $ad - bc \neq 0$. $L$ is called a linear fractional transformation. If $c \neq 0$, $L(\infty) = a/c \neq \infty$. If $a \neq 0$, conjugation by $1/z$ yields the map

$$F(z) = \frac{c + dz}{a + bz}$$

and $F'(0) = (ad - bc)/a^2 \neq 0$. Hence $\infty$ is a regular point for $L$.

More generally, if $R(z)$ is any rational function of the form $P(z)/Q(z)$ where $P$ and $Q$ are polynomials, then $R$ induces an analytic map on the entire Riemann sphere. The point at $\infty$ may either be fixed, as in the case of a polynomial, or periodic, as in the case of $R(z) = 1/z^n$. In fact, $\infty$ may behave as any other point under iteration of an analytic map. We remark that points which are mapped to $\infty$ by a rational map are called *poles*.

**Exercises**

The following exercises deal with rational maps of the form

$$T(z) = \frac{\alpha z + \beta}{\gamma z + \delta}$$

where $\alpha, \beta, \gamma, \delta \in \mathbf{C}$. If $\alpha\delta - \beta\gamma \neq 0$, the map is called a Mobius transformation. Throughout, we assume $\alpha\delta - \beta\gamma \neq 0$.

**1.**  Prove that a Mobius transformation induces an analytic diffeomorphism of the Riemann sphere.

**2.**  Show that the inverse of $T$ exists and is also a Mobius transformation.

**3.**  Calculate $T(\infty)$ and show that $T(-\delta/\gamma) = \infty$.

**4.**  Prove that any Mobius transformation may be written as a composition of translations (maps of the form $z \to z + a$ ), inversions $(z \to 1/z)$, and homothetic transformations $(z \to bz)$. The homothetic transformation may be a contraction $(|b| < 1)$, a dilation $(|b| > 1)$, or a rotation $(|b| = 1)$.

**5.**  Prove that a Mobius transformation maps straight lines in $\mathbf{C}$ to either circles or straight lines. Similarly, prove that circles are mapped to either circles or straight lines. Since straight lines in $\mathbf{C}$ correspond to actual circles in the Riemann sphere, we are justified in calling them (generalized) circles as well. Hence this exercise may be succinctly stated: show that a Mobius transformation maps circles to circles in $\overline{\mathbf{C}}$.

**6.**  Prove that, if $(\alpha - \delta)^2 + 4\beta\gamma = 0$, then $T$ has a unique fixed point at $z = \alpha - \delta$. In this case, $T$ is called a parabolic transformation.

**7.**  Show that a parabolic transformation is analytically conjugate to a translation of the form $z \to z + \mu$.

**8.**  Show that, if $T$ has two fixed points, then $T$ is analytically conjugate (via a Mobius transformation) to a unique linear map of the form $z \to \mu z$. If $|\mu| = 1$, $T$ is called elliptic; if $|\mu| \neq 1$, $T$ is called hyperbolic.

**9.**  Identify each of the following Mobius transformations as parabolic, hyperbolic, or elliptic

   a. $T(z) = 1/z$

   b. $T(z) = 2z + 1$

   c. $T(z) = (z + 1)/(z - 1)$

   d. $T(z) = z/(2 - z)$

   e. $T(z) = iz + 1 - i$

**10.**    Let $C_1$ denote the set of lines through the origin and let $C_2$ denote the set of concentric circles about 0. Note that $z \to \mu z$ preserves the elements of $C_1$ and $C_2$. Under the conjugacy described in exercise 8, both $C_1$ and $C_2$ go over to "circles" in the Riemann sphere. The images of the circles in $C_1$ and $C_2$ are called the *Steiner* circles for $T$. Sketch these circles for the following Mobius transformations

   a.  $T(z) = 1/z$
   b.  $T(z) = 2z + 1$
   c.  $T(z) = z/(2 - z)$

**11.**    Prove that the Schwarzian derivative of a Mobius transformation is identically zero.

## §3.2 QUADRATIC MAPS REVISITED

In this section, we return to an old friend, the family of quadratic maps. This time these maps will be viewed as dynamical systems in the complex plane rather than on the real line. The resulting dynamics will be considerably more complicated, at least for certain parameter values. We choose to study the quadratic maps in the form $Q_c(z) = z^2 + c$ where $c$ is a complex parameter. The reason for this is, for these maps, the critical point is located conveniently at 0. In complex dynamics, the orbit of the critical point plays a dominant role. We remark that for any complex number $\lambda \neq 0$, there is a $c = c(\lambda)$ for which the quadratic map $z \to \lambda z(1 - z)$ studied in the first chapter is analytically conjugate to $z \to z^2 + c$ via a map of the form $z \to az + b$ (see Exercise 7.1 in Chapter One).

We first consider the "simplest" map in this family, $Q_0(z) = z^2$. On the real line, this map has only two fixed points, at 0 and 1, and all other points tend to one of these points or to $\infty$ under iteration. In the complex plane, the dynamics are more complicated, but not much more so.

Observe that, if $|z| < 1$, then $Q_0^n(z) \to 0$, while if $|z| > 1$, then $Q_0^n(z) \to \infty$. Thus the dynamics of $Q_0$ are quite tame off the unit circle. On $S^1$, however, the map is chaotic. Indeed, $Q_0$ reduces to another old friend on $S^1$, the map $\theta \to 2\theta$. In §1.2 we proved that hyperbolic periodic points for this map were dense in $S^1$ and later, in §1.8, we showed that this map was actually chaotic. We recall that the three ingredients of chaotic dynamics were

1. sensitive dependence on initial conditions

2. topological transitivity

3. dense periodic points.

The fact that $\theta \to 2\theta$ expanded arclengths by a factor of two enabled us to prove 1-3 for this map. The unit circle in this example is the *Julia set* for $Q_0$. Most of our efforts in this chapter will be aimed at describing the structure of and the dynamics on this set for other complex analytic maps.

**Definition 2.1.** Let $P : \mathbf{C} \to \mathbf{C}$ be a polynomial. The Julia set of $P$, denoted by $J(P)$, is the closure of the set of repelling periodic points of $P$.

As in previous chapters, a periodic point $z_0 = P^n(z_0)$ is repelling if $|(P^n)'(z_0)| > 1$. For a complex dynamical system, we remark that this derivative is a complex number and hence the absolute value denotes the modulus of $(P^n)'(z_0)$.

Let us note several other properties of $J(Q_0)$. First, $J(Q_0)$ is *completely invariant*. By this we mean $J(Q_0)$ contains all forward images as well as preimages of points in $J$. As a consequence, the complement of $J(Q_0)$ is also completely invariant. We call this set the *stable set* and emphasize again that the dynamics on this set are quite tame.

**Definition 2.2.** The stable set of $P$, denoted by $S(P)$, is the complement of the Julia set.

A more interesting property of the Julia set of $Q_0$ is the fact that neighborhoods of any point in $J$ are smeared over virtually the entire plane by iterates of $Q_0$. More precisely, we have

**Proposition 2.3.** *Let* $|z_0| = 1$ *and suppose* $U$ *is any neighborhood of* $z_0$. *Then for each* $z \in \mathbf{C}$, $z \neq 0$, *there is an* $m$ *such that* $z \in Q_0^m(U)$.

*Proof.* Let $z_0 = e^{2\pi i \theta_0}$. There is an $\epsilon > 0$ such that the portion $W$ of the wedge determined by $|\theta - \theta_0| < \epsilon$ and $|r - 1| < \epsilon$ lies inside $U$. Now $Q_0$ expands $W$ in both the $\theta$- and $r$-directions. In particular, there is an $n$ such that $Q_0^n$ expands the arc $|\theta - \theta_0| < \epsilon$, $r = 1$ sufficiently so that its image covers $S^1$. It follows that $Q_0^n(W)$ contains the annulus $1 - \epsilon < r < 1 + \epsilon$. From this, it follows that $Q_0^{n+k}(W)$ contains the annulus $(1 - \epsilon)^k < r < (1 + \epsilon)^k$. The result then follows immediately.
$$\text{q.e.d.}$$

Another way of viewing this proposition is that every point (except 0 ) has a succession of preimages which converges to the Julia set. Thus, the

Julia set is a chaotic repeller for $Q_0$. Not only do the maps $Q_0^n$ spread points apart on the Julia set but also these maps spread small neighborhoods of points in the Julia set over the entire plane ( 0 is the only point which is missed). This property of the family of maps $\{Q_0^n\}$ will become crucial when we discuss normal families in the next section.

We should remark that, in general, $J(P)$ is not a smooth curve as in the case of $Q_0$. Usually, $J(P)$ is much more complicated geometrically. The following example, again an old friend, shows that $J$ may be a Cantor set.

Consider $Q_c(z) = z^2 + c$. If $c$ is real and $c > 2$, then $Q_c$ is easily seen to be topologically conjugate to a map of the form $z \to \lambda z(1 - z)$ where $\lambda > 4$. Recall that, on $\mathbf{R}$ at least, these maps feature most points tending to $\infty$ under iteration with the exception of a Cantor set on which the map is equivalent to a shift map. The same is true in the complex plane.

**Proposition 2.4.** *Let $|c| > 2$. Suppose $|z| \geq |c|$. Then $Q_c^n(z) \to \infty$ as $n \to \infty$.*

*Proof.* Let $|z| = r \geq |c| > 2$. $Q_c$ maps the circle of radius $r$ centered at 0 to a circle of radius $r^2$ centered at $c$. Since $r^2 > 2r$, it follows that this image circle lies in the exterior of the circle of radius $r$. Consequently, $|P(z)| > |z|$ for all $z$ with $|z| \geq |c|$ and we have $|P^n(z)| \to \infty$.

q.e.d.

As there are no periodic points for $Q_c$ on the exterior of $|z| = |c|$, it follows that all of these points lie in $S(Q_c)$. Moreover, any point which is eventually mapped into this region must also lie in $S(Q_c)$. Let us denote by $\Lambda$ the set of points whose entire forward orbits lie within the circle $|z| = |c|$. The following proposition should come as no surprise.

**Proposition 2.5.** *If $|c|$ is sufficiently large, $\Lambda$ is a $Q_c$-invariant Cantor set on which $Q_c$ is topologically conjugate to the shift on two symbols. All points in $\mathbf{C} - \Lambda$ tend to $\infty$ under iteration of $Q_c$.*

*Proof.* This proof is exactly analogous to that of the real quadratic maps, except that the nested sequence of intervals which defines the Cantor set is replaced by a nested sequence of disks, much as in the case of the solenoid.

Let $\gamma$ be the preimage of the circle $|z| = |c|$. We claim that $\gamma$ is a figure eight curve as depicted in Fig. 2.1. · To see this, we simply observe that 0 is the only preimage of $c$, whereas all other points on $|z| = |c|$ have two preimages. Note that $\gamma$ is contained in the interior of the disk $|z| \leq |c|$. Also note that points between $\gamma$ and the circle $|z| = |c|$ are mapped into the exterior of $|z| = |c|$, and hence these points lie in the stable set.

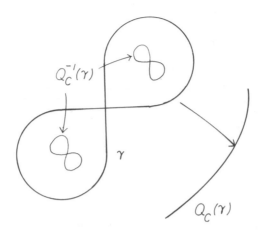

**Fig. 2.1**

Now choose $r < |c|$ so that $\gamma$ is contained in the interior of the disk $D$ given by $|z| \le r$. The preimage of the disk $D$ consists of two simply connected sets, one in each lobe of $\gamma$. Each of these "disks" is mapped diffeomorphically onto $|z| \le r$. Clearly, $\Lambda = \cap_{n=1}^{\infty} Q_c^{-n}(D)$. As in the previous examples, it is easy to check that $Q_c^{-n}(D)$ consists of $2^n$ disks and that $Q_c^{-n}(D) \subset Q_c^{-n+1}(D)$. See Fig. 2.2. The intersection of any nested sequence of these disks is a unique point. To prove this, we need to make one additional assumption at this point (analogous to the derivative condition we used in §1.6 ).

Let $B$ denote the disk of radius $1/2$ centered at the origin. $Q_c(B)$ is the disk of radius $1/4$ centered at $c$. Let us assume that $Q_c(B) \cap \gamma = \phi$. Note that if $|Q_c'(z)| \le 1$, then $z \in B$. Hence our assumption implies that any point with derivative less than one is mapped out of $\gamma$. Hence $|Q_c'(z)| > 1$ for all $z \in \Lambda$ and it follows that $\Lambda$ is a Cantor set.

q.e.d.

**Remarks.**

**1.** Clearly, repelling periodic points are dense in $\Lambda$ and so $\Lambda$ is precisely the Julia set of $Q_c$.

**2.** If $z_0 \in \Lambda$, any small neighborhood $N$ of $z_0$ must contain a disk which is a preimage of $D$. Hence $Q_c$ eventually expands $N$ under iteration so that any point in C lies in some $Q_c^n(N)$.

**3.** We remark that our conditions on $c$ which guarantee that $J(Q_c)$ is a Cantor set may be relaxed somewhat. It can be proved that if $Q_c^n(c) \to \infty$,

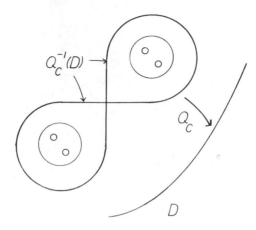

**Fig. 2.2.**

then $J(Q_c)$ is a Cantor set as above.

## §3.3 NORMAL FAMILIES AND EXCEPTIONAL POINTS

In this section, we consider several special properties enjoyed by families of complex analytic maps. It is these properties which give the dynamics of these maps a much richer structure than those previously encountered.

Let $\{F_n\}$ be a family of complex analytic functions defined on an open set $U$. Most often, for our purposes, $F_n$ will be the $n^{th}$ iterate of a given map $F$, but for the moment we will adopt a more general approach.

**Definition 3.1.** The family $\{F_n\}$ is a normal family on $U$ if every sequence of the $F_n$'s has a subsequence which either

    1. converges uniformly on compact subsets of $U$, or

    2. converges uniformly to $\infty$ on $U$.

**Example 3.2.** Let $F(z) = az$ with $|a| < 1$ and set $F_n(z) = F^n(z)$, i.e., the $n^{th}$ iterate of $F$. Then $\{F_n\}$ forms a normal family of functions on any domain in $\mathbf{C}$ since $F_n$ converges uniformly to the constant function 0 on compact subsets.

**Example 3.3.** If $F(z) = az$ with $|a| > 1$, then the above family is normal on any domain which does not include 0, but fails to be normal if the domain includes 0. Indeed, in any neighborhood of 0, there is a point $z$ for which $|F^n(z)|$ is arbitrarily large for some $n$. In particular, we note that any such neighborhood $U$ satisfies

$$\bigcup_{n=1}^{\infty} F^n(U) = \mathbf{C}.$$

The previous examples show that the Julia sets of maps of the form $F(z) = az$, or more generally, $F(z) = az + b$ are quite simple. Therefore, we will exclude these maps from consideration in the sequel. That is, we will consider only polynomials of degree $\geq 2$ from now on.

**Definition 3.4.** The family $\{F_n\}$ is not normal at $z_0$ if the family fails to be a normal family in every neighborhood of $z_0$.

The following proposition gives one of the most important properties of sequences of analytic functions.

**Proposition 3.5.** *Suppose $F_n$ is a sequence of analytic functions which converges uniformly on a domain $U$ to a map $F$. Then $F$ is analytic in $U$ and, moreover,*

$$\lim_{n \to \infty} F_n^{(k)}(z) = F^{(k)}(z).$$

For a proof, we refer to Conway's text. This proposition gives a useful criterion for a family of analytic functions to fail to be normal at a given point.

**Proposition 3.6.** *Let $F$ be analytic and suppose that $z_0$ is a repelling periodic point for $F$. Then the family of iterates of $F$ is not normal at $z_0$.*

*Proof.* We prove this for fixed points. The proof for periodic points is only slightly more complicated. See Exercise 3.1. Assume that $\{F^n\}$ is normal on a neighborhood $U$ of $z_0$. Since $F^n(z_0) = z_0$ for all $n$, it follows that $F^n(z)$ does not converge to $\infty$ on $U$. Thus some sequence of the $\{F^n\}$ has a subsequence $\{F^{n_i}\}$ which converges uniformly to $G$ on $U$. Hence $|(F^{n_i})'(z_0)| \to |G'(z_0)|$. But $|(F^{n_i})'(z_0)| \to \infty$. This contradiction establishes the result.

q.e.d.

**Corollary 3.7.** *Let $F$ be analytic. The family of iterates $\{F^n\}$ fails to be a normal family at any point in $J(F)$.*

One of the most important consequences of the failure to be a normal family at a given point is that the family of functions must then assume virtually every value in any neighborhood of the point. This result is a variant of a theorem known as Montel's Theorem.

**Theorem 3.8.** *Suppose* $\{F_n\}$ *is a family of analytic functions defined on a domain $U$. Suppose there exist $a, b \in \mathbf{C}$ such that $F_n(z) \neq a$ or $b$ for any $n$ and any $z \in U$. Then $\{F_n\}$ is a normal family in $U$.*

For our purposes, the following corollary will be most important.

**Corollary 3.9.** *Let $F$ be an analytic map. Let $z_0 \in J(F)$ and let $U$ be a neighborhood of $z_0$. Then*

$$\bigcup_{n=1}^{\infty} F^n(U)$$

*omits at most one point in* $\mathbf{C}$.

*Proof.* If $F^n(U)$ omitted two points, then $\{F^n\}$ would be a normal family in $U$.

q.e.d.

Thus, the worst possible behavior of the family of iterates of $F$ in a neighborhood of a point in the Julia set is that it omits one value in $\mathbf{C}$. This can happen, as the example $F(z) = z^2$ shows. The Julia set for this map is the unit circle and, if $U$ is an open set which meets $S^1$ but does not meet 0, then

$$\bigcup_{n=1}^{\infty} F^n(U) = \mathbf{C} - \{0\}.$$

See Example 2.1. Such points are called *exceptional points*, and, as their name implies, they rarely occur. Indeed, we may list all possible polynomials which have exceptional points.

**Theorem 3.10.** *Let $P$ be a polynomial. Suppose there is a point $z_0 \in J(P)$ and a neighborhood $U$ of $z_0$ such that*

$$\bigcup_{n=0}^{\infty} P^n(U) = \mathbf{C} - \{a\}.$$

*Then $P(z) = a + \lambda(z - a)^k$ for some $\lambda \in \mathbf{C}$ and some integer $k$.*

*Proof.* Suppose $P(b) = a$. Then $b$ is an exceptional point for $F$, for there is no $z$ in $U$ which maps to $b$ and then to $a$. But then Corollary 3.9 implies that $b = a$. Hence $a$ is fixed for $P$ and, moreover, $a$ is its only preimage.

Thus for some $k$ we may write

$$\frac{P(z) - a}{(z - a)^k} = G(z)$$

where $G$ is a polynomial and $G(z) \neq 0$ for any $z$. Otherwise, we would have an additional preimage of $a$. It follows that $G(z)$ reduces to a constant by the Fundamental Theorem of Algebra.

q.e.d.

These polynomials with an exceptional point at $a$ are in fact topologically conjugate to the simple map $z \to z^k$.

**Proposition 3.11.** *Suppose $P$ is a polynomial of degree $n \geq 2$ which has an exceptional point at $a$. Then $P$ is topologically conjugate to $z \to z^n$.*

*Proof.* Let $Q(z) = z^n$. Theorem 3.10 shows that

$$P(z) = a + \lambda(z - a)^n$$

for some $\lambda \neq 0$. Choose any $(n - 1)^{\text{st}}$ root $\mu$ of $\lambda$, and define $H(z) = \mu(z - a)$. Then a straightforward computation shows that $Q \circ H = H \circ P$.

q.e.d.

It follows that the Julia set of a polynomial with an exceptional value is a circle. This turns out to be a rather special case. Since we may assume that the dynamics of such a map are completely understood, we will henceforth exclude such maps from consideration and study only those polynomials without exceptional points.

**Exercises**

**1.** Prove that the family $\{F^n\}$ is not normal at any repelling periodic point for $F$.

## §3.4 PERIODIC POINTS

There are basically three types of periodic points which may occur for a complex analytic map, attracting, repelling, and indifferent or neutral periodic points. As in previous chapters, a fixed point $z_0$ for $P$ is attracting (resp. repelling, resp. indifferent) if $|P'(z_0)| < 1$ (resp. $|P'(z_0)| > 1$, resp. $|P'(z_0)| = 1$ ). Indifferent periodic points may be quite different from their one-dimensional counterparts, as the rotation map $z \to e^{i\theta} z$ shows. Hence we will concentrate mainly on attracting and repelling periodic points in this section, leaving the difficult neutral case until later.

We may distinguish two distinct types of attracting fixed point: the super-attracting case where $F'(z_0) = 0$ and the attracting case where $0 < |F'(z_0)| < 1$. Analytically, these two cases are quite distinct although they are similar dynamically. We will concentrate on the attracting case for the moment.

Let us first describe the local dynamics near such a fixed point. As we have seen for one-dimensional maps, using a fundamental domain argument, we can set up a topological conjugacy locally between the map itself and the linear map determined by the derivative at the fixed point. For complex analytic maps, we can do much better: the conjugacy can actually be chosen to be complex analytic. Hence we say that a complex analytic map may be linearized near an attracting fixed point. This theorem has a long history, going back to Koenigs in the nineteenth century. The proof we present here is due to Siegel and Moser. We will provide all of the technical details, since they illustrate best what goes wrong in the case of an indifferent periodic point. We remark that it is of no real benefit to assume that the map is a polynomial for local questions like these; hence we will work with a more general analytic function.

**Theorem 4.1.** *Let $P(z)$ be an analytic function satisfying $P(0) = 0$ and $P'(0) = \lambda$ with $0 < |\lambda| < 1$. Then there is a neighborhood $U$ of $0$ and an analytic map $H : U \to \mathbf{C}$ such that*

$$P \circ H(z) = H \circ L(z)$$

*where $L(z) = \lambda z$.*

**Remark.** The functional equation $P \circ H = H \circ L$ is called the Schröder functional equation.

*Proof.* We may write $P$ as a power series in the form

$$P(z) = \lambda z + \sum_{\ell=2}^{\infty} a_\ell z^\ell$$

where this series converges in some neighborhood of 0. We must therefore find an analytic function $H(z)$ that solves the functional equation $P \circ H = H \circ L$. Such an analytic function must be represented as a power series and so we may write

$$H(z) = \sum_{\ell=0}^{\infty} c_\ell z^\ell.$$

We will first determine what the coefficients $c_\ell$ of this power series must be. This is a *formal* solution of the functional equation. Then we must show that the series thus defined actually converges in some neighborhood of 0.

Determining the formal solution is the easy part. Since both $P$ and $L$ fix 0, so too must $H$. This means that $c_0 = 0$. Next, comparing the first order terms of $P(H(z))$ and $H(\lambda z)$, we see that $c_1$ may be chosen arbitrarily (but non-zero). We thus set $c_1 = 1$, so that

$$H(z) = z + \sum_{\ell=2}^{\infty} c_\ell z^\ell.$$

We now proceed to determine the remaining $c_\ell$ recursively. The functional equation may be written

$$\lambda H(z) + \sum_{\ell=2}^{\infty} a_\ell (H(z))^\ell = \lambda z + \sum_{\ell=2}^{\infty} c_\ell \lambda^\ell z^\ell.$$

Therefore we have

$$\sum_{\ell=2}^{\infty} (\lambda^\ell - \lambda) c_\ell z^\ell = \sum_{\ell=2}^{\infty} a_\ell (H(z))^\ell. \tag{*}$$

Suppose that $c_2, \ldots, c_{k-1}$ are known. Then the $k^{th}$ order terms on the right hand side of this equation involve only $c_2, \ldots, c_{k-1}$ since that summation begins with terms of order two. In particular, the coefficient of $z^k$ on the right is a polynomial in $a_2, \ldots, a_k$ as well as $c_2, \ldots, c_{k-1}$ which has positive integer coefficients. We denote this term by $K_k(a_2, \ldots, a_k, c_2, \ldots, c_{k-1})$ and emphasize that, as long as $c_2, \ldots, c_{k-1}$ have been determined, $K_k$ is known.

Then, comparing coefficients, we have

$$c_k = \frac{K_k(a_2, \ldots, a_k, c_2, \ldots, c_{k-1})}{\lambda^k - \lambda}. \qquad (**)$$

Thus, $c_k$ is determined by the previous coefficients, provided $\lambda^k \neq \lambda$, and we have our formal solution.

Before proving that this solution converges, we observe that the expression for $c_k$ shows why this procedure breaks down when $\lambda = 0$ or $\lambda = k^{th}$ root of unity. In either case, the expression for $c_k$ is undefined.

To prove that the power series

$$H(z) = \sum_{\ell=2}^{\infty} c_\ell z^\ell$$

converges, we need several lemmas.

**Lemma 4.2.** *Let $P(z) = z + \sum_{\ell=2}^{\infty} a_\ell z^\ell$. Then $P(z)$ is linearly conjugate to a map $R(z) = z + \sum_{\ell=2}^{\infty} b_\ell z^\ell$ where $|b_\ell| < 1$ for each $\ell$.*

*Proof.* Since the series $\sum_{\ell=2}^{\infty} a_\ell z^\ell$ converges, there is an $A > 0$ for which $|a_{\ell+1}| < A^\ell$ for each $\ell \geq 1$. Let

$$R(z) = z + \sum_{\ell=2}^{\infty} b_\ell z^\ell$$

where $b_\ell = a_\ell / A^{\ell-1}$. Clearly, $|b_\ell| < 1$ for each $\ell$. Let $h(z) = Az$. Then one computes immediately that $h \circ P(z) = R \circ h(z)$.

q.e.d.

**Lemma 4.3.** *There exists $c > 0$ such that $|\lambda^k - \lambda| > c$ for all $k > 1$.*

*Proof.* Since $|\lambda| < 1$, it follows that $|\lambda^k| \leq |\lambda^2| < |\lambda|$ for all $k \geq 1$. Hence $|\lambda^k - \lambda| \geq |\lambda| - |\lambda|^2 > 0$.

q.e.d.

Now observe that the power series in $w$ given by

$$z = w - \frac{1}{c} \sum_{\ell=2}^{\infty} w^\ell$$

converges for $|w| < 1$ and for any $c \in \mathbf{R}$. Indeed, we have

$$z = z(w) = w - \frac{1}{c}\left(\frac{1}{1-w} - 1 - w\right)$$

for $|w| < 1$. Now $z(0) = 0$ and $z'(0) = 1$. Hence it follows that the analytic function $z(w)$ has an inverse which is defined and analytic in some neighborhood of 0. Since $w(0) = 0$ and $w'(0) = 1$, it follows that we may write the inverse map in the form

$$w = w(z) = z + \sum_{\ell=2}^{\infty} \alpha_\ell z^\ell,$$

where we know that the series

$$\sum_{\ell=2}^{\infty} \alpha_\ell z^\ell$$

converges. We will show that this series dominates the formal power series developed for $H$, provided $c$ is chosen small enough.

**Lemma 4.4.** *Choose $c$ as in Lemma 4.3. Then, for each $k \geq 2$, we have*

$$0 \leq |c_k| \leq \alpha_k.$$

*Proof.* Observe that the power series for $w = w(z)$ is a solution of the functional equation

$$cw(z) - cz = \sum_{\ell=2}^{\infty} (w(z))^\ell.$$

Let us solve this equation exactly as we solved the Schröder functional equation. In terms of the $\alpha_k$, this equation reads

$$\sum_{k=2}^{\infty} c\alpha_k z^k = \sum_{\ell=2}^{\infty} (w(z))^\ell.$$

Now compare this equation to equation $(*)$. These equations assume precisely the same form except that $\lambda^\ell - \lambda$ is replaced by $c$ and $a_\ell$ by 1. Hence the solution is determined in precisely the same manner as above and we find that

$$\alpha_k = \frac{K_k(1, \ldots, 1, \alpha_2, \ldots, \alpha_{k-1})}{c}.$$

where the polynomial $K_k$ is given by (**). Now $K_k$ has positive integer coefficients so that, by induction, each $\alpha_k > 0$. Also, we have

$$
\begin{aligned}
|c_k| &\leq \frac{|K_k(a_1, \ldots, a_n, c_2, \ldots, c_{k-1})|}{c} \\
&\leq \frac{1}{c} K_k(1, \ldots, 1, |c_2|, \ldots, |c_{k-1}|) \\
&\leq \frac{1}{c} K_k(1, \ldots, 1, \alpha_2, \ldots, a_{k-1}) \\
&= \alpha_k.
\end{aligned}
$$

q.e.d.

**Remarks.**

**1.** We emphasize that this result has been proved for analytic maps defined locally about an attracting fixed point, not just for polynomials. Indeed, the proof for polynomials is no simpler, as the conjugacy $H$ in general is not a polynomial.

**2.** The same proof works even if $|\lambda| > 1$ by considering the inverse map. Hence repelling periodic points may also be linearized.

**3.** One might be tempted to assume that the Schröder functional equation has a convergent solution if $|\lambda| = 1$ but $\lambda^q \neq 1$ for any integer $q$. After all, it then follows that the term $\lambda^k - \lambda$ is non-zero for all $k$, so that we can determine all of the $c_k$'s. However, the question of convergence of this series is most delicate, since $|\lambda^k - \lambda|$ may be arbitrarily small (which forces the $|c_k|$ to be large see (**)). This is the famous problem of small denominators to which we will return later.

**4.** A similar result holds in the superattracting case. Here $P$ is locally analytically conjugate to a map of the form $z \rightarrow z^k$. The exponent $k$ is determined by the first nonvanishing derivative of $P$ at 0. We omit the details.

We now turn our attention to more global behavior of an analytic map near an attracting periodic point. Let $z_0$ be such a point of period $n$. By the previous results, there is a neighborhood $U$ of $z_0$ in which $F^{jn}(z) \rightarrow z_0$ for all $z \in U$. Consequently, the orbit of any point in $U$ converges to the periodic orbit. We call the set of all points which approach a given attracting periodic orbit the *basin of attraction* of the orbit. Clearly, the basin of attraction is an open set. We call the component of this set which contains the point $z_0$ (and hence, the neighborhood $U$ ) the *immediate attracting basin*. Unlike the map $Q_0(z) = z^2$ for which the immediate attracting basin of 0 is actually the entire basin of attraction, it is very often the case that the basin of attraction consists of infinitely many distinct components, all of which are eventually mapped into one of the immediate attracting basins.

**Remarks.**

**1.**   Let $C$ be any component of the basin of attraction of an attracting periodic point for $F$. Then the family of iterates of $F$ is normal in $C$ (see Exercise 4.1). Hence $C \cap J(F) = \phi$.

**2.**   If $z_0$ is a finite attracting orbit (i.e., $z_0 \neq \infty$ ), then any component of its basin of attraction is simply connected. This fact is an easy consequence of the Maximum Principle (see Exercise 4.2).

**3.**   For a polynomial, $\infty$ may be viewed as an attracting "fixed point." However, the basin of attraction of $\infty$ is not necessarily simply connected. See Proposition 2.5.

One of the distinguishing features of one-dimensional maps with negative Schwarzian derivative was the fact that every attracting periodic orbit attracted at least one critical orbit. This greatly simplified the study of these maps, for it put a bound on the number of attracting orbits. The same result is true for complex analytic maps, but the proof is quite different. Before proving this, we introduce a class of analytic homeomorphisms of the open unit disk $D$.

**Proposition 4.5.** *Let* $|a| < 1$. *Define*

$$T_a(z) = \frac{z - a}{1 - \overline{a}z}.$$

*Then* $T_a$ *is analytic for* $|z| < |a|^{-1}$. *Moreover,* $T_a^{-1} = T_{-a}$ *for* $|z| < 1$ *and* $T_a : D \to D$.

*Proof.* The proof is a straightforward computation and is therefore left as an exercise.

**Theorem 4.6.** *Let* $P$ *be a polynomial and suppose that* $z_0$ *is an attracting periodic point for* $P$. *Then there is a critical value which lies in the basin of attraction of* $z_0$.

*Proof.* Again, for simplicity, we will verify this only for an attracting fixed point, say $z_0$. By our previous results, there is a neighborhood $U$ of $z_0$ and an analytic homeomorphism $H : U \to D$ which linearizes $P$. Let $V$ be an open set containing $U$ and such that $P: V \to U$ is onto. We claim that either $P$ has a critical point in $V$ or else $P$ has an analytic inverse $P^{-1}: U \to V$.

To prove this, let us assume that $P$ does not have an analytic inverse on $U$. Since $P$ is analytic and surjective on $V$, it follows that $P$ must not be one-to-one. Therefore, there exist $z_1, z_2 \in V$ such that $P(z_1) = P(z_2) = q$. Suppose $H(q) = a$ and let $T_a: D \to D$ be as given in the previous proposition.

Consider the circles $C_r$ of radius $r < 1$ centered at 0 in $D$. $T_a^{-1}$ pulls this family of circles back to a nested sequence of simple closed curves surrounding $a$. Since $P: V \to U$ is analytic and $P'(z_i) \neq 0$ for each $i$, it follows that, for $r$ small, $P^{-1}(H^{-1} \circ T_a^{-1}(C_r))$ is a pair of families of nested circles, one family about $z_1$ and the other about $z_2$. Here $P^{-1}$ denotes the inverse image, not the inverse map. Now, as $r$ increases, there is a smallest $r_*$ for which these two families first intersect. Let $p$ be a point common to both simple closed curves of the form $P^{-1}(H^{-1} \circ T_a^{-1}(C_{r_*}))$. Then $p$ is easily seen to be a critical point for $P$.

Thus, we may continue to define $P^{-1}$ on successively larger domains of attraction of $z_0$ until we either meet a critical point or else exhaust the immediate attractive basin by constructing $P^{-k}: U \to \mathbf{C}$ for all positive $k$. Note that, for any $k$, $P^{-k}(U)$ cannot cover all of $\mathbf{C}$; indeed, $P^{-k}(U)$ omits the basin of attraction of $\infty$, which contains $\{z \,|\, |z| > R\}$ for some sufficiently large $R$ (see Exercise 4.4). However, the family of maps $P^{-k}$ is not normal on $U$, as $z_0$ is a repelling fixed point for each. So by Montel's Theorem, we must have that $\cup_{k=0}^{\infty} P^{-k}(U)$ covers $\mathbf{C}$ minus at most one point. This contradiction establishes the result.

$$\text{q.e.d.}$$

**Remarks.**

**1.** The only place where we used the fact that $P$ was a polynomial in this argument was to show that the basin of attraction of $z_0$ omitted at least two points, thereby allowing us to use Montel's Theorem. Eventually, we will show that the Julia set of a complex analytic map contains infinitely many points, all of which lie outside of the basin of attraction of $z_0$. This extends the above theorem to non-polynomial maps.

**2.** This Theorem is more general than our result on maps with negative Schwarzian derivative since it applies to such polynomials as $P(x) = \frac{1}{3}x^3 + x$, which have positive Schwarzian derivative at certain points. Also, complex polynomials may have non-real Schwarzian derivative.

**3.** As we remarked after Theorem 4.1, $\infty$ may be regarded as an attracting fixed point for a polynomial map. The above proof may then be used to show that, if all of the orbits of critical points stay bounded, then the basin of attraction of $\infty$ is simply connected, at least if $\infty$ is included in the basin. We will make this precise later.

**Exercises.**

**1.** Prove that the iterates of an analytic map form a normal family in the basin of attraction of any attracting periodic point.

**2.** Prove that the immediate attracting basin of a (finite) attracting periodic point is simply connected.

**3.** Let

$$T_a(z) = \frac{z - a}{1 - \bar{a}z}.$$

Prove that $T_a$ is an analytic homeomorphism of $D = \{z \,|\, |z| < 1\}$ to itself.

**4.** Prove that if $P$ is a polynomial, there exists $R > 0$ such that if $|z| > R$ then $|P(z)| > |z|$. Conclude that $|P^n(z)| \to \infty$ if $|z| > R$.

## §3.5 THE JULIA SET

The goal of this section is to derive the basic properties of the Julia set of a polynomial map of the complex plane. While some of the arguments we use rely on the polynomial character of the map, most do not. In fact, almost all of the results in this section hold for more general classes of analytic maps such as rational maps or entire functions. Curiously, it is not altogether easy to verify that the Julia set is non-empty. Even for polynomials, most proofs of this fact involve topological rather than algebraic techniques. The following proposition establishes that the Julia set is non-empty for many, but not all, polynomials. We first need a lemma.

**Lemma 5.1.** *Let $R(z)$ be a polynomial of degree $n \geq 2$ with distinct zeroes $\varsigma_1, \ldots, \varsigma_n$. Then*

$$\sum_{i=1}^{n} \frac{1}{R'(\varsigma_i)} = 0.$$

*Proof.* When $n = 2$, the result is a straightforward calculation. For $n > 2$, we use induction. Let $R(z) = (z - \varsigma_n) Q(z)$ where the roots of $Q(z)$ are $\varsigma_1, \ldots, \varsigma_{n-1}$. Suppose all of the $\varsigma_i$ are distinct. By partial fractions we have

$$\frac{1}{Q(z)} = \sum_{i=1}^{n-1} \frac{1}{(Q'(\varsigma_i))(z - \varsigma_i)}.$$

Hence

$$\frac{1}{Q(\varsigma_n)} = \sum_{i=1}^{n-1} \frac{1}{Q'(\varsigma_i)(\varsigma_n - \varsigma_i)}.$$

Now $R'(\varsigma_n) = Q(\varsigma_n)$ and $R'(\varsigma_i) = (\varsigma_i - \varsigma_n)Q'(\varsigma_i)$ for $i < n$. Hence

$$\sum_{i=1}^{n} \frac{1}{R'(\varsigma_i)} = \frac{1}{Q(\varsigma_n)} + \sum_{i=1}^{n-1} \frac{1}{(\varsigma_i - \varsigma_n)Q'(\varsigma_i)}$$

$$= \frac{1}{Q(\varsigma_n)} - \frac{1}{Q(\varsigma_n)}.$$

q.e.d.

**Proposition 5.2.** *Let $P(z)$ be a polynomial. Then either*

    1. *$P(z)$ has a fixed point $q$ with $P'(q) = 1$,*

    2. *$P(z)$ has a fixed point $q$ with $|P'(q)| > 1$.*

*Proof.* Let $R(z) = P(z) - z$. Then the roots of $R$ are precisely the fixed points of $P$. If the roots of $R$ are not all distinct, then there exists $\varsigma$ with $R(\varsigma) = 0$ and $R'(\varsigma) = 0$. But then $P(\varsigma) = \varsigma$ and $P'(\varsigma) = 1$. Hence we may assume that the roots of $R$ are all distinct. Let $\varsigma_1, \ldots, \varsigma_n$ be these roots.

By the lemma, we have

$$\sum_{i=1}^{n} \frac{1}{P'(\varsigma_i) - 1} = \sum_{i=1}^{n} \frac{1}{R'(\varsigma_i)} = 0.$$

Suppose $|P'(\varsigma_i)| \le 1$ but $P'(\varsigma_i) \ne 1$ for all $i$. Then $P'(\varsigma_i) - 1$ lies in the circle $|z + 1| \le 1$ minus the origin. Therefore $1/(P'(\varsigma_i) - 1)$ is well defined and lies in the left-half plane. But since

$$\sum_{i=1}^{n} \frac{1}{P'(\varsigma_i) - 1} = 0,$$

at least one of the $P'(\varsigma_i) - 1$ must lie in the region Re $z \ge 0$. This contradiction establishes the result.

q.e.d.

Thus, either $P$ has a repelling fixed point (and so $J(P) \ne \phi$) or else $P$ has a neutral fixed point with derivative exactly one. In the next section, we will show that such a point is necessarily a limit of repelling periodic points. Assuming this result for the time being, we have

**Proposition 5.3.** *Let $P$ be a polynomial of degree $n \geq 2$. Then $J(P) \neq \phi$.*

**Remark.** A polynomial may not have any repelling fixed points. Indeed $P(z) = z + z^2$ has a unique fixed point at 0 with derivative 1. It is easy to check that $P$ has a repelling periodic point of period two, however.

The proof of the following proposition is straightforward.

**Proposition 5.4.** $J(P) = J(P^n)$.

Our definition of the Julia set as the closure of the repelling periodic points is not the classical one. Since the time of Fatou and Julia, it has been standard to define this set as the set of points at which the family of iterates of $P$ fails to be normal. The following Theorem shows that these two definitions are equivalent.

**Theorem 5.5.** $J(P) = \{z | \{P^n\} \text{ is not normal at } z\}$.

*Proof.* We have already shown that $\{P^n\}$ is not normal at any point in $J(P)$. Hence, it suffices to show that there is a repelling periodic point in any neighborhood of a point where $\{P^n\}$ fails to be normal. Toward that end, suppose $\{P^n\}$ is not normal at $p$ and let $W$ be a neighborhood of $p$. We will produce a repelling periodic point in $W$.

Since $J(P) \neq \phi$, we may find a repelling periodic point $z_0$. By Proposition 5.4, we may assume that $z_0$ is a fixed point for $P$. By the results of the previous section, there is a neighborhood $U_0$ of $z_0$ such that $P : U_0 \to \mathbf{C}$ is a diffeomorphism. Hence $P^{-1}$ is well-defined on $U_0$ and maps $U_0$ inside itself. Let $U_i = P^{-i}(U_0)$ and note that $U_{i+1} \subset U_i$ and $\cap U_i = \{z_0\}$.

Since $\{P^n\}$ is not normal at $p$, there is a point $z_1 \in W$ and an integer $n$ such that $P^n(z_1) = z_0$. Similarly, since $\{P^n\}$ is not normal at $z_0$, there is a point $z_2 \in U_0$ and an integer $m$ such that $P^m(z_2) = z_1$; this uses the obvious fact that $z_1$ is not an exceptional point. Hence $P^{m+n}(z_2) = z_0$. For later use, we note that $z_2$ is a homoclinic point.

We now make the simplifying assumption that $(P^{m+n})'(z_2) \neq 0$. If $z_2$ is a critical point for $P^{m+n}$, then some modifications to the following argument are necessary. We leave these details to the reader. Since $(P^{m+n})'(z_2) \neq 0$, there is a neighborhood $V$ of $z_2$ which is contained in $U_0$ and which is mapped diffeomorphically onto a neighborhood of $z_0$ by $P^{m+n}$. By adjusting $V$, we may assume that $P^m(V) \subset W$ and that $P^{m+n}$ maps $V$ diffeomorphically onto $U_j$ for some $j$. It follows that $P^{m+n+j}$ is a diffeomorphism mapping $V$ onto $U_0$. Consequently, this map has an inverse which contracts $U_0$ onto $V$. By the Schwarz lemma, there is a fixed point for $P^{m+n+j}$ in $V$ which must be repelling. The orbit of this repelling periodic point enters $W$, since $P^m(V) \subset W$.

q.e.d.

This important result has a number of immediate consequences which describe the structure of and the dynamics on the Julia set. For example, the proof shows that any point at which $\{P^n\}$ is not normal is a limit point of repelling periodic points. Hence we have:

**Corollary 5.6.** $J(P)$ *is a perfect set.*

It is clear that the set of repelling periodic points is invariant under $P$. However, it is not so obvious that this is true for inverse images of $P$. Nevertheless, using Theorem 5.5, we have

**Corollary 5.7.** $J(P)$ *is completely invariant.*

*Proof.* If $\{P^n\}$ is not normal at $z_0$, then $\{P^n\}$ is not normal at any inverse image of $z_0$ as well. Hence, by Theorem 5.5, if $z_0 \in J(P)$, then $P^{-1}(z_0) \in J(P)$ as well.

q.e.d.

Recall that a *homoclinic point* $z$ to a repelling fixed point $z_0$ is one for which there exists $n > 0$ for which $P^n(z) = z_0$ and for which there is a sequence of inverse images $P^{-i}(z)$ converging to $z_0$. The proof of Theorem 5.5 also gives

**Corollary 5.8.** *Every repelling periodic point of $P$ admits homoclinic points. Moreover, homoclinic points are dense in $J(P)$.*

**Remarks.**

**1.** It follows that every point in $J(P)$ has a neighborhood on which some sufficiently high iterate of $P$ has an invariant set on which the map is conjugate to the shift. Thus the symbolic dynamics introduced back in §1.6 actually occurs locally near every point in the Julia set!

**2.** It is easy to modify the above arguments to show that, if $z_1$ and $z_2$ are repelling periodic points, then there are heteroclinic orbits connecting them, i.e., there is a point $z$ which eventually maps onto $z_2$ and for which a sequence of backwards iterations of $P^n$ converge to $z_1$.

We have seen that the forward images of a neighborhood of any point in $J(P)$ must eventually cover any non-exceptional point in $\mathbf{C}$, never mind in $J(P)$ alone. If we apply this fact to $J(P)$, we see that the preimages of any point in $J(P)$ must be dense in $J$. That is,

**Proposition 5.9.** *Let* $z_0 \in J(P)$. *Then*

$$J(P) = \text{closure}\left(\bigcup_{k=0}^{\infty} P^{-k}(z_0)\right).$$

This proposition yields a good algorithm for plotting Julia sets graphically. One simply finds a repelling fixed point for $F$ and computes its preimages. One can also use this idea to describe completely certain Julia sets.

Another consequence is the following result.

**Corollary 5.10.** $J(P)$ *has empty interior.*

*Proof.* If $J(P)$ contains an open set, then $J(P) = \mathbf{C}$ minus the exceptional points. But this cannot occur since the basin of attraction of $\infty$ does not lie in the Julia set.

q.e.d.

We remark that this result is the only result in this section which is not true for rational or entire maps.

**Example 5.11.** Let $Q_2(z) = z^2 - 2$. Then $J(Q_2)$ is the closed interval $-2 \le x \le 2$. This may be proved by noting that this interval is closed, completely invariant, and contains a repelling fixed point at $x = 2$ and its preimage at $x = -2$. Thus $J(Q_2)$ is contained in this interval. To see that $J(Q_2)$ actually is this interval, we use the fact that the boundary of the basin of attraction of $\infty$ lies in $J(Q_2)$ (see Exercise 3). Since the orbit of 0 is trapped in the interval, it follows from our remarks after Theorem 4.1 that the basin at $\infty$ is simply connected. Hence $J(Q_2)$ cannot have two disjoint pieces, and so $J(Q_2)$ is the interval.

As a remark, the map $Q_2$ is topologically conjugate to $z \to 4z(1-z)$, so this gives an alternative proof that this map has dense periodic points in the unit interval. See §1.8.

As we mentioned in the introduction to this chapter, the Julia set is precisely the set which carries the chaotic dynamics of $P$.

**Theorem 5.12.** $P$ *is chaotic on* $J(P)$.

*Proof.* As periodic points are dense in $J(P)$ by definition, it suffices to show that $P$ is topologically transitive and also depends sensitively on initial conditions. Let $z_1, z_2 \in J(P)$ and suppose that $U_i$ is a neighborhood of $z_i$. We may assume that $z_1$ and $z_2$ are repelling periodic points. By the remark after Corollary 5.8, there is a heteroclinic orbit connecting $z_1$ and $z_2$. It follows immediately that $P$ is topologically transitive. Since this heteroclinic orbit lies in $J(P)$, it also follows that $P$ has sensitive dependence on initial conditions.

q.e.d.

**Remark.** One can say more: on its Julia set, $P$ is locally eventually onto. This means that if $U$ is any open set meeting $J(P)$, then there is an integer $n$ for which $P^n(U \cap J(P)) = J(P)$. See Exercise 4.

### Exercises

**1.**    Describe the Julia set of $C(z) = z^3 - 3z$.

**2.**    Describe the Julia set of rational maps of the form $R(z) = 1/z^n$ where $n \geq 2$.

**3.**    Prove that the boundary of the basin of attraction of $\infty$ for a polynomial lies in the Julia set.

**4.**    Prove that a polynomial map is locally eventually onto its Julia set.

**5.**    Extend the proof of Proposition 5.2 to the case of rational functions of the form $P(z)/Q(z)$ where $P$ and $Q$ both have degree $n \geq 2$.

## §3.6 THE GEOMETRY OF JULIA SETS

Our goal in this section is to present a variety of examples of Julia sets of polynomials and rational functions. Thus far, the examples of Julia sets that we have encountered (the unit circle, a closed interval, a Cantor set) have been relatively unexciting geometrically. Indeed, one might be fooled into thinking that Julia sets are usually either smooth curves or Cantor sets in C. Actually, this is far from the truth. We hope to illustrate in this section some of the complex and bizarre shapes that a Julia set of a quadratic polynomial may assume.

**Example 6.1.** We first consider the Julia set of $Q_c(z) = z^2 + c$ for $c$ near 0. Recall from §3.2 that $Q_0$ has an attracting fixed point at 0 and that $J(Q_0)$ is the unit circle. This circle clearly bounds both the basins of attraction at 0 and at $\infty$. For $|c|$ small, a similar phenomenon occurs. $Q_c$ is easily seen to have an attracting fixed point near 0 as long as $|c|$ is small. Moreover, it is an easy exercise to show that the boundary of the basin of attraction of

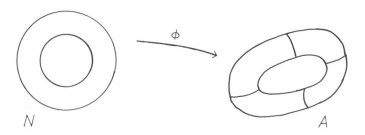

**Fig. 6.1.** The diffeomorphism $\phi: N \to A_1$.

this fixed point again lies in the Julia set of $Q_c$. As before, this boundary is a simple closed curve. However, it is far from being a smooth curve; indeed, it contains no smooth arcs whatsoever!

**Proposition 6.2.** *Suppose $|c| < 1/4$. Then the Julia set of $Q_c$ is a simple closed curve.*

*Proof.* Let $\Gamma_0$ denote the circle of radius $1/2$ about $0$. $\Gamma_0$ contains both the attracting fixed point and the critical point of $Q_c$ in its interior. Moreover $|Q'_c(z)| > 1$ for $z$ in the exterior of $\Gamma_0$.

For each $\theta \in S^1$, we will define a continuous curve

$$\gamma_\theta: [1, \infty) \to \mathbf{C}$$

having the property that $z(\theta) = \lim_{t \to \infty} \gamma_\theta(t)$ is a continuous parametrization of $J(Q_c)$. To define $z(\theta)$, we first note that the preimage $\Gamma_1$ of $\Gamma_0$ under $Q_c$ is a simple closed curve which contains $\Gamma_0$ in its interior and which is mapped in a two-to-one fashion onto $\Gamma_0$. The fact that $\Gamma_1$ is a simple closed curve follows from the fact that both the critical point and its image lie inside $\Gamma_0$.

Hence the curves $\Gamma_0$ and $\Gamma_1$ bound an annular region $A_1$ ($A_1$ may be regarded as a fundamental domain for the attracting fixed point for $Q_c$ ).

Let $N$ be the standard annulus defined by

$$N = \left\{ re^{i\theta} \mid 1 \le r \le 2, \ \theta \text{ arbitrary} \right\}.$$

Choose any diffeomorphism $\phi: N \to A_1$ which maps the inner and outer boundaries of $N$ to the corresponding boundaries of $A_1$. See Fig. 6.1. This allows us to define the initial segment of $\gamma_\theta: [1, 2] \to \mathbf{C}$ by

$$\gamma_\theta(r) = \phi(re^{i\theta}).$$

That is, $\gamma_\theta$ is the image of a ray in $N$ under $\phi$.

For $r \geq 2$, we may extend $\gamma_\theta$ as follows. $Q_c$ has no critical points in the exterior of $\Gamma_1$. Hence there is a simple closed curve $\Gamma_2$ which is mapped in a two-to-one fashion onto $\Gamma_1$. Moreover, $Q_c$ maps the annular region $A_2$ between $\Gamma_1$ and $\Gamma_2$ onto $A_1$, again in a two-to-one fashion. Thus, the preimage of any $\gamma_\theta$ in $A_1$ is a pair of non-intersecting curves in $A_2$. There is a unique such curve which meets the inner boundary $\Gamma_1$. Hence, for each $\theta$, there is a unique curve in $A_2$ which contains the point $\gamma_\theta(2)$. We may thus sew together these two curves in the obvious way at this point, producing a single curve defined on the interval $[1,3]$. Continuing in this fashion, we may extend each $\gamma_\theta$ over the entire interval $[0, \infty)$.

Now recall that $|Q_c'(z)| > k > 1$ provided $z$ lies in the exterior of $\Gamma_1$. Hence the lengths of each extension of $\gamma_\theta$ decrease geometrically. It follows that $\gamma_\theta(t)$ converges uniformly in $\theta$ and that

$$\lim_{t \to \infty} \gamma_\theta(t) = z(\theta)$$

is a unique point in C for each $\theta$.

We claim that $z(\theta)$ parametrizes a simple closed curve in C. Clearly, $z(\theta)$ is continuous, because of the uniform convergence of $\theta$. To show that the image curve is simple, we must prove that if $z(\theta_1) = z(\theta_2)$, then $z(\theta) = z(\theta_1)$ for all $\theta$ with $\theta_1 \leq \theta \leq \theta_2$. However, if this were not the case, then portions of the curves $\Gamma_1$, $\gamma_{\theta_1}(t)$ and $\gamma_{\theta_2}(t)$ would bound a simply connected region containing each $z(\theta)$ in its interior. See Fig. 6.2. This implies that there is a neighborhood of $z(\theta)$ whose images under $Q_c^n$ remain bounded. Hence $z(\theta) \notin J(Q_c)$. But this is impossible.

q.e.d.

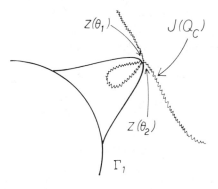

**Fig. 6.2.**

Now suppose that $c$ is complex and satisfies $|c| < 1/4$. $Q_c$ has a repelling fixed point at $z_0 = (1 + \sqrt{1 - 4c})/2$. It is easy to check that $Q_c'(z_0)$ is a complex number which is not pure imaginary. It follows that $z_0$ does not lie in a smooth arc in $z(\theta)$. For if this were the case, then the image of $z(\theta)$ would also be a smooth arc in $J(Q_c)$ passing through $z_0$. Since $Q_c'(z_0)$ is complex, the tangents to these two curves would not be parallel. Therefore, $z(\theta)$ would not be simple at $z_0$. Since the preimages of $z_0$ are dense in $J(Q_c)$ by Proposition 5.9, it follows that $J(Q_c)$ contains no smooth arcs. We have proved

**Proposition 6.3.** *Suppose $c$ is complex and $|c| < 1/4$. Then $J(Q_c)$ is a simple closed curve which contains no smooth arcs.*

Fig. 6.3 illustrates several Julia sets for $Q_c = z^2 + c$ for which $J(Q_c)$ is a simple closed curve.

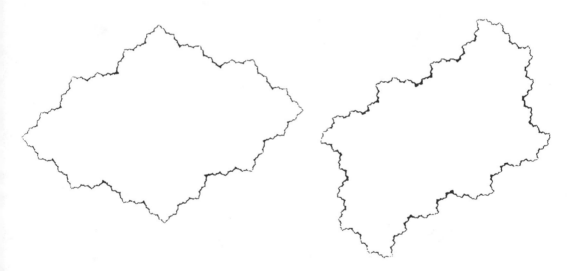

**Fig. 6.3** The Julia sets for $Q_c$ where $c = -\frac{1}{2} - \frac{1}{10}i$, left. and $c = \frac{1}{2}i$.

**Remark.** The condition $|c| < 1/4$ in Propositions 6.2 and 6.3 may be relaxed somewhat. All we need is that $Q_c$ have an attracting fixed point. This occurs for all values of $c$ inside a cardioid in the $c$-plane. See Exercise 1. Moreover, one can actually prove that the Julia set for these $Q_c$ is actually non-differentiable at every point on the simple closed curve.

**Example 6.4.** We now turn to the case of an attracting periodic rather than fixed point. The Julia set is necessarily much different in this case. Let

$P(z) = z^2 - 1$. Note that $P(0) = -1$ and $P(-1) = 0$. Since $P'(0) = 0$, it follows that 0 and $-1$ lie on an attracting periodic orbit of period 2.

The dynamics of $P$ on the real line are relatively straightforward: there are two repelling fixed points at $(1 \pm \sqrt{5})/2$. The fixed point at $(1 - \sqrt{5})/2$ is the dividing point between the basin of attraction of 0 and $-1$. Using arguments as in the proof of Proposition 6.2, one may show that there are two simple closed curves $\gamma_0$ and $\gamma_1$ in $J(P)$ which surround 0 and $-1$ respectively. The curves $\gamma_0$ and $\gamma_1$ meet at the fixed point $(1 - \sqrt{5})/2$. There is much more to $J(P)$ however. Unlike the situation for $Q_c$, the basin of attraction of 0 is not completely invariant. One preimage of the interior of $\gamma_0$ is clearly $\gamma_1$, but there must also be another surrounding the other preimage of 0, namely 1. That is, there is a third simple closed curve in $J(P)$ surrounding 1 as well. Now both 1 and $-1$ must have a pair of distinct preimages, each surrounded by a simple closed curve in $J(P)$. Continuing in this fashion, we see that $J(P)$ must contain infinitely many different simple closed curves. See Fig. 6.4.

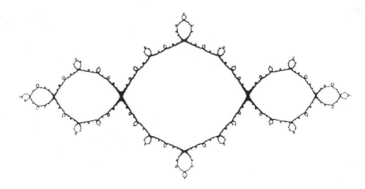

**Fig. 6.4** The Julia set of $P(z) = z^2 - 1$.

The fact that there are infinitely many connected components in the stable set of $P$ is no accident, since we have

**Proposition 6.5.** *Suppose $P$ is a polynomial of degree 2. Then the stable set of $P$ consists of either one, two, or infinitely many connected components.*

The proof of this Proposition is a straightforward exercise (see Exercise 5). We note that all these cases can occur, as we have shown for $Q_c$ when $|c| \geq 2$, $|c| < 1/4$, and $c = -1$ respectively.

**Example 6.6.** Our third example in this section is again a quadratic polynomial. Recall from §1.11 that there is a polynomial of the form $Q_c(z) = z^2 + c$

for which $c$ is real and

1. The critical point at 0 satisfies $Q_c^3(0)$ is a repelling fixed point $-p$, i.e., 0 is eventually periodic.

2. $Q_c$ has dense (repelling) periodic points in the interval $[-p, p]$.

Numerically, the value of $c$ which produces this phenomenon is $c \approx -1.543689$ and $-p \approx -.83928675\ldots$. The graph of $Q_c$ is depicted in Fig. 6.5.

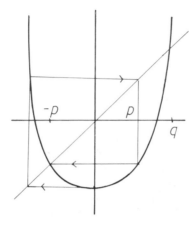

**Fig. 6.5**

Hence the interval $[-p, p]$ is contained in $J(Q_c)$. By backward invariance, all of the preimages of this interval also lie in $J$. Indeed, $J(Q_c)$ is the closure of this set of intervals.

Now $Q_c$ has a second repelling fixed point at $q$ and, using graphical analysis, it is easy to see that $[-q, q] \subset J(Q_c)$. Since $c \in (-q, q)$, the preimage of this interval consists of two intervals, $[-q, q]$ itself and a second interval located symmetrically about 0 but on the imaginary axis. This interval is the preimage of $[-q, c]$. Now the preimage of this pair of intervals consists of four curves as depicted in Fig. 6.6. These curves intersect at 0 and at $Q_c^{-1}(0)$. Continuing, we see that $Q_c^{-n}([-q, q])$ consists of $2^n$ disjoint curvilinear segments. These segments meet one another at the preimages of 0 and the endpoints of these segments are preimages of the fixed point $q$. By the results of §3.5, $J(Q_c)$ is the closure of this set of preimages. Note that, unlike the previous examples, $Q_c^{-n}([-q, q])$ does not bound a region in $\mathbb{C}$. The tree-like structure of

$$\bigcup_{n=0}^{\infty} Q_c^{-n}[-q, q]$$

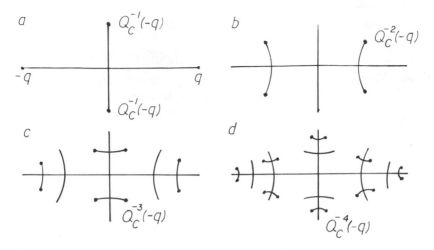

**Fig. 6.6**

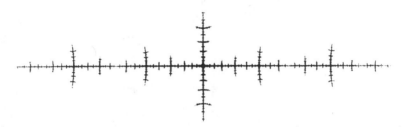

**Fig. 6.7.** The Julia set of $Q_c$.

is called a *dendrite*. The full Julia set of this map is depicted in Fig. 6.7.

These three examples illustrate the dependence of the Julia set on the orbit of the critical point, at least for quadratic maps. When the critical point tends to $\infty$, the Julia set is a Cantor set as illustrated in §3.2. When the critical point tends to an attracting fixed or periodic point, the Julia set is the closure of one or many simple closed curves. And when the critical point is eventually periodic but not periodic, the Julia set is a dendrite as illustrated above.

For the polynomial $Q_i(z) = z^2 + i$, the points $-1 + i$ and $-i$ lie on a repelling periodic orbit of period 2. Note that $Q_i^2(0) = -1 + i$, so 0 is again eventually periodic. The Julia set of $Q_i$ thus shares many of the properties of the previous example. $J(Q_i)$ is depicted in Fig. 6.8.

**Remark.** The quadratic polynomials clearly exhibit a vast array of differ-

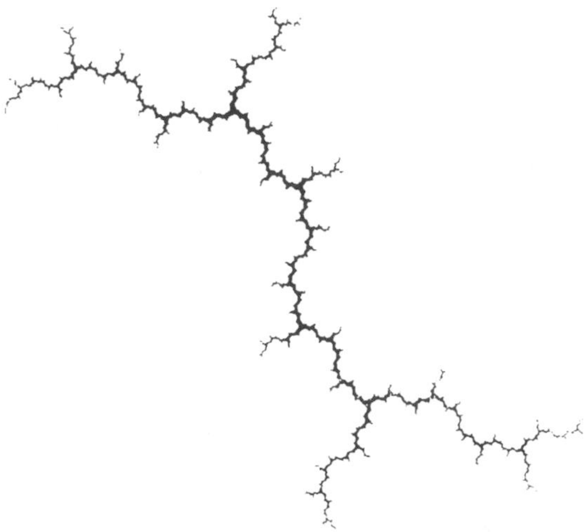

**Fig. 6.8** The Julia set of $Q_i(z) = z^2 + i$.

ent phenomena. One way to catalogue all of this is to sketch the bifurcation diagram for these maps. Since quadratic maps depend on only one parameter, the complex number $c$, this bifurcation diagram lies in the complex plane. The natural way to picture the bifurcation diagram is to indicate all parameter values for which the polynomial has an attracting periodic point. One may also compute the set of parameter values for which the orbit of the critical point remains bounded. These two sets are virtually the same. The resulting locus in the parameter plane is called the Mandelbrot set and is the subject of much contemporary research.

**Example 6.7.** Our final example in this section is intended to show two things. First, this example is a rational function rather than a polynomial, so the dynamics of this map are more efficiently described on the Riemann sphere. Second, the Julia set for this map is the entire Riemann sphere, a phenomenon that cannot occur for polynomials by Corollary 5.10.

To describe this map, we need to use some properties of elliptic functions, and, in particular, the Weierstrass $\wp$-function. This is not an elementary topic in complex analysis. Rather than develop this topic at length, however, we will simply list several properties of the $\wp$-function. We hope that the beautiful geometry associated with this function will motivate the reader to consult any of the standard references such as Ahlfors' text for a complete treatment of this topic.

Let us begin by describing a map on the torus that is very similar to the Anosov or hyperbolic toral automorphisms discussed in §2.4. Let $\omega \in \mathbf{R}$. Regard the torus $T$ as the square in the plane with sidelength $\omega$ and opposite sides identified. Equivalently, $T$ may be regarded as $\mathbf{C}$ modulo the lattice generated by $\omega$ and $\omega i$. By this we mean we identify any two points in $\mathbf{C}$ which differ by a complex number of the form $n\omega + m\omega i$ where $n, m \in \mathbf{Z}$. In particular, the horizontal sides of the square given by $y = \omega i$ and $y = 0$ are to be identified, as are the vertical sides given by $x = 0$ and $x = \omega$. We will regard $\omega$ as a parameter in what follows.

As a remark, we may use any lattice in the plane to define the torus and the resulting map. Let $\alpha, \beta \in \mathbf{C}$ and suppose Im $(\alpha/\beta) \neq 0$. Note that $0, \alpha, \beta$, and $\alpha + \beta$ determine a parallelogram in $\mathbf{C}$, so that $\mathbf{C}$ modulo the lattice generated by $\alpha$ and $\beta$ is also the torus. For simplicity, however, we will restrict our attention to the square lattice generated by $\omega$ and $\omega i$.

Now let $A(z) = 2z$ on $\mathbf{C}$. Since $A$ preserves the lattice points, it follows that $A$ induces a map on $T$ which we also denote by $A$. $A$ is a four-to-one map of $T$ to itself which resembles in many respects the hyperbolic toral automorphisms of §2.4. In particular, repelling periodic points are dense in $T$ (see Exercise 6).

We now wish to project the dynamics of $A$ on $T$ to a map on the Riemann sphere $\overline{\mathbf{C}}$. The projection map $\pi : T \to \overline{\mathbf{C}}$ is easy to describe geometrically. We simply identify each point in $T$ with its "negative" and the result is a sphere. To see this, we first take the fundamental square $S$ in $\mathbf{C}$ with vertices at $0, \omega, \omega i$, and $\omega + \omega i$. Next, draw the diagonal $\Delta$ from $\omega$ to $\omega i$. $\Delta$ divides $S$ into two congruent triangles, $S_1$ and $S_2$. Under the identification of $z$ with $-z$, each point in the interior of $S_1$ is identified with a unique point in the interior of $S_2$. See Fig. 6.9.

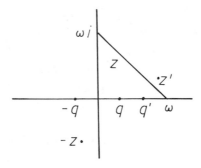

**Fig. 6.9.** The pairs of points $z$ and $z'$
and $q$ and $q'$ are to be identified in $S$

The identifications on the boundaries of $S_1$ and $S_2$ are more complicated. Each point on one of the boundary segments of $S_1$ is identified with a unique point on the same segment with the exception of the endpoints and the midpoints. That is, the points $w/2, wi/2$ and $(w + wi)/2$ have no partners under this identification. Similarly, all of the vertices of $S_1$ are already identified in $T$, and so the point 0 also has no partner under this identification.

Thus we may visualize the sphere as the triangle $S_1$ with all three vertices glued together and each of the sides folded in half and glued together. The result of this operation is a map $\phi: T \to \overline{C}$ which has four critical points at precisely the points $0, w/2, wi/2$, and $(w + wi)/2$.

The map $\phi: T \to \overline{C}$ may be written down explicitly, for $\phi$ is given by the Weierstrass $\wp$- function.

**Definition 6.8.** The Weierstrass $\wp$-function $\wp: C \to \overline{C}$ is given by

$$\wp(z) = \frac{1}{z^2} + \sum_{\nu \neq 0} \left( \frac{1}{(z - \nu)^2} - \frac{1}{\nu^2} \right)$$

where the sum ranges over all of the lattice points $\nu = mw + nwi$ except $\nu = 0$.

We remark that this definition makes perfect sense on any lattice in $C$, not just the square lattice generated by $w$ and $wi$. v Let us list the properties of $\wp(z)$ that we need.

1. $\wp(z)$ is an analytic function with poles at exactly the lattice points in $C$.

2. $\wp(z+w) = \wp(z+wi) = \wp(z)$ so $\wp$ is a doubly-periodic function. Such functions are called *elliptic functions* in complex analysis. Thus $\wp$ may be regarded as a map on $T$.

3. $\wp(z) = \wp(-z)$, so that $\wp$ respects the identification above. Hence we have $\wp: T \to \overline{C}$ as required.

4. $\wp'(z) \neq 0$ as long as $z$ is not one of the half-lattice points $0, w/2, wi/2$, or $(w + wi)/2$.

5. *The Addition Theorem.* $\wp(z)$ satisfies the following relation for any $z, u \in C$

$$\wp(z + u) + \wp(z) + \wp(u) = \frac{1}{4} \left( \frac{\wp'(z) - \wp'(u)}{\wp(z) - \wp(u)} \right)^2$$

where

$$(\wp'(z))^2 = 4(\wp(z))^3 - g_2(w)\wp(z).$$

Here, $(\wp'(z))^2$ means the square of $\wp'(z)$, not the second iterate. Also, the coefficient $g_2$ depends on the parameter $\omega$ only.

6. As a consequence of the Addition Theorem, we have

$$\wp(2z) = \frac{(\wp(z))^4 + \frac{1}{2}g_2(\wp(z))^2 + \frac{1}{16}g_2^2}{4(\wp(z))^3 - g_2\wp(z)}.$$

Consequently, we have the following diagram

$$
\begin{array}{ccc}
T & \xrightarrow{A} & T \\
\wp\downarrow & & \downarrow\wp \\
\mathbf{C} & \xrightarrow{R} & \mathbf{C}.
\end{array}
$$

Because of property 6 above, this diagram commutes. Moreover, the map $R$ is given by

$$R(z) = \frac{z^4 + \frac{1}{2}g_2 z^2 + \frac{1}{16}g_2^2}{4z^3 - g_2 z} = \frac{\left(z^2 + \frac{g_2}{4}\right)^2}{4z(z^2 - g_2)}.$$

That is, $R$ is a rational map of $\overline{\mathbf{C}}$. Now periodic points of $A$ project to periodic points for $R$ and so $R$ has dense periodic points. Consequently, $J(R) = \overline{\mathbf{C}}$, as we claimed.

**Remarks.**

1. It is easy to check that all of the critical points of $R$ are eventually periodic. This is a special case of an important result which asserts that if all critical points of a rational (non-polynomial) map are eventually periodic but not periodic, then $J = \overline{\mathbf{C}}$.

2. All of this works for the general lattice generated by $\alpha, \beta \in \mathbf{C}$ where $\text{Im}\,(\alpha/\beta) \neq 0$. In this case, we have

$$(\wp'(z))^2 = 4(\wp(z))^3 - g_2\wp(z) - g_3.$$

where $g_2$ and $g_3$ are parameters depending only on $\alpha$ and $\beta$. The resulting rational maps assume the form

$$R(z) = \frac{z^4 + \frac{1}{2}g_2 z^2 + 2g_3 z + \frac{1}{16}g_2^2}{4z^3 - g_2 z - g_3}.$$

3. $J = \overline{\mathbf{C}}$ may also occur for entire functions, as we will show in §3.8 for $\exp(z)$.

**Exercises.**

**1.** Let $Q_c(z) = z^2 + c$. Show that $\{c \mid Q_c$ has an attracting point $\}$ is bounded by a cardioid in the $c$-plane. Show that $\{c \mid Q_c$ has an attracting periodic point of period two $\}$ is bounded by a circle in the $c$-plane. This is the beginning of the construction of the Mandelbrot set or the bifurcation set for the family $Q_c$.

**2.** Prove that the boundary of the basin of attraction of an attracting fixed point lies in the Julia set.

**3.** Prove that the boundary of a completely invariant basin of attraction of an attracting fixed point must be the entire Julia set.

**4.** Let $C$ be the immediate basin of attraction of an attracting fixed point. Suppose $D \neq C$ is another component of the stable set which maps onto $C$. Prove that there must therefore be infinitely many components in the stable set.

**5.** Prove that if a polynomial $P$ of degree 2 has a completely invariant component $C$ of its stable set, then $C$ must contain all of the critical points of $P$ (not including, of course, the "critical point" at $\infty$ ). Use this to prove Proposition 6.5.

**6.** Let $T$ be the torus generated by $\mathbf{C}$ modulo the lattice $w, wi$ where $w \in \mathbf{R}$. Let $A_n: T \to T$ be the map induced by multiplication by the integer $n$ with $|n| \geq 2$. Prove that repelling periodic points of $A_n$ are dense in $T$. Show also that $A_n$ is $n^2$- to-one in $T$.

**7.** Construct a lattice which generates a torus $T$ and a map $A: T \to T$ that is induced by multiplication by some nonzero complex number and which is two-to-one on $T$.

**8.** Show that all of the critical points of the map

$$R(z) = \frac{\left(z^2 + \frac{g_2}{4}\right)^2}{4z(z^2 - g_2)}$$

are eventually periodic.

## §3.7 NEUTRAL PERIODIC POINTS

In this section, we take up the difficult problem of the behavior of an analytic function in the neighborhood of a neutral or indifferent periodic point. These points come in two distinctly different varieties. First, there are the rationally indifferent points where the derivative is a rational rotation of the form $e^{2\pi i(m/n)}$ where $m$ and $n$ are integers. We will describe the local behavior completely in this case. The other case, irrational rotations, is much more subtle and difficult and we will confine our remarks to some special cases. As we have noted before, bifurcations often occur when hyperbolicity breaks down. This is the case for complex analytic maps as well and we will describe these local bifurcations in the exercises.

Let us begin with a simple case. Let $F(z)$ have a fixed point at the origin with derivative equal to one. Hence

$$F(z) = z + a_2 z^2 + a_3 z^3 + \ldots + a_n z^n.$$

The first non-zero coefficient $a_k$ plays a crucial role in determining the dynamics of $F$. For the moment, let us assume that $a_2 \neq 0$. By conjugating by $z \to a_2 z$, we may in fact take $a_2 = 1$.

In the simple case of the one-dimensional map $F(x) = x + x^2$, graphical analysis shows that points are attracted to 0 from the left and repelled from 0 on the right. The following proposition shows that a similar phenomenon occurs in the complex plane.

**Proposition 7.1.** *Let* $P(z) = z + z^2 + a_3 z^3 + \ldots + a_n z^n$. *There exists* $\mu > 0$ *such that*

1. *all points in the interior of the circle of radius $\mu$ centered at $-\mu$ are attracted to 0.*

2. *all points in the interior of the circle of radius $\mu$ centered at $\mu$ are repelled from 0.*

*Proof.* The local structure near the fixed point is easiest to understand if we eliminate the fixed point entirely by throwing it to $\infty$. This is accomplished

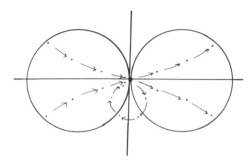

**Fig.7.1.**

by conjugating $P$ with the Mobius transformation $H(z) = 1/z$ as usual. This yields the new map.

$$G(z) = \frac{z^n}{z^{n-1} + z^{n-2} + a_3 z^{n-3} + \ldots + a_n}.$$

Dividing, we may write

$$G(z) = z - 1 + G_0(z).$$

where

$$G_0(z) = \frac{b_2 z^{n-2} + \ldots + b_n}{z^{n-1} + z^{n-2} + a_3 z^{n-3} + \ldots + a_n}.$$

Note that

$$\lim_{|z| \to \infty} G_0(z) = 0.$$

Thus, near $\infty$, $G$ is essentially translation by one unit to the left. In particular, there is a $\delta > 0$ such that, if $\eta > \delta$ and Re $(z) < -\eta$, then Re $(G(z)) < -\eta$ as well. So $G$ maps each half plane Re $(z) < -\eta$ inside itself.

Under $H$, the half plane Re $(z) < -\eta$ is mapped inside the circle of radius $1/2\eta$ centered at $-1/2\eta$. Consequently, points inside these circles are attracted to 0 under iteration of $P$. Thus, if $\mu = -1/2\delta$, part 1 holds. For part 2, we argue similarly, this time using the half-plane Re $z > \eta$.

q.e.d.

Fig. 7.1 illustrates the geometric content of this proposition.

By choosing different regions near $\infty$, one can describe the local dynamics in even more detail. For example, if we choose a portion of a wedge near $\infty$ as illustrated in Fig. 7.2, then the basin of attraction of 0 contains a family of cardioid-like curves.

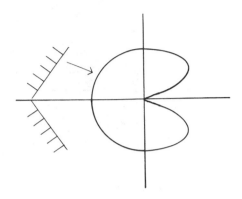

**Fig.7.2.**

We call such a region an attracting *petal* for $P$. More precisely, a simply connected region $C$ is an attracting petal for the indifferent fixed point $z_0$ if $z_0$ is contained in the boundary of $C$ and for each $z \in C$, $P^n(z) \to z_0$. A repelling petal is defined analogously. Note that, if we choose cardoid-like curves for the boundaries of the petals, the attracting and repelling petals can be made to overlap. This means that most, but not all points near 0 simply make a circuit from the repelling to the attracting side of 0. The example $z \to z + z^2$ shows that not all points make this circuit, as points on the positive real axis tend to $\infty$. In fact, it is exactly in the "mouth" of the attracting cardioid that the Julia set slips into a neighborhood of 0, a fact that we will prove below.

When the coefficient $a_2$ vanishes, the situation is more complicated. This is illustrated by the following example.

**Example 7.3.** Let $P(z) = z + z^3$. Note that if $z \in \mathbf{R}$, $z \neq 0$, then $P^n(z) \to \infty$. On the other hand, $P$ preserves the imaginary axis as well and we have $P(iy) = i(y - y^3)$. Graphical analysis of the one-dimensional map $y \to y - y^3$ shows that if $|y| < \sqrt{2}$, then $P^n(iy) \to 0$. The point $\pm i\sqrt{2}$ lies on a repelling periodic orbit of period 2. Thus there are at least two attracting and two repelling petals for $P$. In fact, there are exactly two petals of each type, as we will prove below. See Fig. 7.3.

**Example 7.4.** More generally, consider $P(z) = z + z^{n+1}$. Let $\lambda$ be an $n^{th}$ root of unity. Then the straight line $t \to t\lambda$ is preserved by $P$ since $P(t\lambda) = \lambda t(1 + t^n)$. Graphical analysis allows us to determine the dynamics on each of these invariant lines. For example, if $n$ is even, then $P^j(\lambda t) \to \infty$ for all $t \neq 0$. So all of these lines are repelling. In this case, if $\omega$ is an $n^{th}$

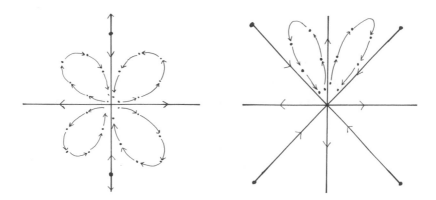

**Fig. 7.3.** The dynamics of $P(z) = z + z^3$ and $P(z) = z + z^5$.

root of $-1$, it is easy to check that the lines $\omega t$ are locally attracting (see Exercise 7.1). See Fig. 7.3.

In the general case, let us suppose that $a_k \neq 0$ but $a_i = 0$ for $i = 2, \ldots, k - 1$. Then we claim that there are $k - 1$ attracting and repelling regions for $P$. This may be seen as follows. Let $H(z) = 1/z^{k-1}$. $H$ is no longer a conjugacy since the inverse of $H$ is not well-defined. However, on the plane $\text{Re } z < -\eta$ used in the proof of Proposition 7.1, we may choose an analytic branch of the inverse, i.e., a well-defined $(k-1)^{st}$ root of $z$. If we fix this inverse map, then one may check that $P$ is conjugate to the map

$$G(z) = \frac{z^n}{z^{n-1} + kz^{n-2} + K(z)}$$

where $K(z)$ involves terms of the form

$$\alpha_{ij} z^i \left( z^{\frac{1}{k-1}} \right)^j$$

for $i < n - 2$ and $0 \leq j < k - 1$. Here we emphasize that the term $z^{1/k-1}$ denotes the fixed branch of the $(k-1)^{st}$ root of $z$ chosen at the outset. The remainder of the proof is now analogous to the above and hence is omitted.

Each different choice of a $(k-1)^{st}$ root yields a different projection from the plane $\text{Re } z < -\eta$ to **C**. Hence we have shown the following.

**Corollary 7.4.** *Let* $P(z) = z + a_k z^k + \ldots + a_n z^n$ *where* $a_k \neq 0$. *Then there are exactly* $k - 1$ *attracting and repelling petals for* $P$ *at* $0$.

The previous discussion allows us to handle the case of rationally indifferent fixed points. Suppose $\lambda^m = 1$ but $\lambda^j \neq 1$ for $1 \leq j < m$. Then the map

$$P(z) = \lambda z + a_2 z^2 + \ldots + a_n z^n$$

has a rationally indifferent fixed point at the origin. Clearly $P^m(z) = z + \ldots$, so that $P^m(z)$ is in the form considered above. Actually, one can prove directly that $P^m$ assumes the form

$$P^m(z) = z + b_{\ell m+1} z^{\ell m+1} + \ldots$$

for some $\ell > 0$. That is, all of the coefficients of $z^k$ vanish (since $\lambda^m = 1$) up to and including the $(\ell m)^{th}$ for some $\ell$ (see Exercise 7.5). We choose to take a different route to this fact by drawing upon our prior work with normal forms.

**Lemma 7.5.** *Let*

$$P(z) = \lambda z + \sum_{k=2}^{n} a_k z^k$$

*where $\lambda^m = 1$ but $\lambda^j \neq 1$ for $1 \leq j < m$. Then for some $\ell > 0$, there is a neighborhood $U$ of $0$ and an analytic map $H : U \to \mathbf{C}$ such that $H^{-1} \circ P \circ H$ assumes the form $\lambda z + b_{m\ell+1} z^{m\ell+1} + \ldots$.*

*Proof.* This proof mimics exactly the reduction to normal form described in §2.9. Hence we will merely begin the reduction process, leaving the remaining details as an exercise.

We try first to eliminate the second order term in $P$. Let

$$P_2(z) = \lambda z + b_3 z^3 + \ldots$$

and set $H(z) = z + A z^2$. The goal is to determine $A$ such that $H \circ P_2 = P \circ H$. One may check easily by comparing second order terms in the above equation that this equality necessitates

$$A = \frac{a_2}{\lambda^2 - \lambda}.$$

Thus, we can only eliminate these terms if either $a_2 = 0$ or $\lambda \neq 0, 1$. One may continue to eliminate successively higher order terms in exactly the same fashion, by determining a conjugacy $H(z) = z + A z^k$ that kills each term in succession.

q.e.d.

Thus we may assume at the outset that $P(z) = \lambda z + z^{m\ell+1} + \ldots$ where we have made a preliminary change of variables that makes the coefficient of $z^{m\ell+1}$ equal to 1. Then one computes readily that

$$P^m(z) = z + m\lambda^{m-1}z^{m\ell+1} + \ldots.$$

This is precisely the form of the maps considered in the first part of this section: hence we conclude that $P^m$ has a total of $m\ell$ attracting and $m\ell$ repelling petals. Thus we have proved

**Theorem 7.6.** *Let $P(z) = \lambda z + \ldots$ where $\lambda^m = 1$ but $\lambda^j \neq 1$ for $1 \leq j < m$. Then there is an $\ell > 0$ such that $P$ admits $m\ell$ attracting and $m\ell$ repelling petals at 0. Each petal is fixed by $P^m(z)$.*

At this point, we can fill the slight gap that we left in our discussion of the Julia set in §3.5. Recall that, in our efforts to prove that the Julia set was nonempty, we showed that a polynomial $P$ map either has a repelling fixed point or a fixed point with derivative one. To complete the proof that $J \neq \phi$, it suffices to show that these latter points are actually limit points of repelling periodic points (establishing, in particular, that there *are* repelling periodic points for $P$). To do this, we will combine the local theory around a neutral fixed point with the ideas of normal families developed in §3.3 to produce a homoclinic point to 0, and thus a nearby repelling periodic point.

As usual, we may assume that the neutral fixed point is at 0. Hence we may write $P(z) = z + \alpha_k z^k + \ldots$ where $\alpha_k \neq 0$. Note first that $\{P^n\}$ is not normal in any neighborhood of 0. This follows since $P^n(z) = z + n\alpha_k z^k + \ldots$ and therefore $(P^n)^{[k]}(0) = n\alpha_k k!$ Consequently,

$$\lim_{n \to \infty} |(P^n)^{[k]}(0)| \to \infty$$

and no subsequence of the $P^n$ can converge to an analytic function at 0.

Since 0 is not an exceptional point, there exists $z_0 \neq 0$ such that $P(z_0) = 0$. Let $U$ be a neighborhood of 0 not containing $z_0$. We may assume that $U$ is contained in the union of attracting and repelling petals, i.e., that if $z \in U$ we either have $P^n(z) \to 0$ as $n \to \infty$ or else as $n \to -\infty$. Since $\{P^n\}$ is not normal in $U$, there is an integer $k$ and $z_1 \in U$ such that $P^k(z_1) = z_0$. Clearly, $z_1$ does not lie in an attracting petal. Hence $z_1$ must lie in a repelling petal. This means that $z_1$ is a homoclinic point for $P$.

The remainder of the proof is exactly the same as the proof of Theorem 5.5. Thus we omit the details.

q.e.d.

We now turn briefly to the behavior of an analytic map whose derivative at a fixed point is an irrational rotation. It turns out that the Schröder functional equation has a solution provided the rotation is sufficiently irrational. This is a celebrated result of C.L. Siegel. The precise statement is

**Theorem 7.7.** *Suppose* $F(z) = \lambda z + a_2 z^2 + \ldots$ *and* $\lambda = e^{2\pi i \alpha}$ *where* $\alpha$ *is irrational. Suppose there exist positive constants* $a$ *and* $b$ *such that* $|\alpha - p/q| > a/q^b$ *for all* $p, q \in \mathbf{Z}$. *Then there is a neighborhood* $U$ *of* $0$ *on which* $F$ *is analytically conjugate to the irrational rotation* $z \to \lambda z$.

For a proof, we refer to the text of Siegel and Moser.

**Remarks.**

**1.** The hypothesis on $\alpha$ requires that $\alpha$ be poorly approximated by rationals. It is a fact that "most" irrationals are of this form.

**2.** A region on which $F$ is conjugate to an irrational rotation is called a Siegel disk. Since there are no other periodic points in a Siegel disks, these regions also lie in the stable set of $F$.

**3.** This theorem does not hold for all irrational rotations, as the following example shows.

**Example 7.8.** Suppose $P(z) = \lambda z + \ldots + z^d$ where $\lambda = e^{2\pi i \alpha}$. Suppose that $\alpha$ is irrational and satisfies

$$|\lambda^n - 1| \le \left(\frac{1}{n}\right)^{d^n - 1}$$

for infinitely many natural numbers. Then we claim that $P$ has a periodic point in every neighborhood of $0$ (and hence cannot be conjugate to a linear irrational rotation).

To see this we simply note that the equation $P^n(z) - z = 0$ assumes the form

$$z^{d^n} + \ldots + (\lambda^n - 1)z = 0.$$

$0$ is one root of this equation. Let $\varsigma_1, \ldots, \varsigma_{d^n - 1}$ denote the remaining roots. Clearly,

$$|\varsigma_1| \cdot \ldots \cdot |\varsigma_{d^n - 1}| = |\lambda^n - 1|.$$

At least one of the $\varsigma_i$ must satisfy

$$|\varsigma_i| \le |\lambda^n - 1| \le \left(\frac{1}{n}\right)^{d^n \quad 1}$$

and so there are periodic points arbitrarily close to 0.

The natural question, of course, is whether there are any irrational numbers $\lambda$ which satisfy the inequality

$$|\lambda^n - 1| \leq \left(\frac{1}{n}\right)^{d^n - 1} \tag{$*$}$$

for infinitely many $n$. To show this, we need to make a brief detour to discuss continued fractions.

Let $a_0, a_1, a_2 \ldots$ be a sequence of positive integers. Define a sequence of rational numbers by

$$\frac{p_n}{q_n} = a_0 + \cfrac{1}{a_1 + \cfrac{1}{a_2 + \cfrac{1}{\ddots \, \cfrac{1}{a_n}}}}$$

From Exercise 8, the denominators of these rationals satisfy

$$q_{n+1} = a_{n+1} q_n + q_{n-1}.$$

As a consequence, this sequence converges to an irrational number, which we denote by $\alpha$ (see Exercise 9). These numbers are the best rational approximations to $\alpha$, and, from Exercise 10, they satisfy the inequality

$$\frac{1}{\left(a_{n+1} + 2\right)q_n^2} < \left|\alpha - \frac{p_n}{q_n}\right| < \frac{1}{a_{n+1}q_n^2}.$$

Now let us construct a $\lambda$ that satisfies the above condition. For any $\lambda = e^{2\pi i \alpha}$ we have

$$|\lambda^n - 1| = |e^{2\pi i n \alpha} - 1|$$
$$= |e^{\pi i n \alpha} - e^{-\pi i n \alpha}|$$
$$= 2|\sin(\pi n \alpha)|$$
$$= 2|\sin \pi(n\alpha - m)|$$

for any integer $m$. For each $n$, we may choose an integer $m_n$ such that $|n\alpha - m_n| \leq 1/2$. If $|x| \leq 1/2$, we have

$$2|x| \leq |\sin(\pi x)| \leq |\pi x| \leq |7x/2|$$

so that

$$4|n\alpha - m_n| \leq |\lambda^n - 1| \leq 7|n\alpha - m_n|.$$

Let us choose the $a_n$ inductively so that

$$a_{n+1} > 7q_n^{(d^{q_n}) - 2} \qquad (**)$$

and let $\alpha$ be the irrational whose continued fraction expansion is determined by the $a_n$. We then have

$$\left(\frac{1}{q_n}\right)^{d^{q_n} - 1} > \frac{7}{a_{n+1}q_n} > 7|\alpha q_n - p_n| > |\lambda^n - 1|$$

for each $n$. Hence $\lambda$ is an irrational rotation which satisfies (*) for infinitely many integers.

We remark that one may choose the first $k$ $a_n$'s arbitrarily and then choose the remainder to satisfy (**). This produces a dense set of irrationals which satisfy (*). Hence there are many irrational rotations which fail to produce Siegel disks.

**Exercises.**

1.   Describe the dynamics of the map $P(z) = z + z^{n+1}$ on the straight lines through the origin given by $\omega t$ where $\omega$ is a $(2n)^{th}$ root of unity.

2.   Show that, if $n$ is even this map admits $n$ repelling periodic orbits of period 2, one orbit on each straight line through 0 and an $n^{th}$ root of $-1$.

3.   Let $\lambda$ be an $n^{th}$ root of unity. Consider the map $P(z) = \lambda z(1 + z^{\ell n})$. Describe the dynamics of these maps on the invariant lines through 0 and an $(\ell n)^{th}$ root of $\pm 1$.

4.   Prove Corollary 7.4.

5.   Let $P(z) = \lambda z + a_2 z^2 + \ldots$ where $\lambda$ is an $n^{th}$ root of unity. Prove directly that $P^n(z)$ assumes the form

$$z + \beta_{\ell n + 1} z^{\ell n + 1} + \ldots.$$

6.   Complete the details of the reduction to normal form in Lemma 7.5.

7.   Let $S(z) = \sin z$. Prove that the real line lies in an attracting petal for $S$ at 0, while the imaginary axis lies in a repelling petal.

**8.**   Let

$$\frac{p_n}{q_n} = a_0 + \cfrac{1}{a_1 + \cfrac{1}{a_2 + \cfrac{1}{\ddots \cfrac{1}{a_n}}}}$$

be the continued fraction expansion of $\alpha$. Let $[\rho]$ denote the greatest integer part of $\rho$. Given the irrational number $\alpha$, show that the $a_i$ can be determined by the following procedure:

Let $a_0 = [\alpha]$ and $r_0 = \alpha - [\alpha]$ so that $\alpha = a_0 + r_0$. We have similarly

$$a_1 = \left[\frac{1}{r_0}\right]$$

$$r_1 = \frac{1}{r_0} - \left[\frac{1}{r_0}\right]$$

so that

$$\alpha = a_0 + \frac{1}{a_1 + r_1}.$$

Continuing in this fashion, show that the $a_n$'s are determined and $a_n \geq 0$.

**9.** Prove that $\alpha$ is irrational.

**10.** Let $p_{-1} = 1, q_{-1} = 0, p_0 = a_0$, and $q_0 = 1$. Use induction to prove that

$$\frac{p_{n+1}}{q_{n+1}} = \frac{a_{n+1}p_n + p_{n-1}}{a_{n+1}q_n + q_{n-1}}.$$

**11.** Use the previous exercise to show that

$$\left| \frac{p_n}{q_n} - \frac{p_{n+1}}{q_{n+1}} \right| = \frac{1}{q_n q_{n+1}}$$

so that $p_n$ and $q_n$ are relatively prime.

**12.**   Prove that the continued fraction expansion of $\alpha$ satisfies

$$\frac{1}{\left(a_{n+1} + 2\right)q_n^2} < \left|\alpha - \frac{p_n}{q_n}\right| < \frac{1}{a_{n+1}q_n^2}.$$

**13.**   Prove that $\alpha = (1 + \sqrt{5})/2$ if $a_i = 1$ for $i = 0, 1, 2, \ldots$.

**14.**   Prove that $\alpha = 10 + 2\sqrt{30}$ if $a_{2k} = 4, \; a_{2k+1} = 5$ for $k = 0, 1, 2 \ldots$.

The following series of exercises is designed to complement our work in §1.12 of one-dimensional bifurcations. There are many different types of bifurcations when we enlarge our viewpoint from the real line to the complex plane. Even the familiar saddle node and period-doubling bifurcations are different in this setting.

**15.**    *The saddle node.* Let $P_c(z) = z^2 + c$ where $c$ is real. On the real line, $P_c$ has two fixed points if $c < 1/4$ and none if $c > 1/4$. Prove that $P_c$ has a pair of repelling fixed points in C when $\lambda > 1/4$. Thus, this complex saddle node features a sink/source pair coalescing to form a pair of repelling fixed points.

**16.**    *The period-doubling bifurcation.* Recall that, when $\mu$ is real, the map $F_\mu(z) = \mu z(1 - z)$ experiences a period-doubling bifurcation at $\mu = 3$: for $1 < \mu < 3$, $F_\mu$ has one attracting fixed point and no period 2 points, but for $\mu > 3$, $F_\mu$ has a repelling fixed point and an attracting period 2 point (at least for $\mu$ close to 3). Prove that, in C, $F_\mu$ has a repelling period 2 orbit for $1 < \mu < 3$. Hence, in the plane, this period-doubling bifurcation features a repelling period 2 and an attracting period 1 point interchanging their qualitative character at the bifurcation.

**17.**    Describe the bifurcation that occurs in C for the family $z \to \lambda e^z$ as $\lambda$ passes through $1/e$ and also as $\lambda$ passes through $-e$.

**18.**    Describe the bifurcation that occurs in C for the family $S_\lambda(z) = \lambda \sin z$ as $\lambda$ increases through 1.

**19.**    Let $t > 0$ and suppose $\lambda$ is an $n^{\text{th}}$ root of unity. Consider the family of maps
$$P_t(z) = t\lambda z + z^{n+1} + \dots.$$
Describe the bifurcation that occurs as $t$ increases through 1.

## §3.8 AN EXAMPLE: THE EXPONENTIAL FUNCTION

Entire transcendental functions are, in many ways, similar to polynomials. There are, however, several major differences. Perhaps the most important difference is the fact that $\infty$ is no longer an attracting fixed point. Unlike polynomials, entire functions cannot even be extended continuously to $\infty$ to give a map on the Riemann sphere. Indeed, $\infty$ is what is known as an essential singularity for the map, and the remarkable theorem of Picard guarantees that the map assumes all but one value infinitely often in every neighborhood of an essential singularity. Thus the dynamics of an entire map are extremely complicated near $\infty$. We will illustrate this point in this section by considering a simple but important family of maps, the complex exponential maps $\lambda e^z$. Although these maps are relatively simple, they nevertheless illustrate a number of phenomena common to many entire functions. For example, we will prove below that the Julia set of these maps may explode as the parameter $\lambda$ is varied from a nowhere dense subset of **C** to the entire plane!

Let us denote the family of maps $\lambda \exp(z)$ by $E_\lambda$. Here $\lambda$ is a real parameter. The dynamics of $E_\lambda$ on the real line are easy to understand. Recall from §1.12 that the graph of $E_\lambda$ assumes three different forms depending upon whether $\lambda > 1/e$, $\lambda = 1/e$, or $0 < \lambda < 1/e$. See Fig. 8.1. It follows that, if $\lambda > 1/e$, then $E_\lambda^n(x) \to \infty$ for all $x$, while if $\lambda < 1/e$, there are two fixed points in **R**, one attracting and one repelling.

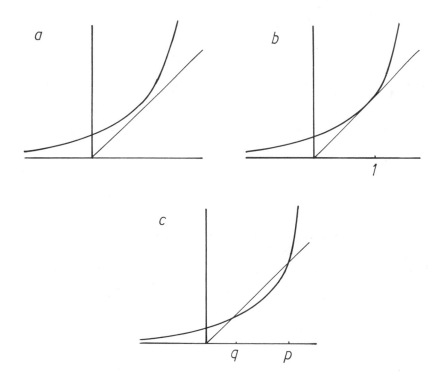

**Fig. 8.1.** The graphs of $E_\lambda(x) = \lambda \exp(x)$ when
a. $\lambda > \frac{1}{e}$, b. $\lambda = \frac{1}{e}$, and c. $\lambda < \frac{1}{e}$.

This, of course, is an example of the saddle node bifurcation. We will show below that this bifurcation is actually much more complicated when regarded globally; the explosion in the Julia set of $E_\lambda$ occurs as $\lambda$ passes through $1/e$.

We first recall some of the elementary properties of the exponential. For simplicity, we will write $E(z)$ for the usual exponential $e^z$. We have $E(x + iy) = e^x e^{iy} = e^x \cos y + ie^x \sin y$. Consequently, $E$ maps vertical lines $x = c$ to circles centered at 0 with radius $e^c$ and horizontal lines $y = c$ to rays $\theta = c$. In particular, horizontal lines of the form $y = (2k + 1)\pi$ for $k \in \mathbf{Z}$ are mapped to the negative real axis, while lines of the form $y = 2k\pi$ are mapped to the positive real axis. See Fig. 8.2. Finally, $E$ is $2\pi i$ periodic and has no critical points, but 0 is a singular "value" in the sense that it is an omitted value. The reader may easily check that any neighborhood of $\infty$ is mapped by $E$ onto $\mathbf{C} - \{0\}$ infinitely often. See Exercise 1.

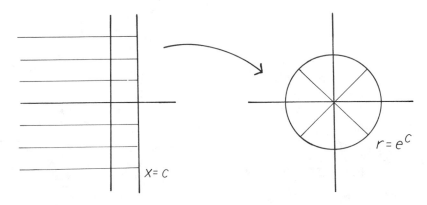

$$X = C$$
$$r = e^C$$

**Fig. 8.2.** The mapping $E(z) = \exp(z)$.

We now turn to a description of the Julia set of $E(z)$. We remark that all of the properties of the Julia set of a polynomial go over to the exponential map with one exception: the Julia set may have interior (see Corollary 5.10). In fact, in 1981, Misiurewicz showed that $J(E) = \mathbf{C}$, answering a sixty year old question of Fatou. We present his proof of this fact below.

The following proposition highlights one of the differences between $E(z)$ and polynomials: points which tend to $\infty$ under iteration of $E$ need not be in the stable set.

**Proposition 8.1.** *The real line is contained in $J(E)$.*

*Proof.* Let $S$ denote the strip $|\text{Im }(z)| \leq \pi/3$. If $z = x + iy \in S$, then since $|e^z \cos y| \geq e^x/2 > x$, it follows that $E(z)$ lies to the right of $z$. The last inequality follows from Fig. 8.1, since $\frac{1}{2} > \frac{1}{e}$. In particular, if $E^i(z) \in S$ for all $i$, we have Re $E^i(z) \to \infty$.

If $z \in S$ with Re $(z) > 1$ and Im $(z) \neq 0$, then we also have

$$|e^x \sin y| > e^x(\frac{2}{\pi}|y|) > |y|.$$

Consequently, if $z \in S$ but $z \notin \mathbf{R}$, then $|\text{Im } (E^i(z))|$ must grow as $i$ increases. Hence there exists $j > 0$ for which $E^j(z) \notin S$. Thus all points in $S$ which do not lie in $\mathbf{R}$ must eventually leave $S$.

Now let $U$ be any neighborhood of $x \in \mathbf{R}$. Recall that $E^j(x) \to \infty$. By the above remarks, there is $N > 0$ such that, for each $j > N$, $E^j(U)$ intersects both $\mathbf{R}$ and the line $y = \pi/3$ at points with real part $> 1$. Consequently, $E^{j+1}(U)$ meets both $\mathbf{R}$ and $y = \pi$, since $y = \pi/3$ is mapped to the ray $\theta = \pi/3$. Hence $E^{j+2}(U)$ meets the negative real axis, and so a portion of $U$ is mapped by $E^{j+3}$ inside the unit disk. Thus, for sufficiently large $j$, there are points $z_1$ and $z_2$ in $U$ for which $E^j(z_1)$ lies in the unit disk and for which $|E^j(z_2)|$ is arbitrarily large. It follows that $\{E^n\}$ is not normal in $U$, and so $x \in J(E)$.

q.e.d.

Thus to show that $J(E) = \mathbf{C}$, it suffices to show that inverse images of the real line are dense in $\mathbf{C}$. For this, we need several lemmas.

**Lemma 8.2.** $|\text{Im } (E^n(z))| \leq |(E^n)'(z)|$.

*Proof.* If $z = x + iy$, we have

$$|\text{Im } (E(z))| = e^x|\sin y|$$
$$\leq e^x|y|$$
$$= |E'(z)||\text{Im } (z)|$$

so that

$$\frac{|\text{Im } (E(z))|}{|\text{Im } (z)|} \leq |E'(z)|$$

if $z \notin \mathbf{R}$. More generally, if $z \notin \mathbf{R}$, we may apply this inequality repeatedly to find

$$\frac{|\text{Im } (E^n(z))|}{|\text{Im } (E(z))|} = \prod_{i=1}^{n-1} \frac{|\text{Im } E(E^i(z))|}{|\text{Im } (E^i(z))|}$$
$$\leq \prod_{i=1}^{n-1} |E'(E^i(z))|.$$

Since $|\mathrm{Im}\ (E(z))| \leq |E(z)| = |E'(z)|$ we may write

$$|\mathrm{Im}\ (E^n(z))| \leq \prod_{i=0}^{n-1} |E'(E^i(z))|$$

$$= |(E^n)'(z)|.$$

q.e.d.

Proposition 8.1 shows that most points must leave the strip $S$ under iteration. The next lemma shows, however, that most points must eventually return.

**Lemma 8.3.** *Let $U$ be an open connected set. Then only finitely many of the $E^n(U)$ can be disjoint from $S$.*

*Proof.* Let us assume that infinitely many of the images of $U$ are disjoint from $S$. If there is an $n$ for which $E^n$ is not a homeomorphism taking $U$ onto its image, then there exist $z_1, z_2 \in U$ for which $E^n(z_1) = E^n(z_2)$. Consequently, there is a $j$ for which $E^j(z_1) = E^j(z_2) + 2k\pi i$ for some $k \in \mathbf{Z} - \{0\}$. But then $E^j(U)$ must meet a horizontal line of the form $y = 2m\pi$ for $m \in \mathbf{Z}$ and so $E^{j+1}(U)$ meets $\mathbf{R}$. Hence $E^{j+\alpha}(U)$ meets $\mathbf{R}$ for all $\alpha > 0$ and only finitely many of the images of $U$ can be disjoint from $S$. We thus conclude that each $E^n$ must be a homeomorphism on $U$.

Now suppose $E^{n_j}(U) \cap S = \phi$. By the previous lemma, $|(E^{n_j})'(z)| \geq (\pi/3)^j$ for each $j$ and all $z \in U$. It follows that, if $U$ contains a disk of radius $\delta > 0$, then $E^{n_j}(U)$ contains a disk of radius $\delta(\pi/3)^j$. See Exercise 2. Hence for $j$ large enough, $E^{n_j}(U)$ must meet a line of the form $y = 2\pi$ and again we are done.

q.e.d.

**Lemma 8.4.** *Let $V$ be an open connected set for which infinitely many of its images are contained in the half plane $H = \{z\,|\,\mathrm{Re}\ (z) > 4\}$. Then there exists $n > 0$ for which $E^n(V) \cap \mathbf{R} \neq \emptyset$.*

*Proof.* Let $W$ denote the set $\{z\,|\,|\mathrm{Im}\ (z)| \leq 2\pi$ and $|\mathrm{Im}\ (E(z))| \leq 2\pi\}$. If a set $A$ satisfies $A \cap W = \emptyset$, then either $A \cap S$ or $E(A) \cap S$ is empty. Consequently, by the previous lemma, only finitely many images of $V$ in $H$ can be disjoint from $W$. Hence almost all images of $V$ are contained in $W$.

Now consider the boundary $|y| = \frac{\pi}{3}$ of $S$ in $H$. If $z$ lies on this boundary, then

$$|\mathrm{Im}\ (E(z))| \geq e^4 \sin\left(\frac{\pi}{3}\right) > 2\pi.$$

Therefore, the boundary of $S$ in $H$ does not lie in $W$. Thus every connected set in $W \cap H$ is either contained in $S$ or disjoint from $S$. Now the image of $S \cap H$ is contained in $H$. Since infinitely many of the images of $V$ are contained in $W \cap H$, it thus follows that infinitely many of them must be disjoint from $S$. This contradicts Lemma 8.3.

q.e.d.

We can now prove

**Theorem 8.5.** $J(E) = \mathbf{C}$.

*Proof.* By Proposition 8.1, it suffices to show that any open set in $\mathbf{C}$ contains some preimage of $\mathbf{R}$. To that end, let $U$ be open and connected and suppose $E^n(U) \cap \mathbf{R} = \emptyset$ for each $n$. By Montel's Theorem, $\{E^n\}$ is a normal family on $U$.

Let $D$ denote the disk of radius $e^4$ about 0. Note that $E(H)$ is the complement of $D$. Hence, by Lemma 8.4, we may assume that infinitely many of the images of $U$ meet $D$.

Now let $F$ denote the limit function of some subsequence of the $E^n$. By the above, $F(U) \cap D \neq \phi$. Choose a point $z_0 \in F(U) \cap D$. If $z_0 \in \mathbf{R}$, then there exists $k > 0$ such that $E^k(U) \cap \mathbf{R} \neq \phi$ and we are done. Thus we assume $z_0 \notin \mathbf{R}$. As we observed in Proposition 8.1, there exists $k > 0$ for which $E^k(z_0) \notin S$. Therefore there exists $w \in U$ and another subsequence of the $E^n$ which converges to a map $F_1$ which satisfies $F_1(w) \notin S$. But then there is an open neighborhood $V$ of $w$ and infinitely many images $E^n(V)$ which do not meet $S$. This contradicts Lemma 8.3 and establishes the Theorem.

q.e.d.

**Remarks.**

**1.** We have actually shown that the family $\{E^n\}$ is not normal at any point in $\mathbf{C}$. To prove that repelling periodic points are dense as well, we may argue as follows. It is easy to check that $E$ has a repelling fixed point; in fact it has infinitely many (see Exercise 3). Then the arguments of §3.5 may be used to produce a repelling periodic point near any point in $\mathbf{C}$.

**2.** It is also true that $J(E_\lambda) = \mathbf{C}$ for $\lambda > 1/e$. One need only modify the sets $S$ and $H$ to prove this. We omit the details.

Now we turn to the bifurcation which occurs at $\lambda = 1/e$. By the above remarks, $J(E_\lambda) = \mathbf{C}$ for all $\lambda > 1/e$. For $\lambda < 1/e$, $E_\lambda$ has two fixed points in $\mathbf{R}$: an attracting fixed point at $q$ and a repelling fixed point at $p$. See Fig. 8.1. Note that $q < 1 < p$ and that $E_\lambda(1) < 1$. Consider the vertical line $x = 1$. $E_\lambda$ maps this line to a circle of radius $\lambda e < 1$ about 0. In particular,

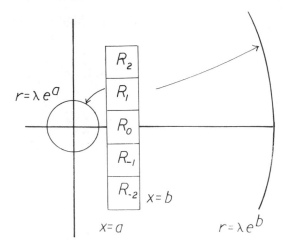

**Fig. 8.3.**

the entire left half plane Re $(z) < 1$ is mapped inside this circle. Since Re $(z) < 1$ is simply connected, and $q$ is a fixed point in this set, it follows from the Schwarz lemma (see Corollary 1.11) that all points in Re $(z) < 1$ tend to $q$ under iteration of $E_\lambda$. Hence there are no repelling periodic points in Re $(z) < 1$ and it follows that the Julia set of $E_\lambda$ lies entirely to the right of the line $x = 1$. Thus, as $\lambda$ increases through $1/e$, the Julia set explodes to cover the entire plane. Obviously, this saddle node bifurcation has global ramifications. This also shows that the Julia set of a complex analytic map need not vary continuously.

A natural question to ask concerns the nature of $J(E_\lambda)$ for $0 < \lambda < 1/e$. By the above results, $J(E_\lambda)$ is contained in the plane Re $(z) \geq 1$. To describe this set, we again invoke symbolic dynamics. For any integer $N$, we first construct a sequence of $2N + 1$ rectangles $R_j$ which we index by $j = -N, \ldots, N$. Each of these rectangles have sides parallel to the $x$- and $y$-axes. Their left boundaries lie on $x = 1$ while their horizontal boundaries lie on the lines $y = (2j+1)\pi$ and $y = (2j-1)\pi$. The right boundaries lie on the vertical line $x = c$ where $c$ depends on $N$ and is chosen so that $\lambda e^c > \sqrt{2}c$. This condition guarantees that $E_\lambda$ maps each $R_j$ onto the annular region $\lambda e < 1 < |z| < \lambda e^c$ which covers each of the $R_k$. See Fig. 8.3. This is a familiar situation. Let

$$R^N = \bigcup_{j=-N}^{N} R_j$$

and define $\Lambda_N = \{z \in R^N | E_\lambda^i(z) \in R^N \text{ for all } i\}$. The following proposition is proved exactly as in preceding sections and hence is left as an exercise.

**Proposition 8.6.** $\Lambda_N$ *is an invariant Cantor set on which* $E_\lambda$ *is topologically conjugate to the shift on* $2N+1$ *symbols. Moreover, if* $z \in \Lambda_N$, *then* $|(E'_\lambda)(z)| > 1$.

The Julia set of $E_\lambda$ is actually more complicated than the above would suggest. The fixed point $p$ is obviously in $J(E_\lambda)$, and, moreover, using Proposition 8.1, the entire interval $[p,\infty)$ is also in $J$. The open interval $(p,\infty)$ consists of points which tend to $\infty$ under iteration of $E_\lambda$ and hence, by the argument in the proof of Proposition 8.1, these points lie in $J(E_\lambda)$. But none of these points are captured by the construction of $\Lambda_N$. We call this curve the *tail* associated to $p$. It turns out that each point in $\Lambda_N$ has a similar tail attached. Let us be more specific.

**Proposition 8.7.** *Let* $w \in \Lambda_N$. *There is a continuous curve* $\psi_w : [0,\infty) \to \mathbf{C}$ *which satisfies*

    1. $\psi_w(t)$ *is one-to-one.*

    2. $\psi_w(0) = w$.

    3. *If* $t \neq 0$, $E^n_\lambda(\psi_w(t)) \to \infty$ *as* $n \to \infty$.

    4. $\psi_w(t) \in J(E_\lambda)$.

We simply sketch the construction of $\psi_w$, leaving most of the details to the reader. Since $w \in \Lambda_N$, there is a sequence $s_0 s_1 s_2, \ldots$ of integers which gives the itinerary of $w$ in the $R_j$. Let $S_j$ denote the semi-infinite strip $\{z \,|\, \mathrm{Re}\,(z) \geq 1 \text{ and } (2j-1)\pi \leq \mathrm{Im}\,(z) \leq (2j+1)\pi\}$. $\psi_w$ is then defined as $\{z \,|\, E^n_\lambda(z) \in S_{s_n}\}$, i.e., $\psi_w$ contains all points which share the same itinerary as $w$ in the $S_j$. One may check that, if the right hand boundary of $R_j$ lies in $x = c$ and if $c$ is large enough, then there is a unique point in this interval which lies in $v_u$. See Exercise 6.

In fact, one can show that the set of points whose orbits lie for all time in one of the above strips is homeomorphic to the Cartesian product $\Lambda_N \times [0,\infty)$, i.e., that the tails vary continuously with the points in $\Lambda_N$. Such a set is called a *Cantor bouquet*. See Exercise 7.

## Exercises

**1.** Prove that $E_\lambda(z) = \lambda \exp(z)$ assumes every value but 0 infinitely often in every neighborhood of $\infty$. This is a special case of the great Picard Theorem.

**2.** Suppose $B$ is a disk of radius $\delta$ and $F$ is an analytic homeomorphism which satisfies $|F'(z)| > \mu$ for each $z \in B$. Show that $F(B)$ contains a disk of radius $\mu\delta$.

**3.** Let $R_k$ denote the open strip $(2k - 1)\pi < \text{Im}\,(z) < (2k + 1)\pi$. Prove that each $E_\lambda$ has a repelling periodic point in $R_k$ if $k \neq 0$.

**4.** Prove that when $\lambda = 1/e$, there are no repelling periodic points in the plane $\text{Re}\,(z) < 1$.

**5.** Prove Proposition 8.3.

**6.** Let $0 < \lambda < 1/e$. Let $w$ be a point in the Cantor set $\Lambda_n \subset J(E_\lambda)$ as described in Proposition 8.6. Prove that $w$ has a tail attached as described in Proposition 8.7.

**7.** Let $\Gamma_n = \{z \,|\, \text{Re}\,(z) \geq 1 \text{ and } |\text{Im}\,(z)| \leq (2n + 1)\pi\}$. Prove that if $0 < \lambda < 1/e$, then $\Gamma_n$ is a Cantor bouquet, i.e., is homeomorphic to $\Lambda_n \times [0, \infty)$.

The following three exercises show that entire functions may differ from polynomials in other respects.

**8.** Let $F(z) = z + e^z$. Prove that $F$ has no fixed points.

**9.** Let $F(z) = z - (\sin z)e^{iz}$. Prove that $F$ has infinitely many attracting fixed points.

**10.** Let $F(z) = z + 1 + e^{-z}$. Prove that if $\text{Re}\,(z) > 0$, then $F^n(z) \rightarrow \infty$, so that entire functions may have partial basins of attraction at $\infty$. Describe $J(F)$.

The following four exercises deal with the structure of the Julia set of the entire function $S_\lambda(z) = \lambda \sin(z)$.

**11.** Prove that the imaginary axis is invariant under $S_\lambda$. If $|\lambda| \geq 1$, prove that it is contained in $J(S_\lambda)$.

**12.** Describe the bifurcation that occurs for $\lambda = 1$ in the family $S_\lambda$.

**13.** Let $|\lambda| < 1$. For $-N \leq j \leq N$ and $\epsilon$ small, let $W_j$ denote the strips given by $\text{Im}\,(z) > \epsilon$ and $(2j - 1)\pi \leq \text{Re}\,(z) \leq (2j + 1)\pi$. Show that $J(S_\lambda)$ contains an invariant Cantor set in the $W_j$ on which $S_\lambda$ is conjugate to the shift on $2N + 1$ symbols.

**14.** Prove that $S_\lambda(z)$ assumes every value in **c** infinitely often in every neighborhood of $\infty$

**15.** Describe the Julia set of $\lambda \exp(z^n)$ for $\lambda$ small.

# FOR FURTHER READING

There are very few accessible references for many of the concepts presented in this section; most of this material is only available in the classical research literature or else has only recently appeared. One survey with an excellent bibliography of the research literature has recently appeared:

Blanchard, P. *Complex Analytic Dynamics on the Riemann Sphere.* Bull A.M.S. **11** (1984), 85-141.

Further details on convergence questions which arise near an indifferent periodic point may be found in:

Siegel, C.L. and Moser, J. *Letures on Celestial Mechanics.* Springer-Verlag, New York, 1971.

Julia sets are also examples of fractals, and the computer graphics work of Mandelbrot has done a lot to bring the subject of complex dynamics back into vogue. Some of this work is contained in:

Mandelbrot, B. *The Fractal Geometry of Nature.* Freeman, San Francisco, 1982.

There are a number of excellent textbooks available for background work in complex analysis. Among them are:

Ahlfors. L.V. *Complex Analysis.* McGraw-Hill Book Co., New York, 1979.

Conway, J.B. *Functions of One Complex Variable.* Springer-Verlag, New York, 1978.

Finally, there is no substitute for going back to the old masters. It is remarkable to see how far Fatou and Julia were able to push the theory of complex analytic dynamics without having access to a computer!

Julia, G. *Memoire sur l'iteration des fonctions rationelles.* J. Math. **8** (1918), 47-245.

Fatou, P. *Sur l'iteration des fonctions transcendantes entières.* Acta Math. **47** (1926), 337-370.

# INDEX